Reinventing Fisheries Management

CHAPMAN & HALL FISH AND FISHERIES SERIES

Amongst the fishes, a remarkably wide range of fascinating biological adaptations to diverse habitats has evolved. Moreover, fisheries are of considerable importance in providing human food and economic benefits. Rational exploitation and management of our global stocks of fishes must rely upon a detailed and precise insight of the interaction of fish biology with human activities.

The *Chapman & Hall Fish and Fisheries Series* aims to present authoritative and timely reviews which focus on important and specific aspects of the biology, ecology, taxonomy, physiology, behaviour, management and conservation of fish and fisheries. Each volume will cover a wide but unified field with themes in both pure and applied fish biology. Although volumes will outline and put in perspective current research frontiers, the intention is to provide a synthesis accessible and useful to both experts and non-specialists alike. Consequently, most volumes will be of interest to a broad spectrum of research workers in biology, zoology, ecology and physiology, with an additional aim of the books encompassing themes accessible to non-specialist readers, ranging from undergraduates and postgraduates to those with an interest in industrial and commercial aspects of fish and fisheries.

Applied topics will embrace synopses of fishery issues which will appeal to a wide audience of fishery scientists, aquaculturists, economists, geographers and managers in the fishing industry. The series will also contain practical guides to fishery and analysis methods and global reviews of particular types of fisheries.

Books already published and forthcoming are listed below. The Publisher and Series Editor would be glad to discuss ideas for new volumes in the series . . .

Available titles

1. **Ecology of Teleost Fishes**
   Robert J. Wootton
   ISBN 0 412 31730 3
2. **Cichlid Fishes**
   Behaviour, ecology and evolution
   Edited by Miles A. Keenlyside
   ISBN 0 412 32200 5
3. **Cyprinid Fishes**
   Systematics, biology and exploitation
   Edited by Ian J. Winfield and Joseph S. Nelson
   ISBN 0 412 34920 5
4. **Early Life History of Fish**
   An energetics approach
   Ewa Kamler
   ISBN 0 412 33710 X
5. **Fisheries Acoustics**
   David N. MacLennan and E. John Simmonds
   ISBN 0 412 33060 1
6. **Fish Chemoreception**
   Edited by Toshiaki J. Hara
   ISBN 0 412 35140 4
7. **Behaviour of Teleost Fishes**
   Second edition
   Edited by Tony J. Pitcher

ISBN 0 412 42930 6 (HB), 0 412 42940 3 (PB)

8. **Genetics and Fish Breeding**
   Colin R. Purdom
   ISBN 0 412 33040 7
9. **Fish Ecophysiology**
   J. Cliff Rankin and Frank B. Jensen
   ISBN 0 412 45920 5
10. **Fish Swimming**
    John J. Videler
    ISBN 0 412 40860 0
11. **On the Dynamics of Exploited Fish Populations**
    Raymond J.H. Beverton and Sidney J. Holt
    (Facsimile reprint)
    ISBN 0 412 54960 3
12. **Sea Bass**
    Graham D. Pickett and Michael G. Pawson
    ISBN 0 412 40090 1
13. **Fish Bioenergetics**
    Malcolm Jobling
    ISBN 0 412 58090 X
14. **On the Sex of Fish and Gender of Scientists**
    Daniel Pauly
    ISBN 0 412 59540 0
15. **Hake**
    Fisheries, products and markets
    Edited by Jürgen Alheit and Tony J. Pitcher
    ISBN 0 412 57350 4
16. **Environmental Biology of Fishes**
    Malcolm Jobling
    ISBN 0 412 58080 2
17. **Electric Fishes**
    History and behavior
    Peter Moller
    ISBN 0 412 37380 7
18. **The Impact of Species Changes in African Lakes**
    Edited by Tony J. Pitcher and Paul B. Hart
    ISBN 0 412 55050 4
19. **Pike**
    Biology and exploitation
    Edited by John F. Craig
    ISBN 0 412 42960 8
20. **Reef Fisheries**
    Edited by Nicholas V.C. Polunin and Callum M. Roberts
    ISBN 0 412 60110 9
21. **Early Life History and Recruitment in Fish Populations**
    Edited by R. Christopher Chambers and Edward A. Trippel
    ISBN 0 412 64190 9
22. **Fish and Fisheries in Tropical Estuaries**
    Stephen J.M. Blaber
    ISBN 0 412 78500 5
23. **Reinventing Fisheries Management**
    Edited by Tony J. Pitcher, Paul J. Hart and Daniel Pauly
    ISBN 0 412 83410 3
24. **Ecology of Teleost Fishes**
    2nd edition
    Robert J. Wootton
    ISBN 0 412 84590 3 (HB), 0 412 64200 X (PB)

Forthcoming titles

Fisheries Ecology
Second edition
Edited by Tony J. Pitcher and Paul J. Hart
ISBN 0 412 33070 9

Tilapias
Biology and exploitation
Edited by Malcolm Beveridge and Brendan McAndrew
ISBN 0 412 80090 X

Reinventing Fisheries Management

Edited by

Tony J. Pitcher

Fisheries Centre
University of British Columbia
Vancouver, British Columbia
Canada

Paul J.B. Hart

Department of Biology
University of Leicester
Leicester
UK

and

Daniel Pauly

Fisheries Centre
University of British Columbia
Vancouver, British Columbia
Canada

KLUWER ACADEMIC PUBLISHERS
DORDRECHT / BOSTON / LONDON

Library of Congress Cataloging in Publication Card Number: 98-070426

ISBN 0 412 83410 3

Published by Kluwer Academic Publishers,
P.O. Box 17, 3300 AA Dordrecht, The Netherlands.

Sold and distributed in North, Central and South America
by Kluwer Academic Publishers,
101 Philip Drive, Norwell, MA 02061, U.S.A.

In all other countries, sold and distributed
by Kluwer Academic Publishers Group,
P.O. Box 322, 3300 AH Dordrecht, The Netherlands.

Printed in Great Britain

Contents

Contributors

Andrew Bakun
FAO, Via delle Terme di Caracalla, 00100 Rome, Italy
andrew.bakun@fao.org

Robert C. Brown
Institute of Fisheries Analysis, Simon Fraser University, Burnaby, BC, Canada V5A 1S6
brown@sfu.ca

Rik C. Buckworth
Department of Primary Industry and Fisheries, GPO Box 990, Darwin, NT, Australia 0810
rik.buckworth@dpif.nt.gov.au

Alida Bundy
Groundfish Division, Department of Fisheries and Oceans, North West Atlantic Fisheries Centre, PO Box 5667, Newfoundland, A1C 5X1 Canada
bundy@athena.nwafc.nf.ca

Anthony T. Charles
Department of Finance and Management Science, Saint Mary's University, Halifax, NS, Canada B3H 3C3
t.charles@StMarys.ca

Kevern L. Cochrane
Fishery Resources and Environment Division, FAO, Via delle Terme di Caracalla, 00100 Rome, Italy
kevern.cochrane@fao.org

Karl K. English
LGL Limited, 9768 Second Street, Sidney, BC, Canada V8L 3Y8

Shane W. Frederick
Department of Social and Decision Sciences, Carnegie Mellon University, Pittsburgh, PA 15213–3989, USA

Jarl Giske
Department of Fisheries and Marine Biology, University of Bergen, High Technology Centre, N–5020 Bergen, Norway
Jarl.Giske@ifm.uib.no

Nigel Haggan
1777 E. 7th Avenue, Vancouver, BC, Canada V5N 1S1
nhaggan@fisheries.com

Rögnvaldur Hannesson
Norwegian School of Economics and Business Administration, Helleveien 30, N–5035 Bergen-Sandviken, Norway
rognvaldur.hannesson@nhh.no

Craig K. Harris
Department of Sociology, Michigan State University, East Lansing, MI 48824–1111, USA
craig.harris@ssc.msu.edu

Paul J.B. Hart
Department of Biology, School of Biological Sciences, University of Leicester, University Road, Leicester LE1 7RH, United Kingdom
pbh@leicester.ac.uk

Trevor Hutton
9 St Clair Road, Plumstead 7800, Cape Town, South Africa
thutton@botzoo.uct.za

Svein Jentoft
Institute of Social Science, University of Tromsø, Breivika, N–9000 Tromsø, Norway
sveinj@isv.uit.no

Michael R. Link
LGL Limited, 9768 Second Street, Sidney, BC, Canada V8L 3Y8
mlink@LGL.com

Thomas R. McGuire
Bureau of Applied Research in Anthropology, Department of Anthropology, University of Arizona, Tucson, AZ 85721, USA
mcguire@u.arizona.edu

William E. Neill
Fisheries Centre, University of British Columbia, 2204 Main Mall, Vancouver, BC, Canada V6T 1Z4
neill@fisheries.com

Daniel Pauly
Fisheries Centre, University of British Columbia, 2204 Main Mall, Vancouver, BC, Canada V6T 1Z4
pauly@fisheries.com

Andrew I.L. Payne
Sea Fisheries Research Institute, Rogge Bay 8012, South Africa
ailpayne@sfri.wcape.gov.za

Randall M. Peterman
School of Resource and Environmental Management, Simon Fraser University, Burnaby, BC, Canada V5A 1S6
peterman@sfu.ca

Calvin N. Peters
ESSA Technologies Ltd, Suite 300, 1765 W. 8th Avenue, Vancouver, BC, Canada V6J 5C6

Tony J. Pitcher
Fisheries Centre, University of British Columbia, 2204 Main Mall, Vancouver, BC, Canada V6T 1Z4
tpitcher@fisheries.com

David Policansky
Board on Environmental Studies and Toxicology, National Research Council, 2101 Constitution Avenue, Washington, DC 20418, USA
dpolican@nas.edu

David B. Preikshot
Fisheries Centre, University of British Columbia, 2204 Main Mall, Vancouver, BC, Canada V6T 1Z4
preikshot@fisheries.com

Jake C. Rice
Fisheries and Oceans Science, 200 Kent Street, Ottawa, Ontario, Canada K1A 0E6
ricej@dfo-mpo.gc.ca

Laura J. Richards
Fisheries and Oceans Canada, Science Branch, Pacific Region, Pacific Biological Station, 3190 Hammond Bay Road, Nanaimo, BC, Canada V9R 5K6
richardsl@dfo-mpo.gc.ca

Christina A. Robb
Lincoln Environmental, PO Box 84, Lincoln University, Canterbury, New Zealand

James P. Scandol
Quantitative Training Unit for Fisheries, Marine Ecology Laboratories A11, University of Sydney, NSW, 2006, Australia
jscandol@bio.usyd.edu.au

Jon T. Schnute
Fisheries and Oceans Canada, Science Branch, Pacific Region, Pacific Biological Station, 3190 Hammond Bay Road, Nanaimo, BC, Canada V9R 5K6
schnutej@dfo-mpo.gc.ca

Anthony Scott
Department of Economics, University of British Columbia, 1873 East Mall, Buchanan Tower 997, Vancouver, BC, Canada V6T 1Z4
adscott@interchange.ubc.ca

John T. Sproul
Fisheries Centre, University of British Columbia, 2204 Main Mall, Vancouver, BC, Canada V6T 1Z4
jsproul@unixg.ubc.ca

Ussif Rashid Sumaila
Chr. Michelsen Inst. and Dept of Economics, University of Bergen, Fantoftvegen 38, N–5036 Bergen, Norway
sumaila@fisheries.com

Michael Sutton
Endangered Seas Campaign, WWF International, Branksome House, Filmer Grove, Godalming, Surrey GU7 3AB, United Kingdom
gmsutton@compuserve.com

Carl Walters
Fisheries Centre, University of British Columbia, 2204 Main Mall, Vancouver, BC, Canada V6T 1Z4
walters@fisheries.com

Meryl Williams
ICLARM, MC PO Box 2631, Makati, Manila 0718, Philippines
williams@cgnet.com

Series foreword

Among the fishes, a remarkably wide range of biological adaptations to diverse habitats has evolved. As well as living in the conventional habitats of lakes, ponds, rivers, rock pools and the open sea, fish have solved the problems of life in deserts, in high altitude torrential streams, in the great deeps of the sea, in the extreme cold of the Antarctic, in warm waters of high alkalinity, in low oxygen conditions, and in habitats like estuaries where such physical factors change relentlessly all the time. Along with these adaptations, we find the most impressive specialisations of morphology, physiology and behaviour. For example, we can marvel at the high-speed swimming of the marlins, sailfish and warm-blooded tunas, air-breathing in catfish and lungfish, parental care in mouth-brooding cichlids, and viviparity in many toothcarps, blennies and sharks.

Moreover, the fish in our oceans, estuaries, lakes and rivers are of considerable importance to the well-being of the human species in the form of nutritious, diverse and delicious food. Sustainable exploitation and responsible management of our global stocks of fishes must rely upon a detailed and precise insight of their biology and their role in aquatic ecosystems. Moreover, there is a need to focus interdisciplinary work in relevant social, economic and historical spheres upon fish and fisheries in order to assess their cultural role and significance, and evaluate the impact of human harvest on aquatic ecosystems.

The Chapman and Hall *Fish and Fisheries Series* aims to present timely volumes reviewing major thematic aspects of the science of fish and fisheries. Most volumes will be of interest to research workers in biology, zoology, ecology, physiology and fisheries, but an additional aim is for the books to be accessible to a wide spectrum of non-specialist readers ranging from undergraduates and postgraduates to those with an interest in industrial and commercial aspects of fish and fisheries.

It has long been the policy of the editorial team that books in the *Fish and Fisheries Series* should adhere to the highest academic standards by being fully peer reviewed – indeed some volumes have inevitably fallen by the wayside in this respect. We want work published in the Series to receive the international recognition it deserves and so citations by the scientific information industry are vital. I am therefore happy to report that, starting in 1998, agreement has been reached that all books, and individual chapters in multi-author edited volumes, will from now on be fully abstracted in Aquatic Sciences and Fisheries Abstracts (ASFA), and this procedure will be applied retrospectively to previous volumes.

Reinventing Fisheries Management comprises the 23rd volume in the Chapman & Hall *Fish and Fisheries Series* since the first volume appeared ten years ago. The main themes and much of the core material in the book started life as a result of discussions at an international symposium held in Vancouver in 1996, but all of the work has been updated, honed and augmented as result of referee and editorial comment since then.

In North America, the meaning of the term 'management' in the title is often restricted to agency control of regulations and allocation decisions. In this book 'management' is given its wider meaning, covering all of the activities with which humans seek to rationalise the exploitation of aquatic resources. Hence, the book covers the principles of the assessment of exploitation status of resources, institutional, societal and economic aspects, and, most important, the establishment of appropriate policy goals. It is surprising how many fisheries around the world are 'managed' from day to day without any clear understanding of what those policy goals may be. Moreover, many of the contributions in this book question the wisdom of the current widely-adopted goal of 'sustainability'.

The chapters include a broad geographical range of examples from such diverse fisheries as those in the African lakes, marine fisheries in Africa, Europe, Australia, Asia, Latin America and Oceana, and fisheries for migratory salmon in North America. Examples also range from small-scale artisanal fisheries to those run by huge commercial international operations: over 90% of the world's trade in seafood is controlled by less than 20 companies. We have seemed powerless to reverse the twin economic evils of huge overcapacity in the catching power of industrial fleets and in the desperately poor coastal peoples of the developing world: both generate positive feedback loops that devastate aquatic resources.

Fisheries science is at a turning point. There is widespread unease with existing science and methods, which has such a poor track record that it provides an increasingly dramatic mismatch with impressive and sophisticated methodology. Overall, this book seeks a new synthesis for fisheries management by examining the elements that are needed for a cognitive shift in the way that we approach fisheries problems.

Professor Tony J. Pitcher
Editor, Chapman and Hall Fish and Fisheries Series
Director, Fisheries Centre, University of British Columbia, Vancouver, Canada

Preface

Why does fisheries management need reinventing? Judged by a dismaying track record, and a consequent downturn in the reputation of fisheries scientists, fisheries management is certainly a candidate for calls for reinvention. It seems that, despite our best efforts, fisheries worldwide have become severely depleted and, along with reductions in the size of fish harvested, fish communities shift towards small, rapidly growing species. These symptoms have been accompanied by a series of fisheries collapses that not only were largely unforeseen by our most advanced assessment methods, but also brought about disastrous economic consequences. Such things have even occurred in Canada, a nation with probably more top-rate fishery scientists per capita than any other. Confidence in our discipline has been eroded at the very time when we need it most if we are to do anything to ensure viable fish harvests in the densely populated world of the coming century.

So fisheries science is now in a state of flux and many feel that we are at a crossroads where new paradigms compete for attention and demand evidence of their utility. Some of the world leaders in our subject have expressed the pessimistic view that no fishery has ever been properly understood or managed. Some consider that we have to conduct experiments with our fishery resources if we are to hope to do any better. Some see a solution in quantifying our ignorance. Others look to the social sciences to bring salvation by trying to understand people as well as manage fish. This book seeks a new paradigm that will place this flux of ideas in perspective and help us choose those that will make management work.

New ways of looking at things often entail interdisciplinary synergy between biological, ecological, social and economic foci. This is a most difficult vision to achieve, not least because exponents of these disciplines find it difficult to step outside their walls as they are invariably rewarded in their careers only for staying securely within them.

The book is organized in six parts:

1. why does fisheries management need reinventing?
2. new policies for a reinvented fisheries management;
3. the role of the social sciences in a reinvented fisheries management;

4. coping with ecology in a reinvented fisheries management;
5. modelling through in a reinvented fisheries management.
6. overview and synthesis.

Issues considered include: understanding of the production base and the ecological impact of harvesting in freshwater and marine ecosystems; assessing fisheries intelligently, quantifying risk and learning to make management adaptive; shaping policy to make fishing responsible and fit both the sustainable limits of the resource and the ambitions of humans; reducing conflict and fostering consensus by understanding fishing communities; mitigating resource depletion through innovative and appropriate economic instruments; and fostering sustainable development and the key research by encouraging the right kind of institutions. By attempting to integrate across these multidisciplinary themes, our ambitious objective is to help create a fresh synthesis and a new paradigm for the management of fisheries.

The book was planned at a symposium of over 100 fishery researchers at the Fisheries Centre, UBC, Vancouver, in February 1996. A number of contributions have been solicited from those who were unable to attend that meeting. All papers have been fully peer reviewed (indeed some fell by the wayside during that process) and revised in the light of comments by referees and discussions at the symposium.

Tony J. Pitcher
Daniel Pauly
Paul J.B. Hart

Fisheries Centre, Vancouver and University of Leicester

Part One

Why Does Fisheries Management Need Reinventing?

Chapter one

World fisheries are in crisis? We must respond!

Rik C. Buckworth

ABSTRACT

World fisheries are reported to be in a crisis. This chapter briefly reviews the evidence for this crisis, underlying problems, and possible responses. Global fishery catches are probably between 130 and 190 million tonnes annually, and are of a similar magnitude to the predicted sustainable catch. A large proportion of fisheries are fully or overexploited. The uncertainty in global limits should not be used to deny the risk that we face. Thus it is imperative that we respond by controlling fishing effort, but in this we are hindered by severe, global overcapacity. This major, structural problem must be addressed.

Many difficulties arise from the uncertainty that pervades the fishery management process, and how it has been dealt with. Uncertainty should be explicitly considered – by imposing safety margins in target fishing levels, by dealing with uncertainty and risk in harvest decisions, and by creating institutional arrangements in which such decisions are transparent and subject to checks and balances. Because there will always be some uncertainty about resource status, and the behaviour of fishers and decision makers, actions must be reversible, management strategies must be safe. We must identify and develop understanding of systems that can sustainably manage resources.

Reinventing Fisheries Management. Edited by Tony J. Pitcher, Paul J.B. Hart and Daniel Pauly. Published in 1998 by Kluwer Academic Publishers, London. ISBN 0 412 83410 3.

1.1 INTRODUCTION

We live in an age of achievement – in technology, in information, and in science. But do we also live in an age where we are reaching the limits of the systems on which we depend, and our skills have outstripped our understanding and ability to manage those systems? In the middle of the last century, the world fish catch was a meagre two million tonnes, and marine resources seemed limitless (McGoodwin, 1990). Just a few generations later, landings from the world's fisheries have increased more than fiftyfold, but we may be in a state of crisis. How bad is the situation, how has it arisen, and what might be done? I argue that world fisheries are near global production limits. Many fisheries are depleted or have collapsed, apparently owing to a mixture of relentless economic pressure and our inability to manage complex, uncertain systems. We must respond: containment of fishing effort on a global scale is thus imperative, and the future lies in finding robust approaches to management and in learning from systems that we have been able to sustain.

1.2 THE FISHERIES CRISIS

Recent issues of magazines for informed readership, such as *Scientific American, National Geographic, The Economist, Nature* and newspapers like *The Australian*, carry articles on the serious problems of the world fisheries. Mistakes of fishery management have been described, for example, as having 'reduced numerous fish populations to extremely low levels, destabilised marine ecosystems and impoverished many coastal communities' (Safina, 1995). This is popular journalism, written to be emotive. But plainly factual publications, produced for example by the US National Marine Fisheries Service (Sissenwine and Rosenberg, 1993), or FAO (Garcia and Newton, 1997), are the more alarming for their sobriety.

There are two main components to the problem. The first is that world catches may be near the limit imposed by oceanic and freshwater productivity. The second component is that many stocks are overfished, ultimately calling into question our ability to manage fisheries sustainably.

World catches

With recorded catches in 1994 near 110 million tonnes, the world fishery supports 200 million people and considerable world trade (Fig. 1.1; Garcia and Newton, 1997). The fishery is of increasing importance to developing nations, with their share of the catch doubling from 35% to 70% between 1960 and 1994 (Fig. 1.1; FAO Fishery Statistics). Most of the increase has

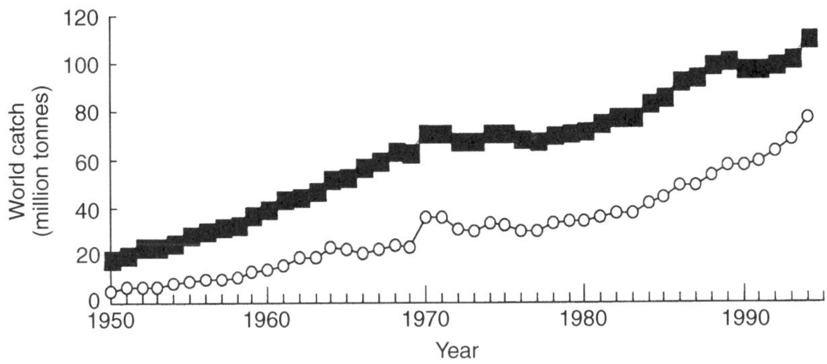

Fig. 1.1 Total world catch (filled squares) and catch from developing countries (circles), 1950 to 1994.
Source: FAO Yearbooks

been in the marine sector, with marine fishery production increasing 300% between 1950 and 1990, from 18.5 million to 82.5 million tonnes (Garcia and Newton, 1997). But the estimates of world catch are not precise. There is also a substantial bycatch – catches of non-target species – from the world's fisheries. In 1992, between 17.9 and 39.5 million tonnes (average 27.0 million) were taken as bycatch and discarded, and so are not represented in the landings figures (Alverson *et al.*, 1994). Thus, the world catch is around 130 million tonnes, at the very minimum. Moreover, the estimate of bycatch did not include many invertebrate, recreational or subsistence fisheries, and data from some regions were sparse. The catch and bycatch estimates are necessarily imprecise, and undeclared catches could be as high as 20–30% of those recorded (Pitcher, 1995). An upper bound on the world catch, allowing for bycatch and undeclared catches, could be around 190 million tonnes.

There are indications in the landings data that the growth in fishery production is near its limit. Strong growth in production through the 1960s to 1980s slowed substantially during the 1990s (Fig. 1.1); the decreased landings in 1990–1992 were cause for alarm (Garcia and Newton, 1997). The increasing catches of the 1960s and 1970s were supported by geographic expansion of fisheries, which has run its course (Alverson and Larkin, 1994). Only minor increases are likely to come from fishing underutilized groups of fish, such as mesopelagics. Most major stocks are now subject to intense fishing. The total world fishing effort in 1989 was near (roughly) estimated sustainable levels (Garcia and Newton, 1997). It can only be concluded that the world fishery is near its sustainable limit.

But how near? Estimates of the potential sustainable yield of the world fishery are common enough, but reviewing a sample of these, Pauly (1996) revealed that many were simplistic, some relying on previous estimates or educated guesses. Only the estimates of 100 million tonnes by Gulland (1971) and 120–150 million tonnes by Moiseev (1994) were based on sufficient data and stratification to be credible. The proportion of global aquatic primary production required to support fisheries has been estimated as 8% (Pauly and Christensen, 1995). This could appear comfortingly small. However, as Pauly and Christensen (1995) note, most primary production is in the extensive but relatively unproductive open ocean (and from which current fisheries appropriate only 2% of primary production); most of the world's fish production is from the freshwater, continental shelf and upwelling systems, where current fishing ultimately appropriates 24–35% of the primary production. These areas are, furthermore, the most subject to environmental damage and reduction of ecosystem diversity (Moiseev, 1994; Pauly and Christensen, 1995). The global catch is thus close to, and may exceed, the more robust estimates of sustainable yield. The very high primary production already required to underpin our fisheries also suggests that the sustainable limit to fisheries must be close. This negates speculation that underutilized groups could provide for substantial increases.

But global assessments disguise the fact that the world fishery is composed of a large number of individual fisheries (Garcia and Newton, 1997). The second component of the fishery crisis is that the potential of the world fishery is being eroded by depletion of individual stocks.

Many stocks are overfished: 22% of the stocks examined by FAO in 1994 were overexploited or depleted; most were fully exploited, while a mere one-third were less than fully exploited, and with this depletion there has been a consequent shift over recent decades to lower-value species (Garcia and Newton, 1997). Production of these lower-value species has disguised the 'slow but steady degradation of the demersal high value resources' (Garcia and Newton, 1997).

The obvious and imperative overall response must be to contain fishing effort. However, the pressures for increase are enormous. Human population growth and economic activity are driving the system. The food fish deficit will grow just as the human population grows: a further 65 million tonnes will be required by the year 2025 (Alverson and Larkin, 1994). Relatively small gains are likely to come through species diversification, aquaculture and direct human use of fishery products now used as animal feed (Pitcher, 1995).

Demand will come from both developed and undeveloped nations. With improving transport and the evolution of global markets, there is opportunity for developing economies to earn foreign exchange and reap the benefits that go with it. The risks are also high. This is exemplified by the huge economic incentives of markets for luxury items, such as live fish, and the damage that responses to these incentives might create. In Hong Kong, a live fish typically attracts a price 400–800% greater than a comparable, dead fish. A fisher in the live fish trade in Indonesia earns 3–10 times the average salary of an artisanal fisher (Erdmann and Pet-Soede, 1996). In the short term, the benefits are obvious. The huge prices for the most favoured species, for example, *Epinephelus*, *Lutjanus* and *Plectropomus* spp., have allowed fishing to remain economic even where fish population densities have already been pushed very low by fishing. The usual fishing method – stunning of fish with sodium cyanide – is not selective, killing non-target species as well as recruits of the target species. In the longer term, sequential, local overexploitations seem almost inevitable. The big losers are in the artisanal fisheries – often the last resort of the landless poor (Pauly, 1994, pp. 112–117).

Underlying problems: effort and uncertainty, and where the risks lie

Containing fishing effort is, of course, a central fishery management objective but attainment of this objective is hindered by overcapitalization: the capability of exerting fishing effort is greater than that needed to use fish resources optimally (Pitcher and Hart, 1982). Overcapitalization is stimulated by competition between fishers for a greater resource share – "more vessels 'racing' to catch fewer fish" (Sissenwine and Rosenberg, 1993). Catching capacity and capital invested in a fishery tend to build up during the good times – the period of fishery development, or periods of good recruitment or prices – in anticipation that better gear or more boats or processing equipment will increase future revenues. But when the fishery stabilizes, or returns to normal, or slumps, government assistance may be provided to prevent unemployment or collapse of investments and infrastructure. On the whole, the overcapacity and subsidies tend to be perpetuated (Hilborn and Sibert, 1988; Ludwig *et al.*, 1993). On a global scale, this phenomenon has become so severe that the cost of fishing exceeds the revenue from harvest, with the deficit effectively offset by government subsidies (Sissenwine and Rosenberg, 1993). Consequently, there is economic pressure to intensify fisheries that may already be harvested above optimum levels, and to create new fisheries. Overcapacity drives rapid redeployment and enables almost instantaneous growth of fishing effort, and has been identified as the most serious factor threatening the continued viability of exploited fish stocks (Mace, 1997).

Containing access to fisheries remains a problem even in larger fishing nations such as the USA (Sissenwine and Rosenberg, 1993; Mace, 1997), and while developed countries move to control access to their fisheries, there is pressure to shift capacity from the developed to developing nations. Furthermore, controlling access to fisheries acts only to reduce rates of increase in capacity: fishers strive to increase their revenues, and compete with each other for the resource. This is well illustrated in Australia's northern prawn fishery (NPF). Declared as a limited-entry fishery in 1980, the NPF has been managed with a series of input controls, including vessel replacement constraints, gear restrictions and seasonal and area closures. Overfishing of tiger prawns (*Penaeus esculentus* and *P. semisulcatus*), which became apparent in 1986, engendered several years of painful adjustments to reduce effort and capacity in the fishery. Despite a 50% reduction of fleet size, and strict imposition of gear and other input controls, steadily increasing fishing power means that further effort reductions will be needed in the near future (Robins *et al.*, 1998).

It is difficult to ascertain exactly what the level of effort or catches in a fishery should be: the maximum potential yield of a resource can be defined only by overexploiting it (Hilborn *et al.*, 1995), and uncertainty pervades all stages of the fishery management process. We cannot predict exactly how the abundance of a resource will vary owing to environmental influences or in response to fishing, our data contain errors, assessment models are gross simplifications, the implementation of management decisions is imperfect, and the behaviour of fishers, industries and decision makers is subject to a variety of external and unaccountable influences. All elements interact. At higher levels of resolution, a fishery system itself is embedded in larger, varying, ecological and economic systems. Our experience and knowledge are always limited. Thus, a fishery is a system that behaves, like any ecological system, in ways that are not always predictable (Ludwig *et al.*, 1993; Walters, 1995).

Two of the more infamous fishery collapses illustrate our difficulty in managing or even acknowledging this web of uncertainty. The Peruvian anchoveta, *Engraulis ringens*, fishery collapsed from more than 12 million tonnes in 1972. Although this was ascribed at the time to unprecedented environmental fluctuation (Boerema and Gulland, 1973), simplistic assessment and management approaches were also major contributors (Glantz, 1981; Hammergren, 1981; Pitcher and Hart, 1982; Hilborn and Walters, 1992).

Although the El Niño phenomenon affects the anchoveta stock by reducing recruitment and survival, and causing stock concentration (Muck, 1989), El Niño events when harvest levels were relatively low, during the 1950s and mid 1960s, had little impact on the fishery. By the early 1970s, however, fishing pressure had grown considerably and age structure of the stock had been eroded, so that combination of the environmental effects and heavy fishing precipitated stock collapse (Glantz, 1981). There had been no previous experience in the fishery of heavy fishing effort combined with the environmental extreme. The fishery was managed under the maximum sustainable yield (MSY) paradigm, which advised the steady increase of effort until a decline in yield signalled that the optimum fishing level had been reached, followed by reduction of fishing effort to the optimum. This simple approach failed to account for an important attribute of the fishery's dynamics: environmental variability in fish production, which implies that the optimum level of fishing effort may change over time. The MSY approach critically failed in that identification of the target fishing level requires fishing beyond optimum levels, where uncertainty as to whether optimum levels have been exceeded can entrench overfishing (Hilborn and Sibert, 1988). Peru became heavily dependent on the fishery: by the time of collapse, the anchoveta fishery provided one-third of Peru's foreign exchange (Glantz, 1981). A government concerned chiefly with boosting economic performance displayed a basic insensitivity to signals from the fishery about the state of the resource (Hammergren, 1981). The risk involved was just not acknowledged. This begs the question, that if the extent of environmental fluctuations affecting Peruvian anchoveta had been known, would the fishery still have collapsed? Entrenched overcapitalization meant that the fishery continued to be overfished even after its collapse (Muck, 1989).

Greater sophistication in assessment approaches and in management structures is, of course, desirable. Nevertheless, it cannot guarantee the safety of the resource, as shown in the collapse of cod stocks of the northwest Atlantic (Hutchings and Myers, 1994; Walters and Maguire, 1996). The fishery was sustainably fished for nearly four centuries, with exploitation limited by available fishing technology. Considerable overexploitation occurred through the 1970s with competition between high-technology, distant-water fleets, when the advent of accurate positioning systems and improved fishing technology made deep-water stocks accessible. The situation could have been redressed with a reduction in harvest rates when or since the fishery came under Canadian control in the early 1980s (Hutchings and Myers, 1994). However, failures of basic assumptions led to optimistic assessments, promoting increased capacity. Difficulties in addressing the uncertainty created by conflicting information in assessments led ultimately to the risky decisions that drove the fishery to commercial extinction by 1992. Despite the large assessment resources available to this fishery, it was unclear even at the time of the collapse, whether environmental fluctuation or overexploitation was at fault (Hilborn *et al.*, 1993). Extensive analyses have since demonstrated that overfishing was certainly the major cause (Hutchings and Myers, 1994; Myers *et al.*, 1996; Walters and Maguire, 1996).

A general objective of fisheries management is that fish resources be harvested sustainably at an optimum level. Under this objective, uncertainty in stock assessments leads to decisions that balance the risk of overfishing, requiring future corrective actions, against the risk of containing a fishery below its economic potential. However, the benefits and risks may not be the same for all stakeholders, or members of a management structure – there is conflict between interests. For example, if the productivity of a fish stock is low, it may be more profitable for operators or other stakeholders to maintain fishing at a high, unsustainable rate over a short period of time – even fishing a stock to extinction – and then to divert their capital investment elsewhere (Clark, 1990). Long-lived, deep-water species, such as orange roughy, *Hoplostethus atlanticus*, can be economically managed only if discount rates are low (Campbell *et al.*, 1993; Smith, 1993). Whales were knowingly overfished (Allen, 1980), simply because there was the economic incentive to do so. Processors involved in the northern cod fishery were probably the least affected by its collapse, yet without pressure from those processors, would the uncertainty in stock assessments have produced greater caution in fishery managers? (Pinkerton and Weinstein, 1995).

Problems arise from the dynamics of responsibility, authority and risk within management structures. In the Atlantic Canadian groundfish fishery of the 1970s, management operated in a top-down framework, vesting most power and responsibility in the Minister of Fisheries, and minimal decision-making power with fishers. This arrangement created pressure to beat the system, rather than a social pressure for adherence to regulations ultimately designed for conservation of the resource (Charles, 1995). A public official may have little personal risk involved if there really is overfishing and, given uncertain scientific advice, inaction and similar risk-prone decisions can be the logical alternative to the vocal opposition that suggestions of catch reductions would create. Thus, there were unheeded warnings of overfishing in the northern cod stock from 1985 until the fishery closed in 1991 (Walters, 1995), and a lack of management response to a decade of scientific warnings in the North Sea herring fishery, before its collapse and closure in 1977 (Saetersdal, 1980).

1.3 RESPONSES AND CHALLENGES

The vital first step to addressing these problems is recognition – that we are near the limits to sustainable fishing, that our assessment and management have not been adequate, and that the uncertainty in this evaluation means that we are in a risky position. There is evidence of recognition: for example, the popular articles described at the beginning of this chapter, the other chapters in this volume, and at the applied level, there are responses such as the FAO guidelines to fishery management (FAO, 1995a,b). Although this is encouraging, we are dealing with a political and social problem: whether such recognition is widely translated into national policies, and subsequently into actions within individual fisheries, will be a test of how much recognition the fishery crisis is really accorded. Only then can there be any optimism for the future of world fisheries.

The problems described above manifest themselves at several levels of resolution. At the level of the individual fisheries, overcapacity is a common result of economic forces acting within a fishery. It impairs economic performance and makes it difficult to reduce fishing effort. It is usually in the interests of fishers, at the individual level, to increase their harvesting power. Overcapacity is most readily managed at the inception of a fishery, rather than at a later stage (Hilborn and Sibert, 1988), within structures that remove that individual economic incentive, such as property rights (Scott, 1989). Containing capacity is thus given prominence in FAO's *Precautionary Approach to Fisheries* (FAO, 1995b), wherein management strategies that include access limitation and capping of capacity are advised. Yet the nasty effect that overcapacity generates – effort cascading from fishery to fishery – acts at a higher level of resolution and has serious consequences at regional and even global levels. Mechanisms to control this problem must be developed; it is important that policies to contain and reduce existing capacity should not merely result in the export of excess capacity. Controlling this problem will remain a major challenge to global fisheries management.

Management of uncertainty also requires recognition of problems at different levels of resolution. Within a well-controlled management system, it is possible to improve fishery management by shrinking the uncertainty about resource status. Thus we can establish more effective information-gathering systems (Schnute and Richards, 1994) and develop more powerful analytic techniques (Richards and Megrey, 1994). Management targets can be made safer by explicitly acknowledging the uncertainty in estimates of exploitation rates or stock size, by using this uncertainty to reduce target harvest rates relative to safety limits (Caddy and McGarvey, 1996), or using the coefficient of variation in a stock size estimate to create safety margins in a total allowable catch (Walters and Maguire, 1996). This approach clearly underlines the level of uncertainty in such estimates. Moreover, the incentive to reduce uncertainty – the value of information – is clearly emphasized to fishing interests and other parties included in the management decision process. The uncertainty and risk can be considered explicitly in decision systems (Hilborn *et al.*, 1993; Rosenberg and Restrepo, 1994), and the 'burden of proof' criterion applied in a precautionary manner: where there is uncertainty, decisions must favour conservation of the resource (Peterman, 1990; FAO, 1995b; Walters, 1995). An adaptive approach (Walters, 1986; Hilborn and Sibert, 1988) would provide for learning from responses of the stock to harvest, and building flexibility and reversibility into fishery development, so that the response of the whole system to different levels of utilization could be observed, yet effort could be reduced as required.

However, no amount of research and monitoring can dispel all uncertainty in assessments. Although the approaches above certainly address elements of the problems that plague fishery management, behaviour of the fishery system is not completely predictable: we are unable to forecast ecological effects beyond our experience, or to predict or control all the motives and influences on decision makers. This irreducible uncertainty (Ludwig *et al.*, 1993) can be addressed only by ensuring that actions are reversible, and by identifying and adopting management systems that are safe. And what approaches could be suitable for fisheries where the scientific and regulatory infrastructure are lacking, or for small fisheries where revenues and sampling power might not support a large investment in research and assessment? Such fisheries support most of the world's fishers (McGoodwin, 1990).

The creation of refuges is a simpler approach that might be suitable for many fisheries. It avoids the need for continuous heavy investment in monitoring and assessment to ensure the sustainability of fished resources. By 'playing it safe' – protecting a substantial proportion of a fished stock from exploitation – the approach may effectively avoid the problem of uncertainty in resource status, and so, perhaps, in institutional behaviour. In some fisheries, areas are closed to fishing to enhance catch value and population reproductive potential – for example, in the prawn fisheries of northern Australia (Die and Watson, 1992; Somers and Wang, 1997). Safety for many stocks has probably been provided by areas of habitat in which fishing has been uneconomical or technically limited. Difficulties in accessing cod offshore and in deep water probably conferred safety on the Newfoundland stocks until technical innovation overcame these barriers (Walters and Maguire, 1996). Research so far suggests that the use of permanently closed areas as refuges is robust to both environmental fluctuation and harvest (Polacheck, 1990; Dugan and Davis, 1993). Refuge-based approaches will require research into the design of reserves with respect to the life history of fished species, and development of assessment methods for fisheries based upon them (Dugan and Davis, 1993; Walters and Maguire, 1996). Management issues would be the location and size of closed areas, and ways in which closures might be enforced.

However, reducing or allowing for uncertainty about fishery status addresses only one or two of the weak links in the chain. We need to develop management systems that are more reliable and, therefore, we must address the uncertainty that arises in the ways that management decisions are made. We are thus dealing with the need for better institutions and approaches to management, rather than the need for better science. Two very different types of responses have been suggested. One response is based on the premise that typical management structures have not ensured that decision makers are necessarily aware of risk entailed in their decisions, yet explicit consideration of risk should be an important aspect of decision making (Hilborn *et al.*, 1993; FAO, 1995a; Caddy, 1997). The approach is, therefore, to place the basis of decisions under scrutiny, by addressing the structure and procedures of management bodies, and including systems that impose checks and balances. This could be accomplished with two-tiered management structures, where the top tier of management is responsible for establishing the long-term objectives and mechanisms of management – creating the decision rules by which total allowable catches (TACs), for example, might be calculated. The next tier addresses the short term, implementing those decision rules (Butterworth *et al.*, 1992; Caddy, 1997). But would this address all the fishery management system problems?

In contrast, another body of thought seeks to recognize that the problems of the behaviour of management systems are fundamental, owing to dislocation of incentive, risk, and responsibility within those systems. The response would be to radically change the fishery management systems that are in place in most developed countries. Hence management would move toward use and property rights systems (Scott, 1989), even toward more traditional community-based systems, in which resource users would be responsible for much of the management function (Hilborn *et al.*, 1995). Such approaches change the underlying economic and social incentives for resource users. Proponents argue that we can learn from community-based systems, where the essence of sustainability is the community view that they have not only the right to manage the resource properly, but also the duty (Pinkerton and Weinstein, 1995). The challenge here is to identify and extensively apply the important features of such systems, maintaining that essence.

1.4 CONCLUSIONS

We are confronted with the classic fisheries problem, on a global scale: the world fishery is near its sustainable limit, under pressure to expand, and our decisions are clouded by uncertainty. The uncertainty in global limits must be understood as a danger, not used to deny the risk that we face. The pressure to feed the world's growing population, and make money, will intensify, not abate. Thus, it is imperative that the need to control fishing effort be recognized. Global overcapacity is the immediate problem to be addressed, not exported.

Despite our improvements in information gathering and assessment methods – our technology, information and science – there will always be uncertainty in the status of fished stocks, and in the responses of fishery control systems. Wherever there is uncertainty, there is risk; we must acknowledge and deal with the risk. We must identify and develop understanding of *systems* that can sustainably manage resources, systems that are robust to uncertainty, and safe. Developing these reliable management systems is the central challenge faced in fishery science and management today.

ACKNOWLEDGEMENTS

This chapter grew from course work and discussions at the Fisheries Centre, University of British Columbia, and I am grateful for the encouragement and discussions with Drs D. Pauly, T. J. Pitcher and C. J. Walters, and graduate students at the Centre. I also thank Mr D. Grey, Ms J. Lloyd, Mr F. O'Gara, Ms D. Pinch and Dr D. Ramm, of NT Department of Primary Industry and Fisheries (DPIF), and Dr R. Brown of Writeway Consulting, Brisbane, for their manuscript suggestions. This work was partly supported by a DPIF Study Award.

REFERENCES

Allen, K.R. (1980) *The Conservation and Management of Whales*, Univ. Washington Press, Seattle, 107 pp.

Alverson, D.L. and Larkin, P.A. (1994) Fisheries: fisheries science and management – century 21, in *The State of the World's Fisheries Resources* (Proc. World Fish. Congr., Plenary Sessions) (ed. C.W. Voigtlander), Oxford and IBH Publishing Co. Pty Ltd, New Delhi, pp. 150–167.

Alverson, D.L., Freeberg, M.H., Murawski, S.A. and Pope, J.G. (1994) A global assessment of fisheries bycatch and discards. *FAO Fish. Tech. Pap.* **339**, 233 pp.

Boerema, L.K. and Gulland, J.A. (1973) Stock assessment of the Peru anchoveta (*Engraulis ringens*) and management of the fishery. *J. Fish. Res. Bd Can.* **30**, 2226–2235.

Butterworth, D.S., Punt, A.E., Bergh, M.O. and Borchers, D.L. (1992) Assessment and management of South African marine resources during the period of the Benguela ecology programme: key lessons and future directions, in *Benguela Trophic Functioning* (eds A.I.L. Payne, K.H. Brink, K.H. Mann, and R. Hilborn) *S. Afr. J. Mar. Sci.* **12**, 989–1004.

Caddy, J.F. (1997) Checks and balances in the management of marine fish stocks: organizational requirements for a limit reference point approach. *Fish. Res.* **30**, 1–15.

Caddy, J.F. and McGarvey, R. (1996) Targets or limits for management of fisheries? *N. Am. J. Fish. Manage.* **16**, 479–487.
Campbell, H.F., Hand, A.J. and Smith, A.D.M. (1993) A bioeconomic model for management of orange roughy stocks. *Mar. Res. Econ.* **8**, 261–267.
Charles, A.T. (1995) The Atlantic Canadian groundfishery: roots of a collapse. *Dalhousie Law J.* **18**, 65–83.
Clark, C.W. (1990) *Mathematical Bioeconomics: The Optimal Management of Renewable Resources*, 2nd edn, Wiley–Interscience, New York.
Die, D.J. and Watson, R.A. (1992) A per-recruit simulation model for evaluating spatial closures in an Australian penaeid fishery. *Aquat. Liv. Resour.* **5**, 145–153.
Dugan, J.E. and Davis, G.E. (1993) Applications of marine refugia to coastal fisheries management. *Can. J. Fish. Aquat. Sci.* **50**, 2029–2042.
Erdmann, M.V. and Pet-Soede, L. (1996) How fresh is too fresh? The live reef food fish trade in Eastern Indonesia. *Naga.* **19**(1), 4–8.
FAO (1995a) *Code of Conduct for Responsible Fisheries*, FAO, Rome, 15 pp.
FAO (1995b) Precautionary approach to fisheries. Part 1: Guidelines on the precautionary approach to capture fisheries and species introductions. Elaborated by the Technical Consultation on the Precautionary Approach to Capture Fisheries (Including Species Introductions). Lysekil, Sweden, 6–13 June 1995 (A scientific meeting organized by the Government of Sweden in cooperation with FAO). *FAO Fish. Tech. Pap.* no. 350, Part 1, 52 pp.
Garcia, S.M. and Newton, C. (1997) Current situation, trends and prospects in capture fisheries, in *Global Trends in Fisheries Management* (American Fisheries Society Symp. 20) (eds E.K. Pikitch, D.D. Huppert and M.P. Sissenwine), AFS, Bethesda, MD, pp. 3–27.
Glantz, M.H. (1981) Consideration of the societal value of an el Niño forecast and the 1972–1973 el Niño, in *Resource Management and Environmental Uncertainty: Lessons from Coastal Upwelling Fisheries* (eds M.H. Glantz and J.A. Thompson), John Wiley and Sons, New York, pp. 449–476.
Gulland, J.A. (1971) Summary, in *The Fish Resources of the Ocean* (ed. J.A. Gulland), Fishing News Books, West Byfleet, Surrey, pp. 246–255.
Hammergren, L.A. (1981) Peruvian political and administrative responses to el Niño, in *Resource Management and Environmental Uncertainty: Lessons from Coastal Upwelling Fisheries* (eds M.H. Glantz and J.A. Thompson), John Wiley and Sons, New York, pp. 317–350.
Hilborn, R. and Sibert, J. (1988) Adaptive management of developing fisheries. *Mar. Policy* **12**, 112–123.
Hilborn, R. and Walters, C.J. (1992) *Quantitative Fisheries Stock Assessment. Choice, Dynamics and Uncertainty*. Chapman & Hall, New York and London, 570 pp.
Hilborn, R., Pikitch, E.K. and Francis, R.C. (1993) Current trends in including risk and uncertainty in stock assessment and harvest decisions. *Can. J. Fish. Aquat. Sci.* **50**, 874–880.
Hilborn, R., Walters, C.J. and Ludwig, D. (1995) Sustainable exploitation of renewable resources. *A. Rev. Ecol. Syst.* **26**, 45–67.
Hutchings, J.A. and Myers, R.A. (1994) What can be learned from the collapse of a renewable resource? Atlantic cod, *Gadus morhua*, of Newfoundland and Labrador. *Can. J. Fish. Aquat. Sci.* **51**, 2126–2146.
Ludwig, D., Hilborn, R. and Walters, C.J. (1993) Uncertainty, resource exploitation, and conservation: lessons from history. *Science* **260**, 17–36.
Mace, P.M. (1997) Developing and sustaining world fisheries resources: the state of the science and management, in *Developing and Sustaining World Fisheries*

Resources: The State of Science and Management (Second World Fisheries Congress, Brisbane 1996) (eds D.A. Hancock, D.C. Smith, A. Grant and J.P. Beumer), CSIRO Publishing, Collingwood, Victoria, pp. 1–20.

McGoodwin, J.R. (1990) *Crisis in the World's Fisheries. People, Problems and Policies*, Stanford Univ. Press, Stanford, CA, 235 pp.

Moiseev, P.A. (1994) Present fish productivity and bioproduction potential of world aquatic habitats, in *The State of the World's Fisheries Resources* (Proc. World Fish. Congr., Plenary Sessions) (ed. C.W. Voigtlander), Oxford and IBH Publishing Co. Pty Ltd, New Delhi, pp. 70–75.

Muck, P. (1989). Major trends in the pelagic ecosystem off Peru and their implications for management, in *The Peruvian Upwelling System: Dynamics and Interactions* (ICLARM Conf. Proc. **18**) (eds D. Pauly, P. Muck, J. Mendo and I. Tsukayama), International Center for Living Aquatic Resources Management (ICLARM), Manila, pp. 386–403.

Myers, R.A., Barrowman, N.J., Hoenig, J.M. and Qu, Z. (1996) The collapse of cod in eastern Canada: the evidence from tagging data. *ICES J. Mar. Sci.* **53**, 629–640.

Pauly, D. (1994) *On the Sex of Fish and the Gender of Scientists. A Collection of Essays in Fisheries Science*, Chapman and Hall, London, 250 pp.

Pauly, D. (1996) One hundred million tonnes of fish, and fisheries research. *Fish. Res.* **25**, 25–38.

Pauly, D. and Christensen, V. (1995) Primary production required to sustain global fisheries. *Nature* **374**, 255–257.

Peterman, R.M. (1990) Statistical power analysis can improve fisheries research and management. *Can. J. Fish. Aquat. Sci.* **374**, 255–257.

Pinkerton, E., and Weinstein, M. (1995) *Fisheries that Work. Sustainability through Community-based Management: A Report to the David Suzuki Foundation*. The David Suzuki Foundation, Vancouver, 199 pp.

Pitcher, T.J. (1995) Stewardship and the sustainability of Pacific fishery resources: the need for critical insight and an encyclopedia of ignorance, in *Our Common Shores & Our Common Challenge. Environmental Protection of the Pacific*. Proc. Fourth Int. Symp. Conf. Asian and Pan-Pacific Univ. Presidents, Anchorage, Alaska, September 1993 (ed. D.G. Shaw), Alaska Sea Grant College Program, Univ. Alaska Fairbanks, Fairbanks, AK, pp. 5–27.

Pitcher, T.J. and Hart, P.J.B. (1982) *Fisheries Ecology*, Chapman and Hall, London, 414 pp.

Polacheck, T. (1990) Year around closures as a management tool. *Nat. Resour. Mod.* **4**, 327–354.

Richards, L.J. and Megrey, B.A. (1994) Recent developments in the quantitative analysis of fisheries data. *Can. J. Fish. Aquat. Sci.* **51**, 2640–2641.

Robins, C.M., Wang, Y.-G. and Die, D. (1998) The impact of global positioning systems and plotters on fishing power in the northern prawn fishery, Australia. *Can. J. Fish. Aquat. Sci.* (in press).

Rosenberg, A.A. and Restrepo, V.R. (1994) Uncertainty and risk evaluation in stock assessment advice for U.S. marine fisheries. *Can. J. Fish. Aquat. Sci.* **51**, 2715–2720.

Saetersdal, G. (1980) A review of past management of some pelagic stocks and its effectiveness. *Rapp. P.-v Réun. Cons. int. Explor. Mer* **177**, 505–512.

Safina, C. (1995) The world's imperilled fish. *Scient. Am.* **273**(5), 46–53.

Schnute, J. and Richards, L.J. (1994) Stock assessment for the 21st century. *Fisheries* **19**(11), 10–16.

Scott, A.D. (1989) Conceptual origins of rights based fishing, in *Rights Based Fishing* (eds P.A. Neher, R. Arnason and N. Mollett), Kluwer, Dordrecht, pp. 11–38.

Sissenwine, M.P. and Rosenberg, A.A. (1993) Marine fisheries at a critical juncture. *Fisheries* **18**(10), 6–13.

Smith, A.D.M. (1993) Risks of over- and under-fishing new resources, in *Risk Evaluation And Biological Reference Points For Fisheries Management* (eds S.J. Smith, J.J. Hunt and D. Rivard), *Can. Spec. Publ. Fish. Aquat. Sci.* **120**, 261–267.

Somers, I.F. and Wang, Y.-G. (1997) A simulation model for evaluating seasonal closures in Australia's multispecies northern prawn fishery. *N. Am. J. Fish. Manage.* **17**, 114–130.

Walters, C.J. (1986) *Adaptive Management of Renewable Resources*, MacMillan, New York, 374 pp.

Walters, C.J. (1995) *Fish on the Line: The Future of Pacific Fisheries. Report to the David Suzuki Foundation, Phase I.* The David Suzuki Foundation, Vancouver, 82 pp.

Walters, C.J. and Maguire, J.-J. (1996) Lessons for stock assessment from the northern cod collapse. *Rev. Fish Biol. Fish.* **6**, 125–137.

Chapter two

Reinventing the tree: reflections on the organic growth and creative pruning of fisheries management structures

Nigel Haggan

ABSTRACT

Fisheries management is seen as the result of a process of organic growth and accretion driven by the realization that human ingenuity has outstripped the productive capacity of fisheries. The evolution of the symbiotic scientific and management communities is traced from an expansionist period in the 1950s and 1960s to today's climate of depletion. Stakeholder trust in the quality of the information that underpins management decisions, and a guarantee of continuing benefits, are proposed as the key elements of sustainability. Inclusivity, perspective and clarity are identified as key contributors to the perception of fairness, which alone can ensure that fishers comply with management systems.

The role of fishing communities, science, government, the public and other interests in policy making is discussed. The reinvention of fisheries management cannot be entrusted to a new partnership, however comprehensive, among those directly involved in fisheries. Fish and fisheries must

Reinventing Fisheries Management. Edited by Tony J. Pitcher, Paul J.B. Hart and Daniel Pauly. Published in 1998 by Kluwer Academic Publishers, London. ISBN 0 412 83410 3.

somehow capture the support of the general public, to whom political decision makers respond. The preservation of fisheries values must become a vital consideration for industry. One way to approach this is to stop selling dead fish and start selling fish as a metaphor for the quality of life. Ecosystem approaches, interdisciplinary cooperation, dialogue with fishers and fishing communities and the incorporation of local ecological knowledge, may yet enable fisheries management to haul itself out of the water like Darwin's fish and peer blearily at the fresh environment. The next evolutionary challenge is to integrate human communities into the model as more than just a component of fishing mortality.

2.1 INTRODUCTION

This chapter takes a slightly irreverent look at the notion of 'reinventing' fisheries management. It argues that fisheries management never was invented, it simply grew like a tree as the realization dawned that human intervention could drive fish stocks to extinction. The first shoot on Canada's west coast sprang with declining salmon catch. It has since sprouted almost as many roots and branches as there are stocks and species, and even more when stock assessment, management, habitat and enforcement are treated separately by organizations such as Canada's Department of Fisheries and Oceans (DFO).

The 'trees' of Europe are much, much older. Indeed, there is a small, but vocal group which claims that the Canadian tree is a European transplant. Seeds from this ancient stock are now sprouting throughout the developing world. Whether to continue to fertilize with Western science, uproot in favour of local varieties or cross-pollinate is a subject of lively debate.

As evidenced by any established church, or indeed, university, institutions are much more resistant to change than the people who created them back in the mists of time. Laws, regulations and beliefs accrete over time like oxides and marine organisms on an ancient seawreck. In Canada, bilateral consultative processes have now started to accrete at an explosive rate. Our collective ability to 'reinvent' fisheries management hinges on retaining elements of value while deconstructing existing management structures and integrating them into new approaches and tools.

The age of innocence

'There are plenty of fish in the sea.' Oral history abounds with tales of abundance. The Hanseatic League, a vast Middle Ages trading empire, was founded on the amazing productivity of the Baltic Sea herring (Cushing, 1988). Old people in Ireland and Elders of British Columbia (BC) First

Nations alike, tell of crossing rivers dryshod on the backs of returning salmon. There was little need for management in those early days because marine productivity was far in excess of catching capacity. Early government documents on the BC fishery do little more than marvel at the abundance, suggest ways to exploit it and present figures on annual catch (British Columbia, 1902). Ironically, in view of their later exclusion, an earlier document (BC, 1875) recommends that 'Indians' take the lead in the commercial fishery.

A good time for scientists

The growth of local and world markets for fish led to rapid and significant improvements in catching and processing ability. Newell (1988, 1989) describes this process in the development of the BC fishery. Newell (1993) also provides a good account of the impact of modern industrial fishing and fisheries legislation on the Aboriginal fishery.

Meryl Williams (Chapter 11, this volume) characterizes this as a time when fisheries scientists were naturalists. They spent their time identifying species and examining life cycle stages. Above all, they worked with fishers. They went out on fishing vessels exploring new grounds, developing fish-finders, testing and developing new gear. Scientists were welcome aboard. A good time was had by all. Government agencies virtually threw subsidies and low-interest loans at fishers and processors. Vessels grew in size and power. Fishing became increasingly lucrative.

The age of experience

Powerful vessels, sophisticated fish-finding and catching gear and onboard processing allowed fishing at great distances and in weather and sea conditions that would have tied up earlier fleets. Fish were found and hunted throughout their range. Walters and Pearse (1996) cite access to offshore spawning aggregations through new technology as a major causative factor of stock collapse. BC abalone were abundant until the 1970s when the subtidal population was opened to a dive fishery. In a few years, spawning biomass was reduced to a point where the entire fishery, including the Aboriginal component, was closed (Parsons, 1993). North Sea plaice stocks, which had recovered during World War II, were fished down. To counter this, a juvenile rearing area or 'plaice box' was set aside off the Dutch coast (Rijnsdorp and Millner, 1996). What this failed to do was to protect adults. As Walters and Maguire (1996) note regarding the northern cod collapse in Atlantic Canada, natural or established refugia are a key component of sustainability. It is clear that these refugia must exist both for adults and for juveniles.

Increasing pressures on fish stocks between the 1970s and early 1990s mark the time when the 'naturalist' was upstaged by the statistician and stock modeller. Computer models, stock assessment, harvest rates and quota setting became the order of the day. In recent years, the mathematical approach has come into question. The collapse of the northern cod and the masking of declines in real abundance by stable or increasing CPUE (catch per unit effort) and incorrect estimates of average age, as in the Pacific Ocean perch fishery (Richards and Schnute, Chapter 30, this volume), are significant factors in this falling off among the faithful. Walters and Pearse (1996) and Walters and Maguire (1996) provide a good account of the problems which arise when fixed quotas or harvest rates are based on inadequate or flawed information on what is happening under the surface. The lack of accountability which allows political pressures to override such scientific advice as does exist is also a significant factor. Walters (1995) discusses these pressures at length.

Community outrage and political heat

One important effect of a major stock collapse is to bring fisheries management and science into disrepute. This is most obvious amongst those fishing communities (local or gear-type) that are affected. The community outrage provoked by loss of livelihood and lifestyle has prompted a move to get social scientists to clean up the mess left by biologists, economists and mathematicians. This chore is rejected by some social scientists, who feel strongly that fisheries scientists have not reciprocated their effort to learn the language and tools of fisheries science. Jentoft, Chapter 12, this volume and Maguire, Chapter 18, this volume, also take a strong position on involving communities in the dialogue. It is less clear, at least to this writer, how one group of scientists learning another's secret language will contribute to grassroots understanding of fisheries management.

To summarize, over the last 30 years, many of the world's great fisheries have gone from abundance to depletion. Over the same period, fisheries science has cycled from the 'generalist' approach of Meryl Williams' naturalist, through a long affair with economics and mathematics, to today's tendency to look to social scientists and fishing communities for answers. Ironically, a bad time for fish is now a good time for social scientists, at least in terms of research funding.

Pruning the tree

Stock depletion affects the livelihood and lifestyle of communities. Equally importantly, it reduces or negates returns to the country as a whole. Mitigating community impact from the northern cod collapse has cost Canada at least $3 billion to date (Canada, 1997). Science and management are most needed when stocks are depleted or endangered. Today, however, many governments are in a lean, tight-fisted mode compared with the expansionist times of the 1960s and 1970s. The days are over when bad news for fish was good news for the fisheries science budget. This vicious cycle, where depletion demands more management and a greater need for science at a time when financial returns are down, is a major consideration for policy makers and scientists.

2.2 FACTORS AFFECTING SUSTAINABILITY

As industry, government and science struggle with the transition from abundance to scarcity, the main forces working against sustainability are:

- lack of information;
- declining benefits and rising costs.

These factors lead inevitably to a breakdown in voluntary compliance. Agreement on conservation and management is unlikely in the absence of quality information. Lack of trust in the source of the information sabotages any chance of broadly based agreement. No agreement will hold when benefits decline and costs rise.

Lack of information

Without information on fish stocks and habitat, there can be no credible management planning. There can be no credible assessment of negative impact by competing human activity. If there is no baseline information on stock strength and habitat quality, there is no yardstick to maintain environmental standards or to protect the quality and quantity of water on which all life depends.

At least in BC, failure to implement a transition phase from rigid government control to cooperative management between government and stakeholders is by far the most immediate threat to sustainable fisheries. Failure to maintain data-gathering capability and confidence in the data is a major roadblock to voluntary compliance among historic and emerging stakeholders. The irony is that much of the information needed to manage any fishery does exist. The problem is that it is compartmentalized in the realm of science, in the databases of government agencies and in the heads and logs of fishers. The integration of local ecological knowledge and mainstream science is one of the major challenges of a new science and practice of fisheries management.

Managing the rate of decline

Pauly (1995) observes that fisheries scientists tend to view the productive potential of a stock as the amount of fish when they started out their scientific career. The perception of productive capacity thus 'ratchets down' with successive generations. This concept is well understood by BC's Aboriginal peoples, whose rich cultures and economic wealth were founded and sustained by abundant fisheries resources. As Percy Walkus of the Oweekeno Nation observed in a 1992 meeting with DFO at which the writer was present, 'All that DFO are doing is managing the rate of decline.'

Take the money and run

The perception that stocks are unlikely to recover, coupled with a lack of political will to face up to the costs of rebuilding, makes it likely that the last of a depleted stock will be 'mined out.' The problem is particularly acute when large sections of a fishery are owned by diversified corporations whose 'stake' in the fishery may well be modest compared with their other business interests. The temptation to catch the last fish and reinvest the money in other, unrelated areas is enormous. The larger problem that many fish stocks provide a lower annual rate of return than other business investment is also an incentive to 'mine them out' and invest the proceeds elsewhere.

Breakdown in the perception of fairness

The commercial sector has been the only significant player in the recent history of the BC salmon fishery. Traditional Aboriginal fisheries were legislated out of existence (Newell, 1993). The sport sector had not emerged as a significant player in the allocation stakes. There was no 'public interest' separate from government for environmentalists to claim as a constituency. The system worked, not so much because DFO had enough of a police presence, as because industry, the *de facto* sole client, self-policed because they perceived the system as fair and equitable to them. To illustrate this, Pearse and Larkin (1992) cite DFO's catch breakdown for the preceeding 5 years as 92.3% commercial, with Aboriginal and sport fisheries at 3.4% each.

The perception of fairness in the BC salmon fishery has broken down within the last decade. The problem of diminishing returns is exacerbated by friction between Canada and the US over interpretation of the 1985 Pacific Salmon Treaty; allocation issues between the three commercial gear types of seiners, gill netters and trollers, increasing allocation to the sportfishing industry; and the re-emergence, through the 1982 *Constitu-*

tion Act (Canada, 1982), the 1990 Sparrow decision of the Supreme Court of Canada (Regina v Sparrow, 1990) and the new Treaty process (British Columbia Claims Task Force, 1991), of a significant, but as yet unquantified, Aboriginal interest in the fishery. The drive to redress past social injustice in South Africa by re-allocating fisheries provides an interesting correlative to the re-emergence of First Nations in the BC fishery (Cochrane and Payne, Chapter 5, this volume). It is evident that, without the perception of fairness, there is little hope that fishers will police themselves.

2.3 TOWARDS SUSTAINABLE FISHERIES MANAGEMENT

Reaching for a common vision

People need security of benefits if they are to work together for any length of time (Scott, Chapter 15; Hart, Chapter 17; Hart and Pitcher, Chapter 16; all in this volume). When stock levels are low, resource rebuilding holds out the best hope because most fish stocks recover much faster than other natural resources such as trees, not to mention oil reserves and ore bodies. The notion of 'primal abundance' (Pitcher and Pauly, Chapter 24, this volume) provides some hope of escape from squabbling over the crumbs on today's allocation table. Success in defining stock production levels before modern industrial fishing began, could provide a common vision. To argue that this is not achievable misses the point that a vision is something worth striving for, something to keep in mind when the going gets tough.

The trick for the players in any fishery is to agree on process and achievable steps. The process and steps cannot ignore the 'politician's dilemma' of balancing long-term conservation against the immediate needs of voters and/or powerful interests. If fisheries management cannot become workable, affordable and acceptable, the late Peter Larkin's axiom 'you can't get there from here' will be its epitaph (Larkin, 1979).

As a first step toward policies for sustainable fisheries, two deceptively simple questions can be posed. First, what are the elements of a sustainable fishery? Second, who are the players who need to be involved in policy making?

What are the elements of a sustainable fishery?

While it is neither possible nor desirable to create a rigid formula for the world's fisheries, the following criteria are put forward for consideration:

- a whole-ecosystem approach to fish, fishers and environment;
- understanding and maintaining biodiversity;
- maintaining aquatic ecosystems at their full productive potential;
- a continuing flow of high-quality benefits to stakeholders;
- public awareness of the nature of the resource and flow of benefits;
- harmonization with 'competing' resource sectors;
- clear understanding of macroeconomic and other external factors.

Who should participate in policy development?

Until recently, policy making has been a prerogative of central government. The fleet expansion/stock depletion cycle was spawned by deliberate government policies. New policies must speak to the hearts and minds of consumptive and non-consumptive interests. The bare essentials of policy making are *inclusivity, perspective* and *clarity*. The previous list of criteria gives some indication of the players who need to be at the policy table. Inclusivity is also the best assurance that all the necessary perspectives will be available. Inclusivity, perspective and clarity are also important contributors to the perception of fairness.

Inclusivity

Fisheries policy decisions involve people, fish and the environment. People can represent themselves. The question of who speaks for fish and environment is much thornier. The old answer was government, but questions persist about the relationship between government and industry. Trade-offs between different resource sectors and other countries are also an issue. Another answer is by one or more of the organizations who have appointed themselves as the conscience of the environment. Not everyone is happy with that. Yet another answer is by the fishers themselves; after all, who has most to lose? In the absence of consensus, the developed world has reverted to the animal trials of mediaeval Europe (where, for example, rats could be hauled into court for damaging a grain crop). The US *Endangered Species Act* and the spotted owl is a case in point. There are two problems with this. Courts are a 'win or lose' option. Court decisions are also pitifully short on direction or resources for implementation. The principle of inclusivity seeks to transcend the bilateral or dualistic processes that have dogged fisheries management – people vs. fish, one gear type vs. another, logging vs. fishing, environment vs. people, and so forth. The principle of inclusivity requires, if not the presence, then at least the input of every interest, whether consumptive or non-consumptive. No doubt this will create cumbersome bodies and tedious processes, but is this any worse than isolated lobbies competing for public support and access to decision makers?

Perspective

The principle of inclusivity provides the best guarantee that considerations such as the public interest, macroeconomic factors and competing resource industry are taken into account. These larger perspectives provide essential context for the more 'local' perspectives of biology, mathematics, social science and the local knowledge of fishers and their communities. Each perspective, whether broad or narrow, can be seen as a lens. The challenge in any particular fishery is to ascertain that all the lenses are present before trying to get the right focus. This will require the participation of skilled facilitators and mediators, at least until the participants get to know and understand each other.

Clarity

If you cross a bureaucrat with a scientist you get a policy you can't understand. Good policy speaks directly to the hearts and minds of those for whom it is designed. The goal is clear statements in simple terms, illustrated, where possible, by relevant examples. The secret languages of science are virtually impenetrable, even between disciplines. The principle of clarity requires the translation of these secret languages. This demystification need not trivialize science, nor talk down to fishers and their communities. It is rather a broadening of the scope of science to include non-traditional data sources and the intuition of those who live by the water.

Workability – the perception of fairness

While inclusivity satisfies some of the perception of fairness, it does not guarantee voluntary compliance amongst fishers. 'Cops and robbers' systems do not work without the will to comply. As a simple example, it is the consensus amongst drivers, pedestrians and cyclists to obey a set of rules that keeps us alive on the roads of the world, not the number of police. Voluntary compliance can exist only in a system that is perceived to be fair. Although fairness is essentially subjective, there can be no doubt that stakeholder participation in the design and delivery of a management regime is a vital contributor. The following elements can be identified:

- stakeholder participation in system design;
- agreement on cost-sharing formulas for management and, where necessary, rebuilding;

- stakeholder participation in information gathering and analysis;
- assurance that the system provides the best guarantee of sustainable benefits;
- a system for ensuring a fair distribution of benefits.

Affordability

One key to affordability is to transfer the responsibility for data collection from government to the fishing community. They know where the fish are. They know the local conditions. They have better boats. Above all, there are more of them. The best of all arguments is that fishers will put faith in data they have a hand in collecting. This should not let government off the hook. Profitable fisheries generate downstream employment and new wealth for the nation. Government plays a role in generating economic development and has a responsibility to re-invest revenue from taxes paid by fishers in the resource. Government also has a responsibility where stocks and/or habitat have been affected by external activity. That said, the relative contributions need to be defined. As a bare minimum, government has a responsibility to lead or participate in the design of science and to audit the results of management.

2.4 NEW CONTEXTS, NEW TOOLS

'When you open a can of worms, the only way to recan it is to use a bigger can' (Caesar, *De Bello Gallico*). This thought prompted Julius Caesar to create the Roman Empire, if not the largest, then at least one of the most durable cans in human history. The Roman Empire is relevant to the reinvention of fisheries management in that it brought diverse elements under one law.

The Roman Empire differs from late 20th century fisheries management in one important respect. Caesar is dead. The late 20th century emphasizes horizontal processes such as collaboration, partnership and community empowerment. The very nature of fish and the water in which they live demands an extraordinary level of dialogue and collaboration between resource users. As an example, BC salmon management involves the logging, mining and power generation industries, farmers, urban development, commercial, sport and Aboriginal users and interaction with the US and other Pacific Rim countries. Although the Pacific salmon is a valuable resource, it pales into insignificance when compared with several of the above-noted sectors.

The reinvention of fisheries management cannot therefore be entrusted to a new partnership, however comprehensive, between those directly involved in fisheries. Fish and fisheries must somehow capture the support of the general public to whom political decision makers respond. The preservation of fisheries values must become a vital consideration for 'competing industry'. One way to approach this is to stop selling dead fish and start selling fish as a metaphor for the quality of life.

Marine protected areas can play an important part in raising human awareness of fisheries values (Sumaila, Chapter 23, this volume). Ecopath II (Christensen and Pauly, 1992) and Ecosim (Walters *et al.* 1997) are important steps towards understanding whole ecosystems. Equally importantly, they are a bridge to local environmental knowledge (Haggan, 1996).

Ecosystem approaches, interdisciplinary cooperation, dialogue with fishers and fishing communities and the incorporation of local ecological knowledge, may yet enable fisheries management to haul itself out of the water like Darwin's fish and peer blearily at the fresh environment. The next evolutionary challenge is to integrate human communities into the model as more than just a component of fishing mortality.

ACKNOWLEDGEMENTS

I would like to thank Tony Pitcher for encouragement and critical comments, Daniel Pauly for showing me the potential for ecosystem modelling, Ron MacLeod for sharing his insights into BC fisheries management, Percy Walkus and other BC First Nation members for their patience and instruction. Likewise my wife Pam Brown of the Heiltsuk Nation for her patience and incisive comment. Any errors or omissions are of course my own.

REFERENCES

British Columbia (1875) Report on the Subject of Indian Reserves by a Committee of the Honourable the Executive Council. Approved by His Honour the Lieutenant Governor the 18th Day of August, 1875.

British Columbia (1902) Fisheries Dept: Report of the Commissioner of Fisheries.

British Columbia Claims Task Force (1991) Report of the British Columbia Claims Task Force. First Nations Summit, N. Vancouver, BC.

Canada (1982) *Constitution Act*. Enacted by the *Canada Act*, 1982, (UK), c. 11, Schedule B.

Canada (1997) Fisheries and Oceans Canada sustainable fisheries framework: Atlantic groundfish. *Report of the Auditor General*, Ottawa, Chapter 14 (Website: http://www.oag-bvg.gc.ca/oag-bvg/rep97/1997e/html/9714e/mp9714e.html).

Christensen, V. and Pauly, D. (1992) ECOPATH II – A system for balancing steady-state ecosystem models and calculating network characteristics. *Ecol. Modeling*, **61**, 169–185.

Cushing, D.H. (1988) *The Provident Sea*, Cambridge Univ. Press, Cambridge, pp. 78–79

Haggan, N. (1996) Integration of local environmental knowledge, in *Mass-balance Models of North-eastern Pacific Ecosystems* (UBC Fish. Centre Res. Rep., **4**) (eds D. Pauly and V. Christensen), UBC Fisheries Centre, Vancouver, p. 88.

Larkin, P.A. (1979) Maybe you can't get there from here: a foreshortened history of research in relation to management of Pacific salmon. *J. Fish. Res. Bd Can.* **36**, 98–106.

Newell, D. (1988) Dispersal and concentration: the slowly changing spatial pattern of the British Columbia salmon canning industry. *J. Hist. Geog.* **14**, 22–36.

Newell, D. (ed.) (1989) *The Development of the Pacific Salmon Canning Industry, A Grown Man's Game*, McGill–Queens Univ. Press, Montreal and Kingston.

Newell, D. (1993) *Tangled Webs of History: Indians and the Law in Canada's Pacific Fishery*. Univ. of Toronto Press, Toronto.

Parsons, T. (1993) *Management of Marine Fisheries in Canada*, National Research Council of Canada and Dept of Fish. & Oceans, Ottawa, 210 pp.

Pauly, D. (1995) Anecdotes and the shifting baseline syndrome of fisheries. *Trends Ecol. Evol.* **10**, 430.

Pearse, P.H. and Larkin, P. (1992) Managing Salmon in the Fraser. Report to the Minister of Fisheries and Oceans on the Fraser River Salmon Investigation. Department of Fisheries and Oceans, Ottawa.

Regina v Sparrow (1990), 4 WWR 410, affirming (1987) 36 DLR (4d) 246 (BC Court of Appeal) and (1975) 52 DLR (3d) 744 (BC Supreme Court).

Rijnsdorp, A.D. and Millner, R.S. (1996) Changes in population dynamics and exploitation of North Sea plaice (*Pleuronectes platessa* L.) since the late 1880s. *ICES J. Mar. Sci.* **53**, 1170–1184.

Walters, C.J. (1995) *Fish on the Line: The Future of Pacific Fisheries*, David Suzuki Foundation, Vancouver.

Walters, C.J. and Maguire, J.-J. (1996) Lessons for stock assessment from the northern cod collapse. *Rev. Fish Biol. Fish.* **6**, 125–137.

Walters, C.J. and Pearse, P.H. (1996) Stock Information requirements for quota management systems in commercial fisheries. *Rev. Fish Biol. Fish.* **6**, 21–42.

Walters, C.J., Christensen, V. and Pauly, D. (1997) Structuring dynamic models of exploited ecosystems from trophic mass-balance assessments. *Rev. Fish Biol. Fish.* **7**, 139–172.

Chapter three

Measuring the unmeasurable: a multivariate and interdisciplinary method for rapid appraisal of the health of fisheries

Tony J. Pitcher, Alida Bundy, David Preikshot, Trevor Hutton and Daniel Pauly

ABSTRACT

This chapter describes a new multivariate, multidisciplinary and interdisciplinary rapid appraisal technique that can be used to classify world fisheries and diagnose their problems. For ecological, technological, economic and social categories, we chose 15–25 attributes that (a) are easily and objectively scored on a ranked scale using readily available data, (b) are likely to discriminate among fisheries, and (c) may be easily related to sustainability. A multivariate ordination using multidimensional scaling (MDS) is performed within each disciplinary set of attributes. The ordination scores are then brought together into an overall interdisciplinary

Reinventing Fisheries Management. Edited by Tony J. Pitcher, Paul J.B. Hart and Daniel Pauly. Published in 1998 by Kluwer Academic Publishers, London. ISBN 0 412 83410 3.

analysis, also using MDS. Reference points for evaluating the relative status of fisheries are provided by constructing hypothetical fisheries that are assigned 'good', 'bad' or 'random' scores, defined in terms of sustainability for each discipline, or that have a random assignment of attribute values.

A pilot analysis of a diverse set of 26 world fisheries from commercial, subsistence, artisanal, and industrial sectors suggests that the technique may be useful in an objective comparison of the 'health' of fisheries. The new technique may provide helpful diagnostics that do not rely on conventional stock assessments; however, the selection of attributes is important because they should remain fixed if future analyses are to be comparable.

Extracting useful information for an interdisciplinary overview entails delving into a range of disciplines, each of which has evolved its own rules and unstated assumptions. Ordination within each category represents these disciplines. This may be approximate, as here, using a simple ranking scale for many attributes, or, after careful surveys have been carried out, made more precise. The hierarchical technique introduced here is designed to withstand robust disciplinary review at this first level of analysis, while re-ordinating the fisheries in interdisciplinary multivariate space at the second stage. This rapid appraisal technique could be useful in diagnosing the scope and nature of problems of a fishery, in providing early warning of impending dangers, or in a triage of fisheries to determine where limited management resources might be focused to greatest effect.

3.1 INTRODUCTION

As part of stock assessment, much effort has gone into determining stock status relative to biological reference points, such as levels of fishing mortality, spawning biomass or age structure (Smith, 1993), where the objective is to obtain diagnostics that may give early warning of collapse or serious depletion. Increasingly, stock assessment in North America and Europe relies upon the estimation of large numbers of stock parameters and requires extensive current and historical data measured from the fishery and biomass surveys. There has been much comment on the mismatch between the complexity of these stock models and the high degree of irreducible uncertainty inherent in fisheries research (Walters, Chapter 21, this volume). At the same time, such extensive data requirements preclude the application of these models to many of the world's fisheries.

While fisheries management is increasingly seen to be as much about managing human behaviour as about fish ecology (Jentoft, Chapter 12, this volume), apart from economics, most analyses of the human aspects of fisheries have been non-quantitative, with little predictive or diagnostic power. Nevertheless, this human dimension is so intertwined with the gear, vessels, markets, biological and economic sustainability, management, allocation and the rebuilding of depleted and collapsed stocks, that the study of fisheries can be regarded as truly interdisciplinary.

This chapter addresses both of these problems by presenting a new multivariate, multidisciplinary, rapid appraisal technique that employs simple, easily scored attributes. We show how it may be used to diagnose emerging problems in fisheries; in other words to evaluate the 'health' of fisheries by making comparisons. For example, the scheme might be useful in a 'triage' of fisheries to determine where limited management resources might be focused to greatest effect (Pauly, 1997). It may also be used to track changes in a single fishery in an attempt to foresee problems before some combination of biological, economic or social effects leads to disaster.

3.2 METHOD

The requirements of a method to evaluate the health of fisheries are that it: should capture the multidisciplinary nature of fisheries; should be robust against disciplinary criticism; should show differences both within and among fisheries; and should clearly indicate changes in a fishery's status with time.

We have based the method on a multivariate ordination technique. This has the advantage that the relative importance of the different measured attributes is determined only by the data rather than by some preconceived model; it is essentially a redescription of data rather than fitting of data to a model. A disadvantage is that the choice of specific attributes can unduly influence the ordination.

Fisheries

One problem that we had to solve at the outset was how to define a fishery. Geographical, jurisdictional, ecological, technical, species, genetic and social factors all provide defensible boundaries to a single fishery. A simple solution is to define a fishery by species, or species complex, and gear type, as this provides a great deal of flexibility. A second problem, only partially resolved here by taking averages of values for numbers of target species, was how to deal with multispecies fisheries.

Sources of information for the diverse set of 26 fisheries are presented in Table 3.1. They include commercial, subsistence, artisanal and industrial

Table 3.1 List of 26 fisheries ordinated in this chapter. Principal sources of data for each fishery are indicated. Sources of information are either in the reference list, for published work, or listed below in acknowledgement for personal communications. Abbreviations for the fisheries are as used in the figures

Location	Fishery/species	Abbreviation	Sources of information*
Peru	Anchovy fishery, drift net, 1969	Peru anch 69	Contributions in Pauly and Tsukayama (1987)
Argentina	Hake, offshore, trawl, 1980s	Arg hake–off	Bezzi *et al.* (1995)
Argentina	Hake, coastal , trawl, 1980s	Arg hake–cst	Bezzi *et al.* (1995)
New Zealand	Hake, trawl, 1990s	NZ hake	Colman (1995)
South Africa	Hake, trawl, 1990s	SA hake	Stuttaford (1994), Payne and Punt (1995)
South Africa	Squid, jigger, 1992	SA squid	Stuttaford (1994)
North Adriatic	Sardines, volante (= pair trawl), 1990	Adtic sard vol	Levi *et al.* (1985), FAO (1987), Esseen (1992), Pitcher *et al.* (1992)
North Adriatic	Anchovy, volante (= pair trawl), 1990	Adtic anch vol	Levi *et al.* (1985), FAO (1987), Esseen (1992), Pitcher *et al.* (1992)
Central Adriatic	Sardines, lampara (= night purse with light), 1990	Adtic sard lamp	Levi *et al.* (1985), FAO (1987), Esseen (1992), Pitcher *et al.* (1992)
Central Adriatic	Anchovy, lampara (= night purse with light), 1990	Adtic anch lamp	Levi *et al.* (1985), FAO (1987), Esseen (1992), Pitcher *et al.* (1992)
North Minch, Scotland	*Nephrops* trawl fishery, 1990	Scot Neph trawl	Strange (1981); Bailey *et al.* (1986); Anon. (1989); Bundy (1990)
North Minch, Scotland	*Nephrops* creel (= trap) fishery, 1990	Scot Neph creel	Strange (1981); Bailey *et al.* (1986); Anon. (1989); Bundy (1990)
San Miguel Bay, Philippines	Municipal trawl fishery	SMB trawl	ICLARM (1995)
San Miguel Bay, Philippines	Gillnet fishery	SMB gillnet	ICLARM (1995)

San Miguel Bay, Philippines	Minitrawl fishery	SMB minitrawl	ICLARM (1995)
San Miguel Bay, Philippines	Push net fishery	SMB push net	ICLARM (1995)
San Miguel Bay, Philippines	Fixed gear fishery	SMB fixed gear	ICLARM (1995)
San Miguel Bay, Philippines	Filter net fishery	SMB filter net	ICLARM (1995)
Atlantic Canada	Lobster fishery, traps, 1995	Atlantic lobster	Anon. (1977), Kearney (1984), Canada (1989, 1993, various), Davis (1991), A. Davis[1]
BC, Canada	Herring roe fishery, seine net, 1995	BC herring	D. Hay[2], S. Mackinson[3]
BC, Canada	Black cod fishery, traps, 1995	BC sablefish	Parsons (1993), Anon. (1994), C. Walters[4]
Papua New Guinea	Shrimp fishery, trawl, 1996	PNG shrimp	Evans *et al.* (1995)
Federated States of Micronesia	*Trochus* harvest, 1990s	FSM trochus	Clarke and Ianelli (1995)
Ghanaian Sakumo lagoon	Tilapia, 1971	Ghana tilapia	Pauly (1975)
Bolinao, Philippines	Reef flat fishery, 1996	Bolinao reef	McManus *et al.* (1992)
BC, Canada	First Nations Fraser River sockeye, 1996	BC FN sock	Pearse (1992), Anon. (1994), McDaniels *et al.* (1994)

*Addresses for sources of personal communications: [1]Dr A. Davis, Francis Xavier University, Antigonish, Nova Scotia, Canada; [2]Dr D. Hay, Pacific Biological Station, Department of Fisheries and Oceans, 3190 Hammond Bay Road, Nanaimo, British Columbia, Canada V9R 5K6; [3]S. Mackinson, Fisheries Centre, University of British Columbia, 2204 Main Mall, Vancouver, British Columbia, Canada V6T 1Z4; [4]Dr C. Walters, Fisheries Centre, University of British Columbia, 2204 Main Mall, Vancouver, British Columbia, Canada V6T 1Z4.

fisheries from many areas of the world, and were chosen for this pilot analysis on the basis of personal experience, available data and a mix of similar and contrasting natures so that similarities and differences might be reflected in the ordinations.

For example, the fishery for sardine and anchovy in the Adriatic is divided by two gear types that overlap only partially in space: pair trawls in the north and purse seines operated with lights in the central Adriatic. Adriatic anchovies went through a collapse and rebuilding in the mid 1980s, but sardines did not (Esseen, 1992; Pitcher *et al.*, 1992), and hence we are justified in considering these as separate fisheries despite their being prosecuted from the same vessels. Ordination allows those features in common (such as the location, environment, gear and vessel) to cluster these fisheries, while allowing those attributes where they differ (such as exploitation status, price) to spread them out.

As a second example, six fisheries from San Miguel Bay, Philippines, are included in the 26 fisheries (ICLARM, 1995). One fishery represents the commercial trawl sector, and the other five are small-scale gears which include a large and important gillnet sector, a small-scale trawl, fixed gear such as fish corrals, filter nets and a simple push or scissor net. The six fisheries vary in features such as scale, species targeted, operation and income, while they share the fact that they are all in a tropical multispecies habitat in a developing country. The ordination method can be used to explore how these fisheries are separated for each discipline. Furthermore, using several fisheries from one geographical location allows the differences within the group to be explored.

Disciplines

The method ordinates fisheries in four disciplinary areas that we consider critical to long-term viability of a fishery:

- ecological (including fish population parameters and environment);
- technological (including gear and fishing characteristics);
- economic (including both micro and macroeconomic factors);
- social (including social and anthropological factors).

The interdisciplinary ordination combines these areas.

Ordination attributes

Within each ordination, a set of attributes were defined and scores for each fishery determined from the literature or interviews with experts. The selection of the attributes is important because they should remain fixed if future analyses are to be comparable. For this analysis, attributes were defined in a series of eight rounds of meetings. Criteria for choosing attributes were that they were easily and objectively scored, that they were available for all fisheries, that scores could be easily refined without disruption to the analysis, and that extreme values of attribute scores could be assigned unequivocally as either 'good' or 'bad' in terms of their relationship to sustainability. Many attributes, such as vessel size, that might be appropriate for distinguishing fisheries in taxonomic sense, turned out not to be so useful according to this criterion. Values for certain economic and social aspects were obtained from the CIA *World Factbook* (Central Intelligence Agency, 1995).

Full details of the attributes within each of the four disciplines are given in Table 3.2, together with the scores we used as 'good' or 'bad'. Many attributes are scored on a three- or four-point ranked scale. This makes it relatively easy to obtain a score in the absence of precise surveys and for a group of experts to agree on a score. Moreover, ranked scores can easily be replaced later by precise values without disrupting the main features of the ordination. Actual scores used in this analysis are available in an Excel file from the Web site 'fisheries.com'.

Ordination method

We used principal components analysis and canonical correspondence analysis in pilot work, but found that multidimensional scaling (MDS) was the most suitable for our sets of attributes, which differ in scale and include a mixture of ordinal and nominal measures (Stalans, 1995). MDS was implemented using the SPSS statistical package (SPSS, 1997), using squared Euclidean distance measures. Each attribute was normalized using Z-scores. Scores on the first two MDS axes were plotted against each other (axis one is always horizontal) for each of the four disciplinary analyses. These two MDS scores from each analysis, making eight scores in all, were used in the final interdisciplinary ordination. Loading of each MDS dimension on the original attributes was estimated using multiple linear regression.

To distinguish favourable from unfavourable directions in the ordination, we constructed three additional types of fisheries, 'good', 'bad' and 'random'. The 'good' fishery was given attribute scores that we considered most likely to lead to sustainable, 'healthy' fisheries within each of the disciplines. The 'bad' fishery had the converse scores for each attribute. The 'good' and the 'bad' fisheries, which always occupy opposite quadrants, are plotted on each ordination, and the direction of the axes chosen by convention so that 'good' is in the top left quadrant and 'bad' in the lower right quadrant. We also constructed a set of 20 'random' fisheries with scores chosen randomly from the data range for each attribute. For the

Table 3.2 Definitions of fishery attributes, scoring system used in the ordinations and values ascribed for each attribute to the 'good' and 'bad' constructed fisheries

Attribute	Scoring	'Good'	'Bad'	Notes
Ecological attributes				
1 Catch/fisher	t/fisher/year	1	6	Catch/total fisher crew (logged)
2 Catch trend	0; 1; 2	0	2	No significant, medium, large changes; past 5 yearrs (+ or –)
3 Catch variability	0; 1; 2	0	2	Low COV < 40%; medium COV 40–100%; high COV >100%
4 Exploitation status	0; 1; 2	0	2	Low; fully; overexploited (half FAO scale)
6 Egg production	0; 1; 2	2	0	Low; medium; high
8 Recruitment variability	0; 1; 2	0	2	Low COV < 40%; medium COV 40–100%; high COV > 100%
7 Catch < maturity	0; 1	0	1	Significant numbers caught before maturity
5 Target species	0; 1	1	0	Is catch dominated by a target species (1), or not (0)
9 Discarded bycatch	0; 1; 2	0	2	Low 0–10%; medium 10–40%; high > 40% of target catch
10 Landed bycatch	0; 1; 2	0	2	Low 0–10%; medium 10–40%; high > 40% of target catch
11 Species caught	0; 1; 2	0	2	Low 1–10; medium 10–100; high > 100 species
12 Migration	0; 1; 2	0	2	Low; medium; high overlap with another jurisdiction
13 Primary production	0; 1; 2	2	0	In g C/m^2/year; low = 0–50; medium = 50–150; high = 150+
14 Upwelling	0; 1; 2	2	0	None; seasonal; constant
15 Trophic level	2 to 4	4	2	Of target species
16 Habitat status	0; 1; 2	0	2	Undetectable; some; serious impact
Technological attributes				
1 Commercial	0; 1	0	1	Commercial or industrial fishery, large scale (1)
2 Artisanal	0; 1	1	0	Small boat fishers, small scale (1)
3 Subsistence	0; 1	0	1	Catch important food for fisher's family (1)
4 Recreational	0; 1	1	0	Sport fishery (1)
5 Auxiliary vessels	0; 1	0	1	Supply, factory, service in addition to fishing vessels (1)
6 Trip length	Days	1	35	Average days at sea per fishing trip
7 Landing sites	0; 1; 2	0	2	Dispersed (0), medium(1); centralized (2)
8 Preservation on board	0; 1; 2	2	0	None (0); ice (1); freezer/sophist. (champ./live tank) on vessel (2)
9 Onboard processing	0; 1; 2	0	2	None (0); some gutting etc. (1); filleting/surimi plant on board (2)
10 Geopositioning	0; 1; 2	0	2	Traditional (0); radar, radio, astronavigation (1); GPS (2)
11 Acoustic device	0; 1; 2	0	2	None (0); some (1); sophisticated (2)
12 Gear	0; 1	0	1	Active (1) or passive (0)
13 Selective gear	0; 1; 2	2	0	Device(s) in gear to increase selectivity (1)

14 Power gear	0; 1	0	1	Power assistance to gear (1)
15 FADS	0; 1	0	1	Fish attraction devices used (1)
16 Interaction	0; 1	0	1	Direct physical interaction with another sector/fishery (1)
Economic attributes				
1 Price	US$/tonne	9000	1	US$/tonne of landed product for analysis time
2 Fisheries in GDP	0; 1; 2	2	0	Importance of fisheries sector in country: low; medium; high
3 GDP/person	US$/capita	23	0.4	In country of fishery, thousand US$ per capita
4 Subsidy level	0; 1; 2	0	2	< 15%; 15–50%; > 50% of fishery value
5 Limited entry	0; 1; 2	2	0	Almost none (0); some (1); strong (2) – includes informal limitation
6 Marketable right	0; 1; 2	2	0	Marketable (transferable) quota/share/property right in harvest
7 Other income	0; 1; 2	2	0	Mainly casual (0), part time (1) or full time (2) fishers
8 Unemployment income	0; 1	0	1	Unemployment benefit for fishers (1)
9 Boat origin	0; 1; 2	0	2	Local (0); mixed (1); imported (2)
10 Ownership	0; 1; 2	0	2	Local (0); mixed (1); foreign (2); profits from fishery
11 Interest rates	0; 1; 2	0	2	Low (0); medium (1); high (2) return on govt bonds
12 Market	0; 1; 2	0	2	Principally local (0); national (1); international (2)
Social attributes				
1 Owner driver	0; 1	1	0	No (0); yes (1), on average
2 Kin crew	0; 1; 2	2	0	None (0); weak (2); strong (2) kin in crew of a vessel
3 Crew welfare organization	0; 1	1	0	Yes (1) or no (0)
4 Share system	0; 1	1	0	In proceeds of catch (1)
5 Skipper experience	Years	30	1	Average – to nearest 5 years
6 Fishing community age	0; 1; 2	2	0	< 20 years, 20–50 years, > 50 years
7 Fishing community growth	0; 1; 2	0	2	Growth of fishing community over past 10 years
8 Size fishery sector	0; 1; 2	0	2	Proportion of households fishing in community, < 1/3; 1/3–2/3; > 2/3
9 Infant mortality	Per thousand	0	63	National figure
10 Education level	0; 1; 2	2	0	Fishers below; same; above population average
11 Literacy	Per cent	100	0	In the population–national average
12 Conflict status	0; 1; 2	0	2	Level of conflict with other sectors
13 Information sharing	0; 1; 2	2	0	None; a little; a lot
14 Fisher influence	0; 1; 2	2	0	Strength of fishers' direct influence on fishery regulations
15 Fishing income	0; 1; 2	0	2	Fishing income, % total family income: < 50%, 50–80%; > 80%
16 Kin involvement	0; 1	1	0	Kin sell catch/ process/maintain/accounting

ordination scores of the 'random' fisheries, we plotted the mean and 95% distribution limits on each axis.

3.3 RESULTS

The results of the ordinations in the four disciplinary areas are provided in Figs 3.1 to 3.8. In general, it is encouraging that similar fisheries, such as the two Adriatic species and the San Miguel Bay fisheries, cluster together in most of the ordinations.

In the ecological ordination (Fig. 3.1), most fisheries lie outside the zone of random attribute scores. We can see a trend for small-scale or fixed-gear fisheries, such as the BC trap fishery for sablefish and San Miguel Bay fixed-gear fishery, to lie in the upper left quadrant and larger-scale, trawl fisheries below and to their right, a trend reflected in the separation of the two Scottish *Nephrops* fisheries. Hake fisheries, most of which appear to be in reasonable shape (Pitcher and Alheit, 1995), lie together in the top left quadrant. Among the San Miguel Bay fisheries, the medium-scale trawl fishery lies in the 'worst' position. In the Adriatic, the two purse seine fisheries (*lampara*) ordinate in better positions that the pair trawls (*volante*) for the same species, and moreover, the anchovy fishery, which went through a collapse in the mid 1980s, ordinates in a 'worse' location than the sardines, which did not. In 1969, the Peruvian anchovy was about to go through a major collapse, and it certainly ordinates in an extreme position at bottom centre. The Bolinao reef fishery ordinates nearest to the constructed 'bad' ecologically, as befits the very fishery that inspired the concept of 'Malthusian overfishing' (Pauly, 1989) through its environmental destruction. Figure 3.2 shows that the first dimension extracted from the ecological ordination derives from 7 of the 16 attributes: egg production; trophic level; number of species caught; number of target species; amount of catch less than maturity; catch trend; and discards. Dimension two is linked significantly with 12 of the 16 attributes, including six (exploitation status, catch per fisher, habitat status, migration, recruitment and catch variability), that were not significant for axis one.

In the technological ordination (Fig. 3.3), along axis one (the horizontal axis), fixed gears, such as traps, and small-scale gears, locate in better positions than large-scale and trawl fisheries. But the ordination exhibits relatively little dispersal of most of the fisheries along axis two, and this is surprising considering the diversity of gears involved. Hake fisheries and the squid fishery score well on axis two but not on axis one. Figure 3.4 shows that axis one is linked significantly to only three attributes (auxiliary vessels, artisanal and recreational fisheries), while axis two is linked to five (on-board processing and preservation, gear interactions, and sub-

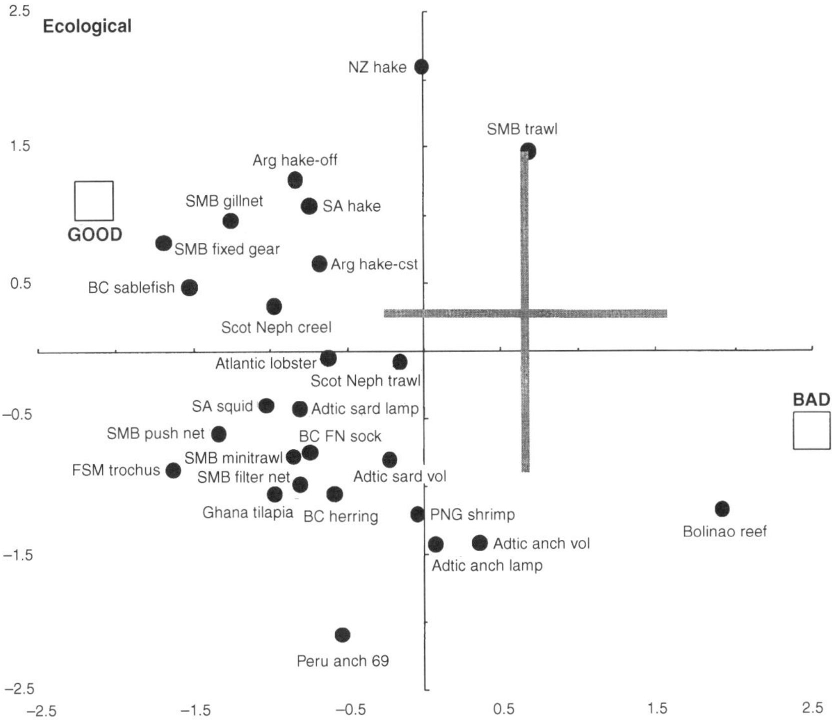

Fig. 3.1 Ordination of the 26 fisheries using multidimensional scaling of 16 ecological attributes. Locations labelled as 'good' and 'bad' represent locations of constructed, hypothetical fisheries all of whose attribute scores represent the best or worst possible scores in terms of sustainability. Shaded cross shows the 95% limits on the mean scores of 20 constructed fisheries whose attribute scores were chosen at random from the ranges available. For identities of fisheries labels see Table 3.1. In this and similar figures, the horizontal axis is designated axis one.

sistence). In general this ordination is less satisfactory and harder to interpret than the ecological.

The economic ordination (Fig. 3.5), produces clearer comparisons. High-value fisheries (e.g. PNG shrimp) with marketable harvest rights (e.g. BC sablefish) locate near to the constructed 'good' fishery while heavily subsidized fisheries (e.g. BC sockeye) lie near to the 'bad'. Small-scale fisheries (e.g. the San Miguel Bay fisheries) tend to lie to the right on axis one, while higher-technology fisheries (e.g. NZ hake, BC herring) lie to the left. The two gears in the Adriatic fisheries ordinate in the same place, and the two species are close, the anchovy, with higher market value, falling in a slightly 'better' position. A number of the San Miguel Bay fisheries are not

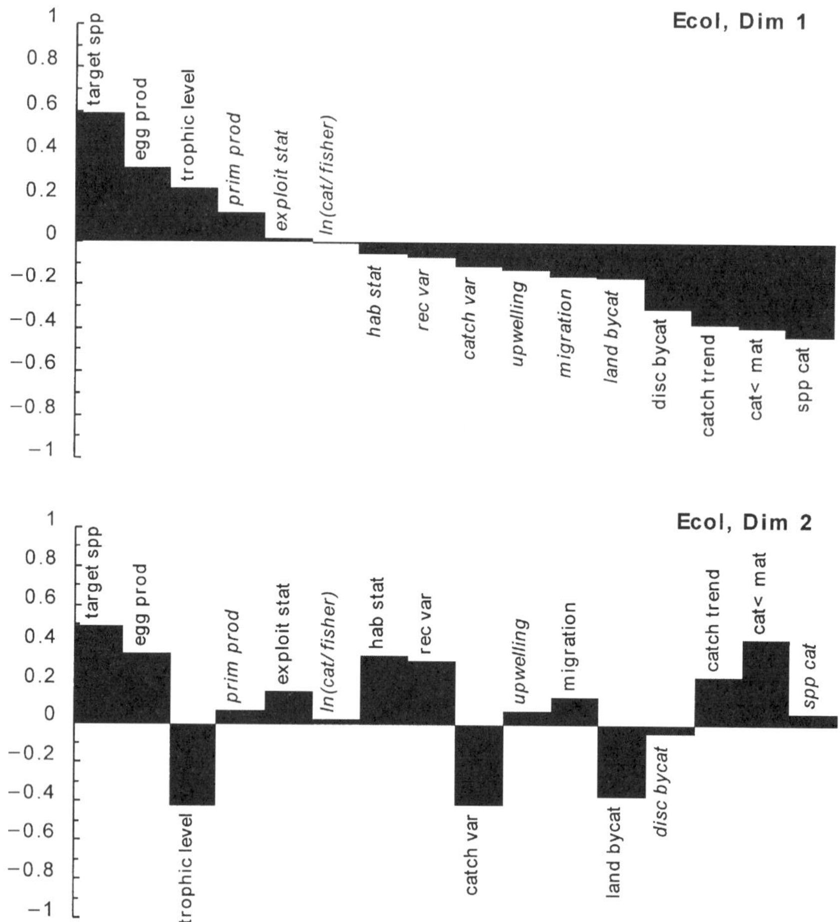

Fig. 3.2 Loadings of the original variables onto the first two dimension axes of the multidimensional scaling for the ecological ordination. Loadings were estimated using multiple linear regression, with an intercept of zero, using the untransformed variables scores. Loadings that are statistically non-significant at the 0.05 level are indicated in italics.

distinguishable from the random scores. Axis one in the economic ordination is related to only three attributes (ownership location, limited entry and importance of fisheries in the national economy, Fig. 3.6), while axis two relates to nine of the twelve attributes, principally, marketable harvest rights and limited entry in the positive sense, and in the opposite sense, unemployment pay, boat origin, subsidies, and interest rates.

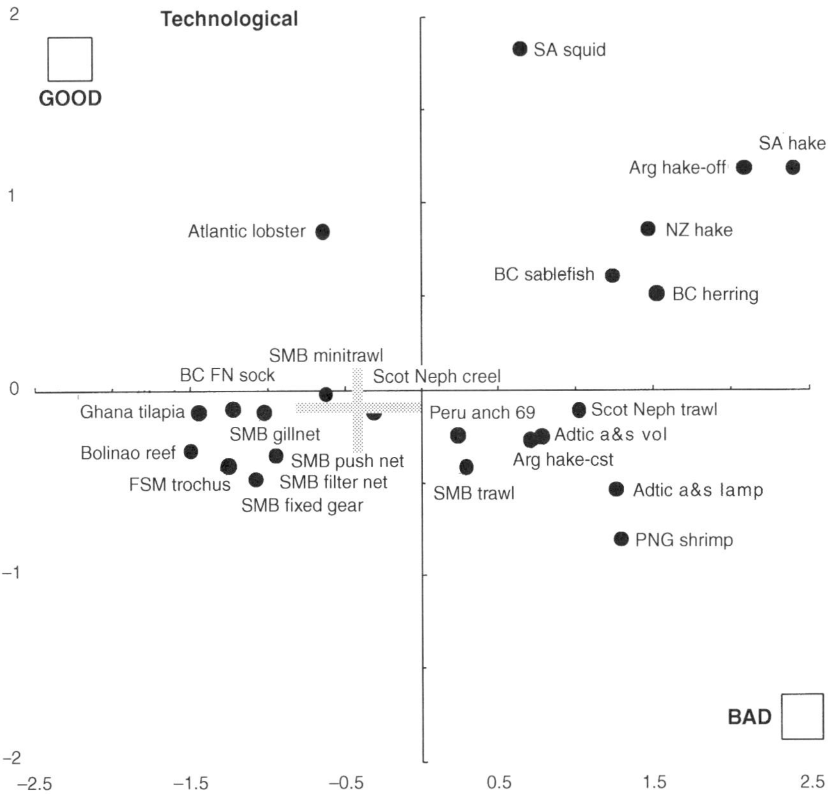

Fig. 3.3 Ordination of the 26 fisheries using multidimensional scaling of 16 technological attributes. Locations labelled as 'good' and 'bad' represent locations of constructed, hypothetical fisheries all of whose attribute scores represent the best or worst possible scores in terms of sustainability. Shaded cross shows the 95% limits on the mean scores of 20 constructed fisheries whose attribute scores were chosen at random from the ranges available. For identities of fisheries labels see Table 3.1; the label 'Adtic a&s' designates superimposed points for Adriatic anchovies and sardines. Note that dots that are not clearly visible are overlapped by nearby dots, for instance 'SMB filter net' and 'SMB fixed gear' overlap.

In the social ordination (Fig. 3.7), axis two appears to separate small-scale fisheries (lower quadrants) from the rest (top), with fixed gears to the left, nearest to the 'good' and other gears to the right. The Atlantic lobster fishery lies close to small-scale tropical fisheries in the social ordination. Figure 3.8 shows that the first social dimension axis is related to seven of the 16 attributes; positively with both kin factors, share systems, skipper experience, and negatively with growth of the community and fishing

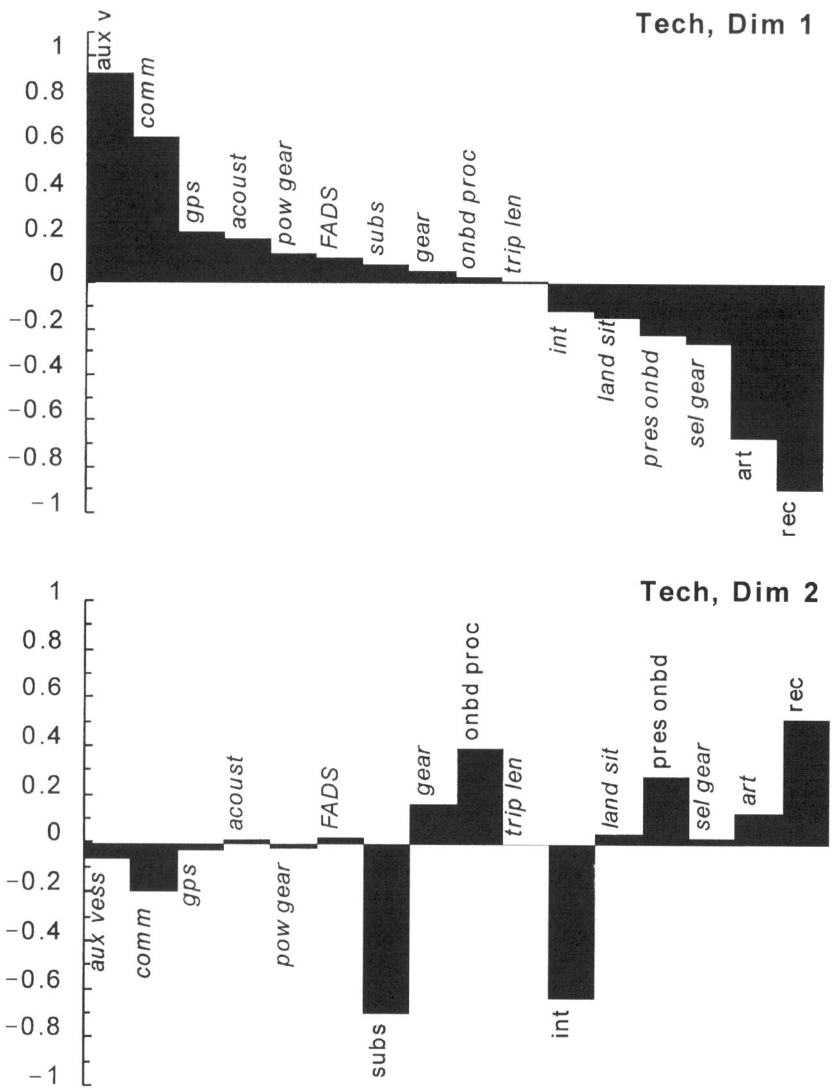

Fig. 3.4 Loadings of the original variables onto the first two dimension axes of the multidimensional scaling for the economic ordination. Loadings were estimated using multiple linear regression with an intercept of zero, using the untransformed variable scores. Loadings that are statistically non-significant at the 0.05 level are indicated in italics.

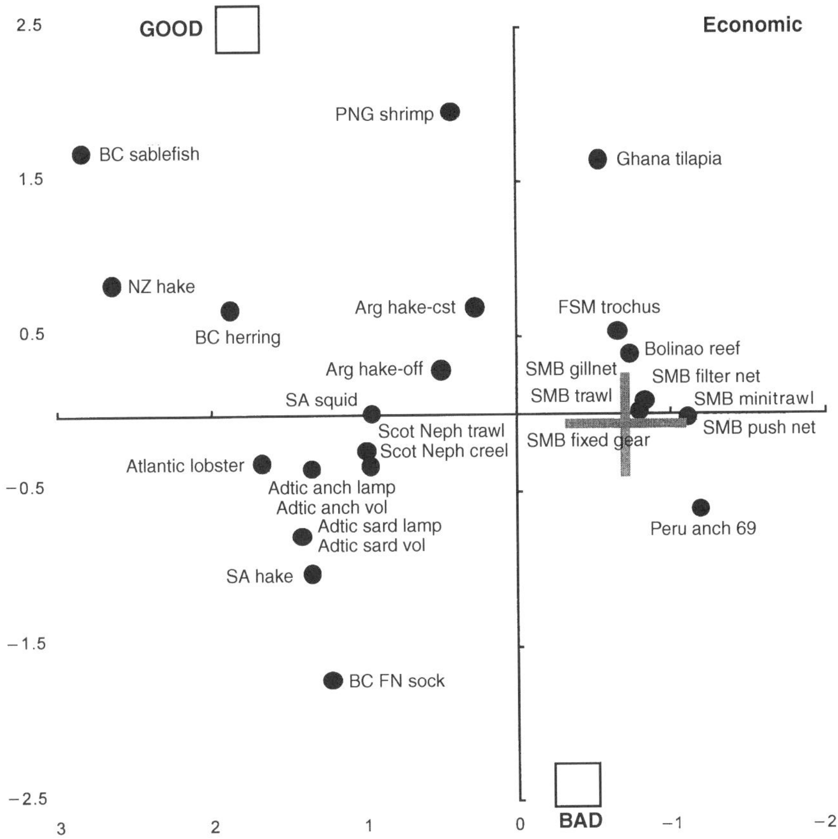

Fig. 3.5 Ordination of the 26 fisheries using multidimensional scaling of 12 economic attributes. Locations labelled as 'good' and 'bad' represent locations of constructed, hypothetical fisheries all of whose attribute scores represent the best or worst possible scores in terms of sustainability. Shaded cross shows the 95% limits on the mean scores of 20 constructed fisheries whose attribute scores were chosen at random from the ranges available. For identities of fisheries labels see Table 3.1.

income. Axis two is related to six of these, but also to crew welfare, education level and, negatively, with conflict status.

The interdisciplinary ordination is illustrated in Fig. 3.9; most fisheries lie outside the 'random' scores area. In general, fixed-gear and small-scale fisheries lie nearest to the 'good' location. Most of the trawl fisheries cluster in the lower left quadrant, indicating features that score well on axis one but poorly on axis two. The 'worst' fisheries overall, falling in the lower right quadrant, are the PNG shrimp and Peruvian anchovy. Figure

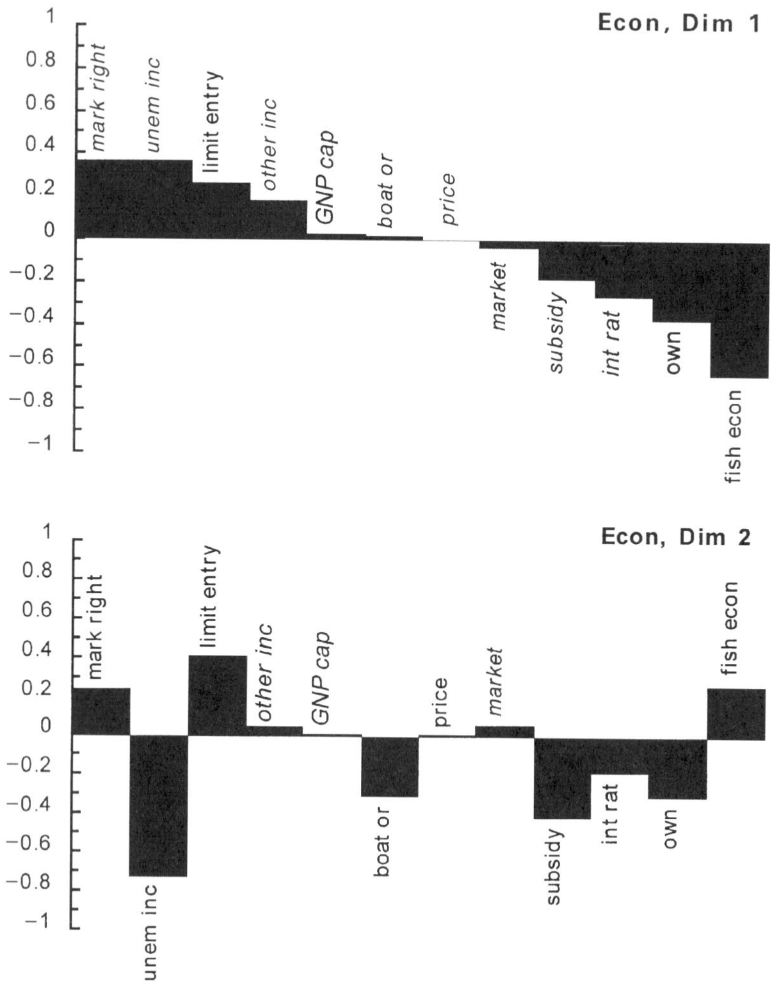

Fig. 3.6 Loadings of the original variables onto the first two dimension axes of the multidimensional scaling for the economic ordination. Loadings were estimated using multiple linear regression with an intercept of zero, using the untransformed variable scores. Loadings that are statistically non-significant at the 0.05 level are indicated in italics.

3.10 shows that all eight disciplinary axes are significantly related to dimension one, and five of the eight to dimension two of the interdisciplinary ordination. The Bolinao reef fishery scores almost as poorly as the Peruvian anchovy on axis one, related to all the disciplines, whereas

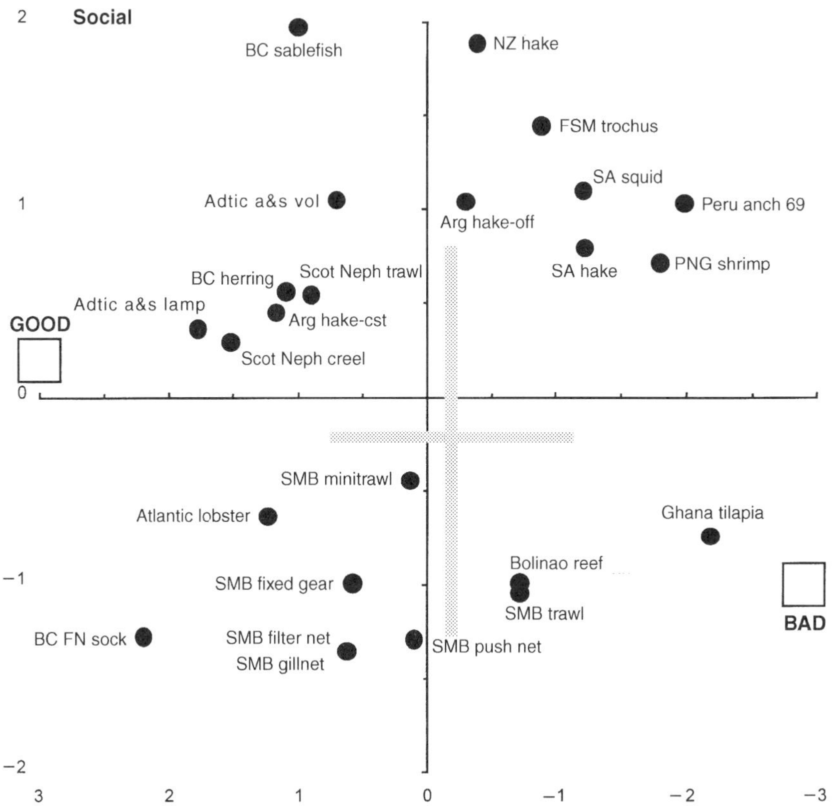

Fig. 3.7 Ordination of the 26 fisheries using multidimensional scaling of 16 social attributes. Locations labelled as 'good' and 'bad' represent locations of constructed, hypothetical fisheries all of whose attribute scores represent the best or worst possible scores in terms of sustainability. Shaded cross shows the 95% limits on the mean scores of 20 constructed fisheries whose attribute scores were chosen at random from the ranges available. For identities of fisheries labels see Table 3.1.

axis two, related mainly to social and economic factors, places it in the top right quadrant, indicating 'good' scores on this axis.

3.4 DISCUSSION

Alternative techniques that address the same problem as this work are multiparameter stock assessment models (e.g. stock synthesis, Methot,

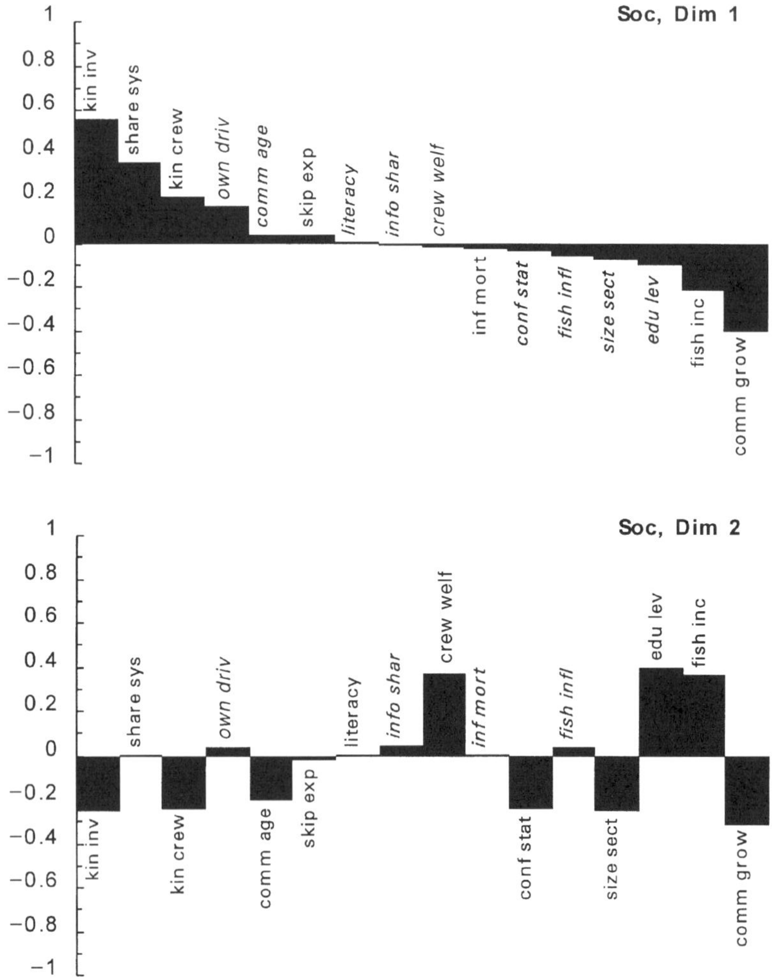

Fig. 3.8 Loadings of the original variables onto the first two dimension axes of the multidimensional scaling for the social ordination. Loadings were estimated using multiple linear regression with an intercept of zero, using the untransformed variable scores. Loadings that are statistically non-significant at the 0.05 level are indicated in italics.

1990), ecosystem harvest models (e.g. Ecosim, Walters *et al.*, 1997), multi-attribute utility analysis (e.g. Boutillier *et al.*, 1988; McDaniels *et al.*, 1994), and its recent offshoot, statistical decision analysis (Peterman *et al.*, Chapter 29, this volume). Lane and Stephenson (1990, 1997) describe an

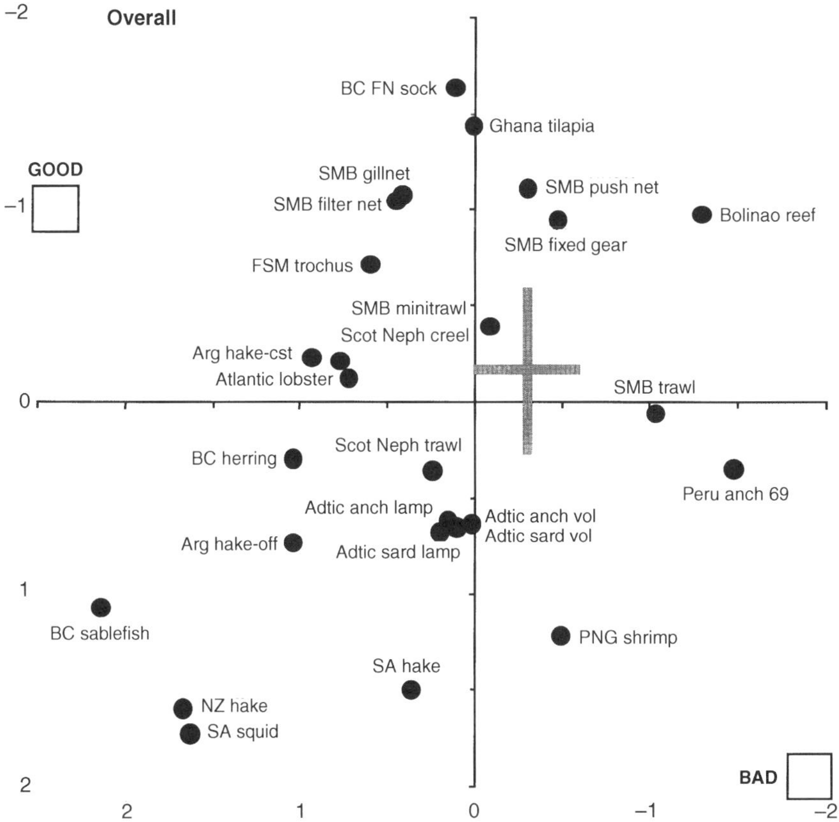

Fig. 3.9 Interdisciplinary ordination of the 26 fisheries using multidimensional scaling of the first two dimensions of the four disciplinary ordinations. Locations labelled as 'good' and 'bad' represent locations of constructed, hypothetical fisheries all of whose attribute scores represent the best or worst possible scores in terms of sustainability. Shaded cross shows the 95% limits on the mean scores of 20 constructed fisheries whose attribute scores were chosen at random from the ranges available. For identities of fisheries labels see Table 3.1.

integrated analysis method for fisheries management that enables managers to examine sophisticated forecasting scenarios as well as socio-economic evaluations. But the method can be used only in data-rich fisheries where stock recruitment and population dynamics relationships are known, and where detailed economic information, such as tax rates on each gear sector, monthly catches and prices by market sector are available. In contrast, the ordination technique presented here provides a

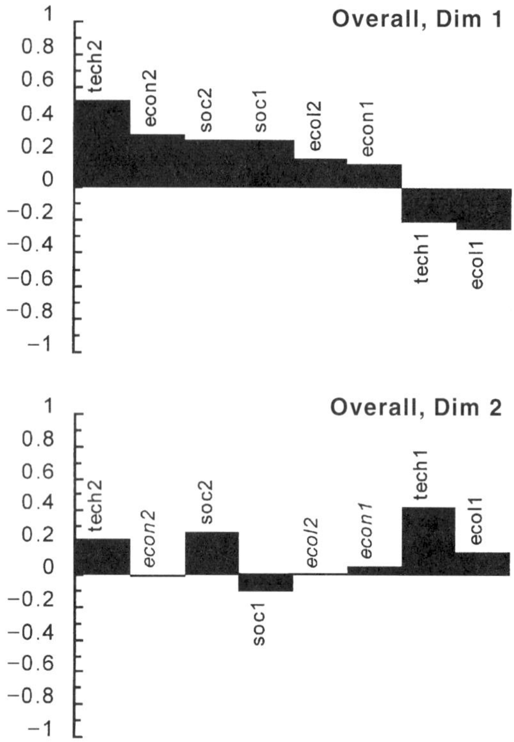

Fig. 3.10 Loadings of the original variables onto the first two dimension axes of the multidimensional scaling for the 'overall' ordination. Loadings were estimated using multiple linear regression with an intercept of zero, using the first two dimension scores from each of the disciplinary ordinations. Loadings that are statistically non-significant at the 0.05 level are indicated in italics.

'statistical common currency' for multivariate measures that are free both of the underlying models of stock assessment or modelling, and of the underlying assumptions that relate utility scores to measured attributes. Provided that enough simple attributes are used and enough fisheries, or points from a single fishery, are documented, the ordination should be relatively unbiased.

The absence of an underlying model in the non-parametric ordination technique of MDS is an advantage for interdisciplinary work, as the quantitative models used by, say, fisheries ecologists and fisheries economists are rarely compatible, while some social scientists are suspicious of any quantitative model.

Two straightforward extensions to the present work may be envisaged.

First, further multivariate classification techniques may be employed to identify more formally groups of 'healthy' and 'less healthy' fisheries (Preikshot and Pauly, 1998), and the attributes culled to those that are most valuable in discriminating among fisheries. Secondly, following the trajectory of a historically documented collapse (Pitcher *et al.*, 1998) will provide insight of which attributes or combinations of attributes might be more valuable in this respect.

By defining fisheries with 'good' and 'bad' values for all attributes and including these invented fisheries in the ordinations, we have imposed a reference direction on the ordination. We found that using the criterion of sustainability reduced disagreement about the value to give the 'good' and 'bad' scores. The technique also distinguishes actual fisheries from those constructed with random attribute scores.

Unfortunately, until recently, most workers in the different fisheries disciplines have not communicated well with one another: a sad example was a 1994 American Fisheries Society meeting in Halifax, where the Newfoundland cod fishery collapse was discussed simultaneously and separately in three adjacent rooms by ecologists, anthropologists and economists. The leaders within a discipline have a tendency, rewarded by improvements to their status, to defend fiercely the sole legitimacy of their view. Our new method is robust in the sense that disciplinary improvements resulting from this criticism can only serve to improve the power of the ordinations within each field.

The lists of attributes within each of the four disciplinary areas need to be sifted and those that are not important in defining 'health' eliminated. We have made no attempt to do this in the present chapter. Attributes identified by other workers, such as Charles (1994) or the Marine Stewardship Council (Sutton, Chapter 8, this volume), should be examined alongside those used here. In our experience, it is very easy to suggest long lists of attributes, but less easy to obtain data for many fisheries or to agree on which are the most valuable. The number of attributes may be reduced more objectively by examining the leverage of individual attributes on the relative position of ordinated points and removing those with little influence. The resulting subset of easily scored attributes with high leverage could be used in a new version of this rapid but powerful appraisal of fisheries.

ACKNOWLEDGEMENTS

We are grateful to Dr Rashid Sumaila for advice concerning social and economic attributes, and we thank Dr Jake Rice for discussion of suitable ordination techniques.

REFERENCES

Anon. (1977) Costs and earnings of selected fishing enterprises Nova Scotia, 1975–1976. Department of Fisheries, Ottawa.

Anon. (1989) Report of the working group on *Nephrops* stocks. International Council for the Exploration of the Sea, Shellfish Committee, C.M. 1989 Assess:18, 99 pp. (mimeo).

Anon. (1994) Pacific coast commercial licensing handbook. Fisheries and Oceans Commercial Licence Unit, Vancouver, 57 pp.

Bailey, N., Howard, F.G. and Chapman, C.J. (1986) Clyde *Nephrops*: biology and fisheries. *Proc. R. Soc. Edinburgh* **90B**, 501–518.

Bezzi, S.I., Varazay, G.A. and Dato, C.V. (1995) Biology and fisheries of Argentine hakes (*M. hubbsi and M. australis*), in *Hake: Fisheries, Ecology and Markets* (eds J. Alheit and T.J. Pitcher), Chapman & Hall, London, pp. 239–268.

Boutillier, J., Noakes, D., Heritage, D. and Fulton, J. (1988) Use of multi-attribute utility theory for designing invertebrate fisheries sampling programs. *N. Am. J. Fish. Manage.* **8**, 84–90.

Bundy, A. (1990) A study of the North Minch *Nephrops* fishery, MSc thesis, University of Wales, Bangor, UK, 97 pp.

Canada, Government of (1989) The Scotia–Fundy Lobster Fishery. Phase 1 – Issues and Considerations. Working Committee Draft Report for Industry Consultation.

Canada, Government of (1993) Report of the task force on incomes and adjustment in the Atlantic Fishery. Minister of Supply and Services, Ottawa.

Canada, Government of (various years) Annual Statistical Review of Canadian Fisheries. Minister of Supply and Services, Ottawa.

Central Intelligence Agency (1995) *World Factbook 1995–1996*. Brassey, Washington, DC, 512 pp.

Charles, A.T. (1994) Towards sustainability: the fishery experience. *Ecol. Econ.* **11**, 201–211.

Clarke, R.P. and Ianelli, J.N. (1995) Current paradigms in *Trochus* management and opportunities to broaden perspectives, in South Pacific Commission and Forum Fisheries Agency workshop on the management of south Pacific inshore fisheries. Manuscript collection of country statements and background papers, Vol. 1 (eds P. Dalzell and T.J.H. Adams), South Pacific Commission, Nouméa, pp. 371–415.

Colman, A. (1995) Biology and fisheries of New Zealand hake (*M. australis*), in *Hake: fisheries, ecology and markets* (eds J. Alheit and T.J. Pitcher), Chapman & Hall, London, pp. 365–388.

Davis, A. (1991) Dire straits: the dilemmas of a fishery. The case of Digby Neck and the Islands. Inst. Social Econ. Stud., Memorial University, St Johns, Newfoundland, no. 43.

Esseen, M. (1992) Analysis of Adriatic pelagic fish stocks and an investigation of fishing power in part of the Adriatic pelagic fishing fleet, MSc thesis, University of Wales, Bangor, UK, 153 pp.

Evans, C., Opnai, L. and Kare, B. (1995) Research and management of the industrial prawn fishery of the gulf of Papua, in South Pacific Commission and Forum Fisheries Agency workshop on the management of south Pacific inshore fisheries. Manuscript collection of country statements and background papers, Vol. 1 (eds P. Dalzell and T.J.H. Adams), South Pacific Commission, Nouméa, pp. 653–689.

FAO (1987) Conseil general des peches pour la Mediterrane. FIPL\Report 395.
ICLARM (1995) *The San Miguel Bay Story*, ICLARM, Manila, Bureau of Fisheries and Aquatic Resources and Fisheries Sector Program, Quezon City, Philippines (CD-ROM).
Kearney, J.F. (1984) Working together: a study of fishermen's responses to government management of the district 4A lobster fishery. University Sainte Anne, Nova Scotia.
Lane, D.E. and Stephenson, R.L. (1990) Saturn: a framework for integrated analysis in fisheries management. *Infor.* **34**, 156–179.
Lane, D.E. and Stephenson, R.L. (1997) Fisheries management science: integrating the roles of science, economics, sociology and politics into effective fisheries management, in *Developing and Sustaining World Fisheries Resources: the State of Science and Management* (eds D.A. Hancock, D.C. Smith, A. Grant and J.P. Beumer), CSIRO, Collingwood, Australia, pp. 177–182.
Levi, D., Cingolani, N., Giannetti, G.F., Scorcelletti, R., Peloni, E. and Frontinin, N. (1985) Stock assessment of *Sardina pilchardus* and *Engraulis encrasicolus* of the northern-central Adriatic Sea: some preliminary results. *Quad. Inst. Ric. Pesca Maritima* **4**, 1–10.
McDaniels, T.L., Healey, M. and Paisley, R.K. (1994) Cooperative fisheries management involving First Nations in British Columbia: an adaptive approach to strategy design. *Can. J. Fish. Aquat. Sci.* **51**, 2115–2125.
McManus, J.W., Nañola, C.L. Jr, Reyes, R.B. Jr and Kesner, K.N. (1992) *Resource ecology of the Bolinao Coral Reef System*. (*ICLARM Stud. Rev.* **22**), ICLARM, Manila, 117 pp.
Methot, R.D. (1990) Synthesis model: an adaptable framework for analysis of diverse stock assessment data. *Int. North Pac. Fish. Comm.* **50**, 259–277.
Parsons, L.S. (1993) *Management of Marine Fisheries in Canada*, National Research Council of Canada and Department of Fisheries and Oceans, Ottawa, 763 pp.
Pauly, D. (1975) On the ecology of a small West African lagoon. *Meeresforschung / Rep. Mar. Res.* **24**, 46–62.
Pauly, D. (1989) On Malthusian overfishing. *Naga, ICLARM Q.* **10**(4), 11–12.
Pauly, D. (1997) When is fisheries management needed? in *South Pacific Commission and Forum Fisheries Agency Workshop on the Management of South Pacific Inshore Fisheries* (eds T.J.H. Adams, P.J. Dalzell and S.A. Robert), Vol. 3, South Pacific Commission, Nouméa, New Caledonia, pp. 103–113.
Pauly, D. and Tsukayama, I. (eds) (1987) *The Peruvian Anchoveta and its Upwelling Ecosystem: Three Decades of Change*. (*ICLARM Stud. Rev.*, **15**), ICLARM, Manila, 351 pp.
Payne, A.I.L. and Punt, A.E. (1995) Biology and fisheries of South African Cape hakes (*M. capensis and M. paradoxus*) in *Hake: Fisheries, Ecology and Markets* (eds J. Alheit and T.J. Pitcher), Chapman & Hall, London, pp. 15–48.
Pearse, P.H. (1992) Managing salmon in the Fraser. Report to the minister of fisheries and oceans on the Fraser River salmon investigation, 36 pp.
Pitcher, T.J. and Alheit, J. (1995) What makes a hake? A review of the critical biological features that sustain global hake fisheries, in *Hake: Fisheries, Ecology and Markets* (eds J. Alheit and T.J. Pitcher), Chapman & Hall, London. pp. 1–14.
Pitcher, T.J., Cingolani, N., Arneri, E., Giannetti, G. and Bombace, G. (1992) Analysis of Adriatic pelagic fish stocks, MRAG, London, 34 pp.
Pitcher, T.J., Mackinson, S., Vasconcellos, M., Nøttestad, L. and Preikshot, D. (1998) Rapid appraisal of the 'health' of fisheries for small pelagics using multivariate, multidisciplinary ordination, in *Fishery Stock Assessment Models for the*

21st Century, Lowell Wakefield Symposium, Alaska (ed. B. Baxter), Alaska Sea Grant, Anchorage, AK (in press).

Preikshot, D. and Pauly, D. (1998) A multivariate interdisciplinary assessment of small scale tropical fisheries, in *Fishery Stock Assessment Models for the 21st Century. Lowell Wakefield Symposium, Alaska* (ed. B. Baxter), Alaska Sea Grant, Anchorage, AK (in press).

Smith, S.J. (1993) Risk evaluation and biological reference points for fisheries management: a review, in *Management Strategies for Exploited Fish Populations. Lowell Wakefield Fish. Symp. 10, Anchorage.* (eds G. Kruse, D.M. Eggers, R.J. Marasco, C. Pautzke and T.J. Quinn), Alaska Sea Grant, Anchorage, AK, pp. 339–353.

SPSS Inc. (1997) SPSS for Windows, release 7.5. SPSS Inc., Chicago, IL.

Stalans, L.J. (1995) Multidimensional scaling, in *Reading and Understanding Multivariate Statistics* (eds L.G. Grimm and P.R. Yarnold), Am. Psychol. Ass., Washington, DC, pp. 137–168.

Strange, E.E. (1981) An introduction to commercial fishing gear and methods used in Scotland. *Scottish Fisheries Information Pamphlet*, No. 1 (2nd edn, revised), 35 pp.

Stuttaford, M. (ed.) (1994) *South African Fishing Industry Handbook: South Africa and Namibia*, 22nd edn, Marine Information cc, Stellenbosch, South Africa.

Walters, C.J., Chistensen, V. and Pauly, D. (1997) Structuring dynamic models of exploited ecosystems from trophic mass-balance assessments. *Rev. Fish Biol. Fish.* 7, 139–172.

Part Two

New Policies for a Reinvented Fisheries Management

Chapter four

Science and decision making in fisheries management

David Policansky

ABSTRACT

Many fishery-resource controversies are couched in scientific terms, although often they are not scientific disputes. I discuss controversies involving several topics recently studied by the US National Research Council, including anadromous salmon in the north-western United States, sea turtles off the US south-east and Gulf of Mexico coasts, tuna and dolphins in the eastern tropical Pacific, and the Bering Sea ecosystem. In each case, the scientific aspects were not obvious to at least one party in the dispute; in some cases, they were not clear to any of them. The lack of scientific information is sometimes a consequence of the inherent complexity of the situation and the difficulties of collecting good data. Often, however, much better information would have been available if managers had taken the time to design data collection into their management regimes. Common barriers to resolving fishery-resource controversies include mutual mistrust, the amount of money and time needed, the lack of a coherent management authority or multiple jurisdictions, and scientific uncertainties. The last is strongly related to the others; good scientific understanding reduces mistrust, saves time and money, and encourages coherent management authority. Scientific uncertainty increases mistrust

Reinventing Fisheries Management. Edited by Tony J. Pitcher, Paul J.B. Hart and Daniel Pauly. Published in 1998 by Kluwer Academic Publishers, London. ISBN 0 412 83410 3.

and incoherent management, and often increases expenditure of time and money. Understanding these relationships is necessary for effective management and decision making.

4.1 INTRODUCTION

Controversies surrounding natural resources – including fisheries – are bitter and increasingly common. Living and nonliving resources are involved, and the causes of, and potential solutions to, the controversies have been discussed in scientific papers (Gordon, 1954; Hardin, 1968; Ludwig *et al.*, 1993) and the popular press (e.g. in the United States, almost any issue of *The National Fisherman* or *Saltwater Sportsman*). Often, the disputants couch their arguments in scientific terms, even when the true sources of disagreement concern policy matters. Scientists are often called on in such situations, and they often become disillusioned when their advice is not taken or does not even seem relevant to the problem at hand. Nonetheless, many scientists and most scientific associations remain convinced that providing scientific advice to decision makers and managers is a useful role and a responsibility.

The US National Research Council (NRC) was established in 1916 as the operating arm of the private, nonprofit National Academy of Sciences. It is often called on by the US Congress or executive-branch agencies to help resolve such controversies by providing an objective review of their scientific aspects. In the course of directing some of those studies, I have observed many of the factors that affect the outcome of natural-resource controversies; four of the cases are the focus of this chapter. The cases discussed here were all topics of NRC reports. They include the decline of endangered sea turtles (NRC, 1990), the tuna–dolphin problem (NRC, 1992), changes in the Bering Sea ecosystem (NRC, 1996a), and the problem of declining salmon populations in the Pacific Northwest (NRC, 1996b). All the cases were complex to varying degrees. They involved scientific uncertainty, unarticulated goals, and politics, among other factors, and they involved mistrust between government officials and users of the resource, and among various private-sector organizations and groups.

In each of the cases, the scientific reports helped to reduce the scientific uncertainties, or at least to delineate them. The effectiveness of the reports in decision making is difficult to assess precisely, and two of the reports were published quite recently, so there has been less time for them to have any effect. Nonetheless, reasonable inferences can be made about their usefulness to decision makers and some general conclusions can be drawn. For each case, I briefly describe the report and its conclusions and the events that followed the report's publication. I then try to relate them to various political and economic factors. In the final section, I attempt to draw some general conclusions about the roles of science in decision making in fishery management.

4.2 ENDANGERED SEA TURTLES

The NRC study

Several species of large marine turtles are listed as threatened or endangered under the US *Endangered Species Act*. The populations of two species in particular, the Kemp's ridley (*Lepidochelys kempi*) and the loggerhead (*Caretta caretta*), have been declining* off the south-eastern and Gulf of Mexico coasts of the United States. Studies by the National Marine Fisheries Service (NMFS), the US Fish and Wildlife Service, and state agencies had led to the conclusion that trawling for shrimps (shrimping) was a major cause of turtle deaths, and therefore NMFS had issued a regulation requiring shrimp trawlers to use turtle-excluder devices (TEDs) in their trawl nets. Shrimpers – especially in Louisiana and Texas – argued that they rarely caught turtles, sometimes fewer than one per year, and that they couldn't be responsible for the declines. In reauthorizing the *Endangered Species Act* in 1988, Congress asked the NRC to examine the scientific data on the extent and causes of the declines of sea turtle populations and to evaluate options for reducing or reversing the declines.

The committee examined and reanalysed the available data. As a result it concluded that shrimp trawling probably killed more turtles than all other human causes combined. It recommended a combination of the use of TEDs and limited trawling times as the most effective conservation methods available, with closures in those few places and times where no other conservation measures were likely to be adequate.

Science's role in management and decision making

The controversy over this issue was quite bitter. The first administrator of the National Oceanic and Atmospheric Administration, Dr Robert White, often refers to his effigy being burned by shrimpers. Opposition to the TED regulations was often expressed in scientific terms: people would question the NMFS conclusions that shrimp trawling killed many turtles by pointing to other sources of mortality, especially beach development, pollution, explosive removal of oil platforms, and other fisheries, and by pointing out that most shrimpers catch no turtles in a year of trawling. Although the

*Kemp's ridley populations appear to have increased since 1990 (Gallaway *et al.*, 1995).

NRC report recognized those factors as significant causes of turtle deaths, its analysis of the data convincingly affirmed the NMFS conclusion that shrimping was the largest single human-caused source of turtle deaths. Indeed, shrimping probably caused more turtle deaths than all other human causes combined; the committee estimated that shrimping accounted for 86% of human-caused deaths of juvenile and adult turtles. While it is true that most shrimpers catch no turtles in a year of shrimping, there is so much trawling effort – the equivalent of 345 000 24-hour days of trawling per year during 1987 and 1988 – that most trawls obviously catch no turtles. However, the cumulative effect of so much effort is significant, even though each individual trawl usually catches no turtles.

Since the report's publication, there has been no significant challenge to the conclusions that shrimp trawling kills turtles and that TEDs are effective in preventing turtle deaths. Indeed, a recent study (Gallaway *et al.*, 1995) reaffirmed those major conclusions. The debate since 1990 has focused more on the effects of TEDs on shrimp catches, on who should bear the costs of protecting turtles, and on the economic effects on shrimp prices of imported, farmed shrimp and appropriate responses to that competition. I believe this example represents an excellent relationship between science and decision making. Once the scientific uncertainties were reduced, the appropriate policy choices could be debated by the appropriate stakeholders.

4.3 THE TUNA–DOLPHIN PROBLEM

The NRC study

Yellowfin tuna (*Thunnus albacares*) are large, pelagic tunas (up to about 100 kg, occasionally larger) of high food value. In the eastern tropical Pacific Ocean (ETP), they associate with dolphins (family Delphinidae). Because they are mammals, dolphins must often surface to breathe, and thus they are easily visible from a boat. Fishers chase schools of dolphins and set large nets around them, because to catch dolphins is to catch tuna. Although the fishers attempt to free all the dolphins unharmed, some die. In the 1960s and early 1970s, it is possible that 500 000 dolphins were killed each year by tuna fishers in the ETP; certainly mortality then was high (NRC, 1992). These deaths aroused much concern, and were in part responsible for the passage of the US *Marine Mammal Protection Act* in 1973.

Estimated deaths in 1989, when the committee started its work, were about 97 000. By 1991, they had declined to 27 292 and by 1996, to 2738 (Marine Mammal Commission, 1997). Despite these reductions in the numbers of dolphins killed, the number was not zero. As a result of the continuing mortality, the US Congress asked the National Research

Council to review the available scientific information, to make recommendations on new and promising techniques for catching yellowfin tuna without killing dolphins, and to identify knowledge gaps and the research needed to achieve zero mortality.

The committee concluded that the most important variable affecting the number of dolphin deaths was the skill of the captains, and therefore recommended that attention be paid to improving that skill through training and certification programmes. The committee judged that no new and promising techniques were likely to have a large effect on reducing dolphin deaths in the near future as long as the primary fishing method included setting nets on dolphins, although it did suggest some possible improvements in gear and fishing techniques and some promising research directions. The committee also concluded that because many nations are involved in catching, processing, and marketing tuna, unilateral actions by the United States might have unintended effects and would probably be less effective than cooperative multilateral actions. Finally, the committee estimated that an annual kill rate of 40 000 dolphins would allow their populations to stabilize and that a kill rate of 20 000 would allow for substantial growth, as long as particularly vulnerable populations were not disproportionately killed. Thus, a complete ban on killing dolphins is not required to conserve their populations; in other words, although zero mortality may properly be a policy goal, it is not scientifically required for conservation. As mentioned above, recent death rates of fewer than 5000 dolphins per year are clearly consistent with conservation of larger dolphin populations than are present today.

Science's role in management and decision making

The effectiveness of the NRC report on the tuna–dolphin problem in influencing policy is much harder to assess than it is for the sea turtles report. Events since its publication are in many ways consistent with its predictions and recommendations, but it does not necessarily follow that the report influenced those events, even though the report was initially requested by Congress. However, it seems likely that if the report had been used, we would be about where we are today, and perhaps the pathway would have been shorter and smoother.

Shortly after the report's publication in 1992, Congress held hearings on the tuna–dolphin problem and then passed the *International Dolphin Conservation Act* of 1992†. Despite its title, some of its provisions involved unilateral actions by the United States. In particular, the act prohibited

† For a fuller discussion of this history, see Marine Mammal Commission (1997).

after 1 June 1994 the sale, purchase, transport, or shipment in the United States of tuna that is not 'dolphin-safe', which was defined as tuna caught by setting nets on dolphins, even if no dolphins were killed. Also in 1992, the governments of all countries that fish for yellowfin tuna in the eastern tropical Pacific (ETP) met at La Jolla, California (home of the Inter-American Tropical Tuna Commission, a nonprofit, international organization). They adopted an agreement to work multilaterally to reduce dolphin mortality 'to levels approaching zero' and to continue and strengthen the international observer programme, among other items.

Under the provisions of the *Marine Mammal Protection Act*, the United States had placed embargoes on non-dolphin-safe tuna, which continued despite the La Jolla agreement. Six signatories to the agreement protested these embargoes in 1995 and argued that using only dolphin-safe methods to catch yellowfin tuna (i.e. not setting nets on dolphins) would harm marine ecosystems by increasing discards of juvenile tuna and increasing bycatch of non-target species such as swordfish, marlins, and sharks.

After the 1994 congressional elections in the United States, new bills were introduced that would ease the United States' embargoes and redefine 'dolphin-safe' to mean tuna caught without killing dolphins. This change would allow tuna caught in nets set around dolphins to be imported as long as no dolphins were killed. One of these bills, the *International Dolphin Conservation Program Act*, was passed in August 1997 and became Public Law 105–42‡ It codifies the Panama Declaration, signed in 1995 by 12 fishing nations, including the United States. The declaration reaffirms most aspects of the La Jolla agreement and continues to express the goal of reducing dolphin mortality to levels approaching zero through annual mortality limits (quotas), with the eventual elimination of all dolphin deaths by finding ways of catching large yellowfin tuna not associated with dolphins.

These events have confirmed two of the NRC committee's conclusions: that multilateral actions are more likely to be effective than unilateral actions (although the threat of a unilateral US boycott certainly had an effect in the short run); and that dolphin mortality can be significantly reduced without major changes in fishing technology. Another conclusion of the committee – i.e. that no effective alternative to setting nets on dolphins is available for catching large yellowfin tuna in the ETP – has not been falsified to date. After the passage of the *International Dolphin Con-*

‡For more details of this law, see *Congressional Greensheets Weekly Bulletin*, 21 July 1997, and The Library of Congress's Legislative World Wide Web site (http://thomas.loc.gov). Only in the ETP could tuna caught in nets set around dolphins be certified as 'dolphin-safe', and only if no dolphins had died. Outside the ETP, 'dolphin-safe' would require certification that the nets were not intentionally deployed around dolphins.

servation Program Act, the situation with respect to the tuna–dolphin problem has become very similar to the conclusions and recommendations of the NRC report, although that report was not explicitly used as the basis for most of the actions taken by the Congress.

4.4 CHANGES IN THE BERING SEA ECOSYSTEM

The NRC study

The Bering Sea ecosystem is a productive, semi-enclosed basin lying between the Arctic Ocean and the Gulf of Alaska. It has been a source of fish, marine mammals, and shellfish for human use for many centuries; in recent years, total landings from the eastern Bering Sea – mainly by the United States – have been around 2 million tonnes per year. Recent declines of some species of marine mammals (particularly northern fur seals, harbour seals, and Steller sea lions) and some birds (particularly common and thick-billed murres and some populations of black- and red-legged kittiwakes) have caused concern, and some people blamed the pollock (*Theragra chalcogramma*) fishery for the declines. At the request of the United States Department of State, the NRC established a committee to examine the information on changes in the Bering Sea, their possible causes, and their management and institutional implications.

The committee concluded that there have been substantial changes in the ecosystem over the past 50 years, notably the declines described above, but also declines in some invertebrate species and small fish such as two species of smelts, capelin (*Mallotus villosus*) and eulachon (*Thaleichthys pacificus*). There have also been increases in some species such as flatfishes, pollock, and some bird populations. There is considerable evidence suggesting that juveniles of the declining mammal and bird species are short of food and are surviving poorly, thus causing the populations to decline. Adults and young animals still receiving parental care appear to be surviving normally.

Because much information is lacking – especially for the western (Russian) side of the Bering Sea – and because of the complexity of a large marine ecosystem such as the Bering Sea, it is not possible to be sure what has caused the changes. However, the most plausible explanation seems to be that environmental changes, in particular a climatic 'regime shift' that occurred in the mid or late 1970s, combined with effects of human exploitation of the ecosystem, in particular the removals of many large mammals and some predatory fish species, have allowed pollock to become dominant. Associated with this domination by pollock was a decline in forage fishes such as capelin, eulachon, and sandlance, which seem to be nutritionally important for some species of young marine mammals and birds. As a result, the young birds and mammals are short of food, and populations of those species are declining.

If this explanation is correct, then reducing pollock catches is unlikely to lead to recovery of the mammal and bird populations in the short term. The committee recommended an increased emphasis on an ecosystem view of the fishery and increased adaptive management. For example, the committee suggested experimental overfishing of pollock in some places and experimentation with the size and placement of exclusion zones around sea-lion rookeries. It also noted the recent concentration of trawling effort for pollock in time and space and suggested that a broader spatial and temporal distribution of fishing effort, with no overall increase in catches, might be helpful to the mammals and birds.

Science's role in management and decision making

This study was requested by the US State Department, and thus is the only one of the four not requested by the US Congress. As a result, it is not surprising that no legislative action has occurred since its publication. But perhaps more important than that is the report's conclusion that the Bering Sea ecosystem is so complex, and so many factors are acting there, that it is not possible to make accurate, quantitative cause–effect predictions. This report, therefore, had a role to play that was somewhat different from the other three. Instead of providing a solution to an immediate problem or resolving the science underlying a controversy, the report provided a general background, recommended a series of broad approaches to a fishery problem (adaptive management, awareness of ecosystem effects, and understanding and resolving institutional arrangements) and helped to identify research priorities. Nonetheless, the report provides a detailed description of many facets of the ecosystem, including institutional arrangements and difficulties; it considers both the Russian and US areas of the Bering Sea; and it analyses the available information. Thus, the report should help various decision makers by providing background documentation and information, by outlining the steps needed to get the information needed for additional decisions, and by helping to provide a broad framework for decision making.

4.5 SALMON DECLINES IN THE PACIFIC NORTHWEST

The NRC study

Pacific salmon (*Oncorhynchus* spp.) have disappeared from 40% of their historical breeding ranges in the Pacific Northwest of North America. In other areas, many populations have declined and many populations are supported mainly by hatcheries. Many changes have occurred in the ecosystems the salmon use; rivers have been dammed, fishing occurs, industrial and residential developments have occurred, logging and grazing continue, and the climate has fluctuated, among other changes. Because so many factors have affected salmon and because so many groups, organizations, and government agencies are interested in them, it has been difficult for people in the region to agree on actions to halt or reverse the declines. As a result, the US Congress asked the NRC to assess the status of salmon populations, analyse the causes of decline, and evaluate options for intervention.

The committee concluded that the salmon problem has been caused by many interacting factors (most but not all related to human activities), and that therefore there is no simple or inexpensive way to reverse or even stop the declines. The report emphasized five major points: the importance of genetic and population diversity in maintaining salmon runs, the failure of current institutions to protect salmon, the importance of natural environmental fluctuations, the importance of reconciling differing values among various groups in addressing the problem, and the importance of efforts to rehabilitate salmon habitats by enhancing and restoring natural structures and processes. Options for intervention that the committee considered to be useful were: changes in the operation of the hydropower system on major rivers and transportation of young salmon around the hydropower projects (dams); changes in fishery management to give higher priority to genetic and population diversity by protecting less-productive populations and focusing on escapement as a primary goal rather than on catch; increased cooperation between the Pacific Northwest, British Columbia, and Alaska; a change in the operation of hatcheries so that they are used as tools to help rebuild natural populations rather than as substitutes for natural production; various approaches to rehabilitation of freshwater habitats, including urban habitats; cooperation among institutions that reflects watershed and regional scales and greater emphasis on adaptive management; and the establishment of an independent, objective science advisory board to provide advice on research plans and to evaluate scientific information.

Science's role in management and decision making

The decline of salmon in the Pacific Northwest has provoked an enormous amount of controversy and acrimony. Many people, groups, organizations, agencies, regions, states, and even countries are involved including the following:

US Government National Marine Fisheries Service;
US Army Corps of Engineers;
US Fish and Wildlife Service;
US Forest Service;
Canadian Government Department of Fisheries and Oceans;
British Columbia Government Ministry of Environment, Lands and Parks;
British Columbia Government Ministry of Agriculture, Fisheries and Food;
Alaska Department of Fish and Game;
Oregon Department of Fish and Wildlife;
Washington Department of Fish and Wildlife;
other US states, especially Idaho and California;
Bonneville Power Administration;
Northwest Power Planning Council;
Columbia River Intertribal Fish Commission;
Oregon Trout;
Northwest Forestry Association;
Pacific Salmon Trollers' Association;
American Sportfishing Association;
American Fisheries Society;
Federal District Court;
United States Congress;
Greenpeace;
individual tribes of Native Americans in the US;
individual tribes of First Nations in Canada;
electric power industry;
universities;
consultants;
private individuals.

Many blame each other for the declines. The truth is that almost no activity is blameless and a comprehensive programme to restore salmon runs will affect most people and groups. It is also true that groups' blaming each other has not been effective in restoring salmon. The NRC (1996b) report *Upstream* emphasized the complexity of the problem, including the multiplicity of factors and the importance of people's values in deciding on desired end points and courses of action. It also emphasized biological (e.g. genetic) and physical (e.g. oceanic) aspects of the problem.

Since the committee's findings were made public late in 1995, the National Marine Fisheries Service and the Northwest Power Planning Council have established an Independent Science Advisory Board to focus on the Columbia River salmon, and the Senate has included an amendment in the FY 1997 Appropriations Bill that would establish an independent scientific committee to review the approximately $100 million annual research budget of the Columbia Fish and Wildlife Authority, an organization of agencies that fund research on salmon and salmon management in the Columbia River. In addition to those formal actions, many newspaper editorials and articles have discussed the report, mostly using it or recommending it be used as a starting point for additional analyses and action.

Perhaps the greatest contribution of the report was in describing a scientific consensus, making clear where scientific uncertainties lie, and distinguishing between scientific uncertainties and conflicts among various values and institutional goals. It was helpful that other scientific activities – most notably the reports of the Snake River Salmon Recovery Team (1994), the Northwest Power Planning Council's Independent Science Group (ISG, 1996) and Stouder *et al.* (1997) – came to many similar conclusions and some of the same recommendations. By emphasizing and documenting the importance of genetic considerations, the importance of natural environmental fluctuations, and the importance of reconciling differing values and institutional goals, the NRC report was also able to suggest some aspects of the problem that cannot be overlooked. But effective solutions to the salmon problem will need a large investment of human and financial resources – and some good fortune – so it will be a long time before the role that science has played in this matter can be properly evaluated.

4.6 CONCLUSIONS

In each of the four cases discussed, scientific uncertainty aggravated controversy. In some cases, notably the sea turtles and salmon in the Pacific Northwest, the uncertainty was to some degree a matter of poor communication among interested parties rather than disagreement among scientists. In each case, science was recognized by at least some of the parties as an important part of solving the problem. However, science was and should be only part of the solution. Although policy decisions and fishery management should be informed by science, nevertheless values, economics, politics, and other considerations should go also into them. Why, then, was the sea turtle case more successful than the others?

There are several answers to that question, including the comment that success is in the eye of the beholder. But clearly the sea turtle case was simpler than the salmon and Bering Sea cases. Only two main species in one environment were under consideration; there was an abundance of useful and relevant data on the lives of the turtles and on the shrimp fishery; only one major agency (NMFS) and one major user group (shrimp trawlers) were involved, although other agencies, particularly the US Fish and Wildlife Service, and other interest groups, particularly the Center for

Marine Conservation, were also involved. There was considerable animosity and mistrust among some of the players, although many of the individuals were known to each other, which seemed to limit the degree of open hostility. There were no significant miscommunications about people's values and desires; everyone knew that NMFS was complying with the *Endangered Species Act* to protect the turtles and that the shrimpers did not want to lose their small margin of profitability by incurring additional expenses and reducing their catches. Finally, the scientific conclusions were so clear and strongly documented that it was obvious that the debate should focus on other topics.

I believe that a large change in any one of those factors would have changed the outcome. If many jurisdictions and agencies had been involved, as is the case for salmon (see list, p. 66), the problem would have been much harder to solve, even if the science were clear. It is likely that if many different goals and values had been involved, or several species in several environments, especially with complex interactions, or many user groups, or a large degree of scientific uncertainty, the problem would also have been much harder. The remaining three cases provide examples of this increased difficulty.

Perhaps the most instructive is the tuna–dolphin problem. In that case, some people were talking at cross-purposes. Many of those promoting the *Marine Mammal Protection Act* provisions and the *International Dolphin Conservation Act* were opposed to the killing of any dolphins, and objected as well to any activity that harassed dolphins. That is a policy-based goal. The membership of the IATTC, in the face of increasing concern among many people and groups about the enormous number of dolphins being killed by tuna fishing in the 1970s, was trying to reduce dolphin mortality to the point that their survival and population increases were assured – a science-based criterion. Although few if any people want to kill dolphins, it appears to me that the IATTC membership would accept an annual dolphin-kill rate small enough to ensure population growth but larger than zero.

In this case, science was able to make estimates of the effects of fishing-caused deaths on dolphin populations, including an estimate of a mortality rate that would ensure survival and growth of dolphin populations. It could say that 'a complete ban on dolphin fishing or the purchase of tuna caught on dolphins is not required to ensure the survival and even increase of dolphin populations' (NRC, 1992). However, for science to be useful in decision making and management (and indeed as a general principle), people's agendas, values, and goals must be explicit.

Indeed, the question of different goals and values is a difficult one in fishery management. Most fishery managers have to balance several, often poorly articulated goals in doing their job. Examples of such, sometimes incompatible, goals for commercial and recreational fisheries include stabilizing yield, maximizing yield, increasing employment, maintaining long-term sustainability of the fishery, preserving a lifestyle, minimizing risk of stock collapse, maximizing accessibility, maximizing the average size of fish caught, and so on (Policansky, 1993). Subsistence fisheries have additional goals. The salmon problem in the Pacific Northwest abounds with such examples, including differing views of what nature should look like and differing values placed on the existence values (i.e. the values placed on a resource by the fact of its existence rather than from its use) of salmon (NRC, 1996b). As long as the various goals are not articulated and discussed openly, fishery managers and the scientists who advise them will often be unable to provide appropriate advice and take appropriate action, and debates over management policies will be more acrimonious and less productive than they need be. In a case like the Pacific salmon, there are so many values and interests that the problem will never be easy to solve.

Finally, it is important for scientists to recognize that science is only a part of the solution to fishery problems (Fig. 4.1), although it is an essential part. Much of fishery management and decision making (and resource

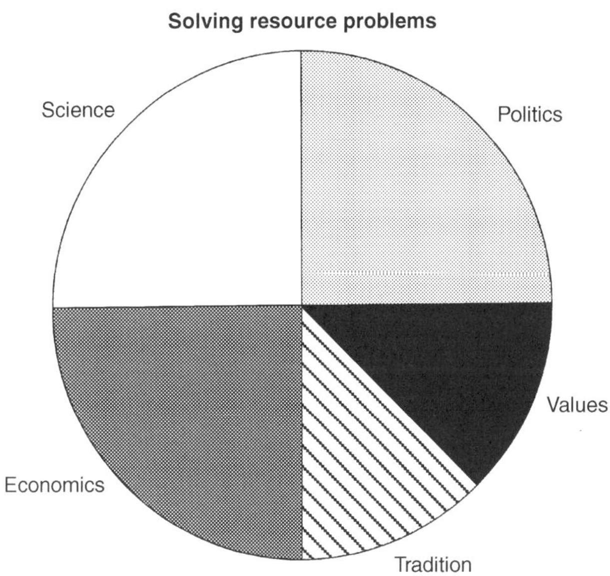

Fig. 4.1 Pie diagram of some major factors involved in fishery management and decision making. The relative sizes of the segments are only approximations, but the science segment rarely is, nor should it be, much larger than depicted here.

management in general) lies outside science, involving economics, values, politics, and other factors. (Yet the poor fishery manager is expected to be expert in all those disciplines and to make everybody happy while ensuring a supply of fish!) It is easy for scientists to complain that they have been ignored in decision making, when in fact they have not been ignored, but other factors have been considered in addition to their advice.

Deciding how much to say about policy can be a delicate issue for scientists, one that is difficult to get right. Scientific advice should be separate from policy advice, but not disconnected from it. A scientist who advocates policy too strongly can lose scientific credibility, but many people see such advocacy as a civic duty. What scientists can and must do is to ensure that our advice is understood and not misrepresented. While it is often proper for decision makers and managers to weigh other factors as heavily as scientific advice, it is never proper for them to claim scientific support for a decision when there is none.

ACKNOWLEDGEMENTS

The volunteers who served as NRC committee members taught me a great deal. The committee chairs in particular – Bob Francis (Bering Sea, tuna–dolphin) and John Magnuson (sea turtles, salmon) – provided many insights in our conversations. James Reisa and Paul Gilman of the NRC have consistently asked hard questions and provided wise advice in all these activities. The Vancouver workshop discussants helped to clarify several issues. I thank them all.

REFERENCES

Gallaway, B.J., Bryan, J.D., Martin, L.R. and Cole, J.G. (1995) Sea Turtle and Shrimp Fishery Interactions – Is a New Management Strategy Needed? LGL Ecological Research Associates, Inc., Bryan, TX.

Gordon, H.S. (1954) The economic theory of a common property resource: fish. *J. Polit. Econ.* **62**, 124–142.

Hardin, G. (1968) The tragedy of the commons. *Science* **162**, 1243–1248.

ISG (Independent Scientific Group) (1996) Return to the River: Restoration of Salmonid Fishes in the Columbia River Ecosystem, prepublication copy, Northwest Power Planning Council, Portland, OR.

Ludwig, D., Hilborn, R. and Walters, C. (1993) Uncertainty, resource exploitation, and conservation: lessons from history. *Science* **260**, 17 and 36.

Marine Mammal Commission (1997) *Annual Report to Congress, 1996*, Marine Mammal Commission, Washington, DC.

NRC (National Research Council) (1990) *The Decline of the Sea Turtles: Causes and Prevention*. National Academy Press, Washington, DC.

NRC (National Research Council) (1992) *Dolphins and the Tuna Industry*, National Academy Press, Washington, DC.

NRC (National Research Council) (1996a) *The Bering Sea Ecosystem*, National Academy Press, Washington, DC.

NRC (National Research Council) (1996b) *Upstream: Salmon and Society in the Pacific Northwest*, National Academy Press, Washington, DC.

Policansky, D. (1993) Evolution and management of exploited fish populations, in *Proceedings of the International Symposium on Management Strategies for Exploited Fish Populations* (eds G. Kruse, D.M. Eggers, R.J. Marasco, C. Pautzke and T.J. Quinn II), Alaska Sea Grant College Program Report no. 93–02, University of Alaska, Fairbanks, AK, pp. 651–664.

Snake River Salmon Recovery Team (1994) Snake River Salmon Recovery Plan Recommendations, Final Report, National Marine Fisheries Service, Seattle, WA.

Stouder, D.J., Bisson, P.A. and Naiman, R.J. (eds) (1997) *Pacific Salmon and their Ecosystems: Status and Future Options*, Chapman & Hall, New York.

Chapter five

People, purses and power: developing fisheries policy for the new South Africa

Kevern L. Cochrane and Andrew I.L. Payne

ABSTRACT

At present, arising in part from the former policy of apartheid, many of the major fisheries in South Africa are in the hands of a limited number of quota holders, dominated by the white sector of the population. Several of the major stocks are either in a productive state or are recovering, under biologically conservative management, from earlier depletions. At the same time, with the democratization of South Africa, the current access rights system is being challenged by many who had previously been excluded from access. The policy makers are faced with the task of introducing a fair, equitable and widely accepted system of user rights to a mature fishery, while taking cognizance of the important social and economic role of the existing industry and without overexploiting the marine resources.

The search for suitable approaches has highlighted some of the strengths and weaknesses of existing South African approaches to fisheries management. Analysis of these features, and consideration of future approaches, has relevance to fisheries management in many areas elsewhere in the world.

This analysis suggests that far greater emphasis than in the past needs to be placed on rigorous analysis of the social and economic dynamics of

Reinventing Fisheries Management. Edited by Tony J. Pitcher, Paul J.B. Hart and Daniel Pauly. Published in 1998 by Kluwer Academic Publishers, London. ISBN 0 412 83410 3.

fisheries, and incorporation of the results from this analysis into management procedures and approaches. This will inevitably also require greater participation by users in the management of fisheries. It is suggested that similar changes are happening, and need to be encouraged, in other fisheries around the world. Developing the capacity for adequate consultation, co-management and human-orientated research in fisheries management agencies will generally require substantial restructuring. Of overriding importance is that this restructuring must be accompanied by the political will to ensure that only sustainable utilization is performed.

5.1 INTRODUCTION

Crisis in world fisheries

The state of fisheries around the world is cause for considerable concern. Rosenberg *et al.* (1993) reported that about 45% of the stocks in the USA for which data are available are overexploited and that 59% of stocks in European waters are considered to be in a similar state. In their review of the state of world marine fishery resources in 1993, the Food and Agriculture Organization of the United Nations (FAO) pointed to a number of disturbing signs of overexploitation of living marine resources, including declines in the world total annual catch in recent years and growing contributions of low-value small pelagic fish to the global total (FAO, 1993; Buckworth, Chapter 1, this volume).

This crisis has led to extensive criticism and scrutiny of current approaches to stock assessment and management of living marine resources. Walters and Maguire (1996) suggested that the blame for the collapse of the cod stocks can be attributed to a combination of gross overestimation of stock sizes by scientists, misreporting and dumping, and government policies which subsidized increasing fishing effort. Rosenberg *et al.* (1993) included uncertainty in resource status and ecological processes, failure of resource managers to accept the advice offered by scientists, and systems of open access to resources as causes for failures of fisheries management. Charles (1995a) suggested that a failure to develop a holistic, integrated approach to fisheries management is responsible for this alarming state of affairs.

Perhaps synthesizing many of these and similar theories, Stephenson and Lane (1995) listed six limitations of current approaches to fisheries management. These can be paraphrased as follows:

1. problems generated by environmental, biological and economic variability;

2. the absence of longer-term management objectives;
3. an absence of short-term accountability for fisheries managers or management agencies;
4. historical, and still prevailing, dominance of biological input to, and corresponding influence on, decision makers, and corresponding deficiency of advice on economic, social and operational considerations;
5. lack of involvement by 'stakeholders and vested interest groups' in fisheries management;
6. reduced efficiency in regulatory agencies as a result of economic rationalization.

New fisheries policy in South Africa

Democratic elections were held in South Africa for the first time in April 1994, marking the end of white minority rule and the beginning of a new democracy. This event was a political watershed, but the process of bringing about real change to all aspects of life will still take some time. To bring about this change, virtually all aspects of South African economic, social and political life are undergoing intense scrutiny, analysis and discussion to identify the problems within the new political context and their optimal solutions. This is reflected in the Reconstruction and Development Programme (RDP), which underlies much of the strategic thinking and planning that is taking place in the country now (African National Congress, 1994). Similar principles are to be found in the country's new Growth Employment and Redistribution programme.

With regard to fisheries, the RDP identified the priorities for the country's marine fisheries as:

> Marine resources must be managed and controlled for the benefit of all South Africans, especially those communities whose livelihood depends on resources from the sea. The fishing stock must be managed in a way that promotes sustainable yield.... The democratic government must assist people to have access to these resources. Legislative measures must be introduced to establish democratic structures for management of sea resources.

(African National Congress, 1994). It is significant that this strategy identifies the following key features: the need for management and control; the beneficiaries; sustainable utilization; clarification of access to the resources; and the nature of the management structures. In the eyes of a, then, government-in-waiting, with little experience of fisheries management but a clear view of the needs of the country, these five points were seen as the cornerstones of a future fisheries policy.

Within this context, this chapter discusses the existing approaches to utilization and management of South Africa's living marine resources and fisheries and the process of developing a policy that includes the above five goals. It examines the key issues that are emerging from this process, and attempts to identify principles that may be applicable to the problems facing fisheries throughout the world.

5.2 MAJOR COMMERCIAL FISHERIES AND THEIR MANAGEMENT

In terms of the magnitude of its total catch, South Africa can be classified as a middle-order fishing nation and is usually ranked about 30th in the list of the world's fishing nations (Fishstat database: FAO, 1996). The total wholesale value of the national catch was nearly US$400 million in 1994 and the fisheries employ some 25 000 people (Table 5.1). The total value of the catch is equivalent to approximately 0.5% of the gross domestic product of the country. Hence fishing in South Africa is of only limited national importance. Nevertheless, it does provide jobs, food and income for tens of thousands of people, and in a country where 'poverty is the single greatest burden' (African National Congress, 1994), no sources of income or employment can be lightly discarded. In addition, in some coastal areas, where poverty is particularly high (Issel, 1995), fishing provides a major source of income and employment.

Table 5.1 Social (as indicated by number employed in 1991) and economic value (1994) of the major South African fisheries. Information on employees from Branch *et al.* (1996), and on wholesale value from Stuttaford (1996); US$1 = SA Rand 3.57 (exchange rate on 31 December 1994)

Fishery	Employees (1991)		Wholesale value – processed (1994) (US$ millions)
	Fishers	Onshore	
Demersal and midwater	2 800	5 900	201
Pelagic	1 000	3 700	81
Rock lobster	2 200	1 600	47
Linefish, squid, tuna	6 200	2 000	46
Abalone	140	80	15
Others			7
Total	>12 340	>13 280	397

Demersal fishery

The single most valuable fishery in South Africa is the demersal or groundfish fishery, accounting for about half of the total earnings from the fishing sector (Table 5.1) and employing nearly 9000 people. This fishery is based largely on two species of hake, *Merluccius capensis* and *M. paradoxus*, which together account for more than 70% of the catch (Payne and Cochrane, 1994). At its peak production in the early 1970s, this fishery yielded about 300 000 t in a year (Table 5.2). However, this heavy exploitation was above the sustainable yield of the stocks, the biomass of which declined dramatically. In 1978, a number of conservative management measures began to be implemented, including a total allowable catch (TAC) and limited entry (Payne, 1989). The status of these stocks is now considered to be recovering (Table 5.2), although recent analyses of the CPUE data throw some doubt on the extent of this recovery (Cochrane *et al.*, 1997).

Table 5.2 The peak annual catch of key resources in the South African fishery, the 1996 TAC, and a qualitative description of the probable status of the resources. The 1996 TACs were obtained from Stuttaford (1996) and peak annual catches were obtained from Stuttaford (1996), SFRI (unpublished data), Penney *et al.* (1989) and Tarr (1989). MSYL, maximum sustainable yield level; TAC, total allowable catch

Resource	Peak annual catch (t) and (year)	1996 TAC (t)	Likely status
Hakes	295 000 (1972)	151 000	< MSYL, recovering
Pilchard	410 000 (1962)	64 500	Dynamic, increased this decade
Anchovy	596 000 (1987)	200 000	Dynamic, decreasing probably for environmental reasons
West Coast rock lobster	>18 000 (1951)	1 500†	Depleted
South Coast rock lobster	973 (tail mass, 1974)	427 (tail mass)†	> MSYL, stabilized
Abalone	2 800 (1965)	615†	Uncertain
Linefish	18 000 – 20 000 (late 1960s – early 1970s)	9026‡	Generally depleted

†1995/96 season.
‡Total catch = 9026 in 1993 (non-quota species).

The fishery is currently dominated by deep-sea trawlers, which caught about 90% of the reported demersal catch in 1993 (SFRI 1995a). The deep-sea fleet targets primarily hake. This fishery shows considerable concentration of quota holdings and 71% of the quota for hakes, 151 000 t, was held by two major companies in 1996 (Table 5.3). However, in recent years there has been growing interest in longlining for hakes. This operation can take place at a variety of scales, and can be much less capital intensive.

The response to this growing interest in small-scale fisheries, and the resulting threat to the existing trawling concerns and the thousands of people dependent on them, has been, in many respects, a model for future development in all South Africa's fisheries. A carefully planned scientific and socio-economic experiment has been devised, with the participation and support of both the existing fishing concerns and the aspirant newcomers. This experiment is investigating and comparing the relative biological, social and economic impacts of trawling and longlining. It commenced in 1994 and is scheduled to run for up to five years. The intention is that, with the results from the experiment, the final decision on the future of this fishery can be based on rigorous information pertaining to its potential to meet the immediate and long-term needs of the country.

Other species targeted by the demersal fishery are Agulhas sole, *Aus-*

Table 5.3 Number of quota holders in the major fisheries sectors in South Africa and the percentage of quota held by the top 2–5 quota holders in each fishery in 1996. Information from Stuttaford (1996). Total number of quota holders shows all quotas in all different sectors. Statistics for major quota holders refer to the sum of quotas held by the parent company in each case

Resource	Total number of quota holders	Number of major quota holders	% Quota held by major holders
Hakes	53	2	71
Pilchard (directed + bait)	56	4	71
Anchovy (reduction)	17	5	78
West Coast rock lobster*	103	5	60
South Coast rock lobster	8	4	94
Abalone	16†	5	91
Linefish‡	3223	n/a	n/a

*1995/96 season.
†1995/96 season, increased from 6 in 1994/95.
‡Not regulated by quota: number given is the sum of licences for category A and B handline, tuna and squid.

troglossus pectoralis, of which a TAC of some 850 t is taken annually by inshore trawlers, and horse mackerel, *Trachurus trachurus capensis*, for which a precautionary upper limit of 58 000 t of adults on the South Coast is set but, for reasons of limited commercial interest, never taken.

Pelagic fishery

The second major fishery, in terms of commercial value, is usually the pelagic fishery (Table 5.1), dependent mainly on anchovy, *Engraulis capensis*, and sardine, *Sardinops sagax*. The commercial fishery started in 1943 (Crawford *et al.* 1987), exploiting mainly *S. sagax*, but a probable combination of overfishing and unfavourable recruitment led to dramatic declines in the biomass of this species, and in the mid 1960s *E. capensis* became the dominant species in the catch. In terms of total catch, anchovy remained the dominant species until 1995, yielding peak landings of nearly 600 000 t in 1987 and 1988 (Table 5.2). However, in 1994 and subsequently, hydroacoustic biomass estimates indicated that sardine was the more abundant species in South African waters, in terms of adult biomass. This reversal has come about through poor recruitment in recent years leading to a decline in anchovy biomass, and several years of good recruitment accompanied by conservative management in sardine, leading to an increasing biomass of this species (Fig. 5.1). As a result, the anchovy fishery was closed in 1997, while the sardine TAC was set at approximately 100 000 t (SFRI, unpublished information). Subsequently, in June 1997, the anchovy fishery was reopened following good recruitment estimates at mid-year. Anchovy are used in a reduction fishery, while sardine are used mainly for canning for human consumption and hence yield a substantially higher return per unit mass caught.

In 1991, nearly 5000 people were employed in this fishery (Table 5.1), although many of them were employed on a seasonal basis. In 1996 over 70% of the TAC in the sardine fishery was held by four companies and over 75% of the quota in the anchovy fishery was held by five companies (Table 5.3). Several quota holders hold quotas for both species. The pelagic fishery is also capital intensive and not readily open to small-scale new entrants. There could, however, be some scope for such entrants in the bait fishery, using smaller vessels and nets and operating close inshore.

A complication in this fishery, which exemplifies a typical fisheries management problem, has been bycatch of juvenile sardine in anchovy-directed catches. Uncontrolled bycatch has the potential to reduce substantially the potential yield in the directed fishery for adult sardine. Clearly, it is impossible to maximize yields from both fisheries simultaneously and one or other needs to be given priority to the detriment of the other. Nevertheless, the various sectors of the fishery have been unable to reach agree-

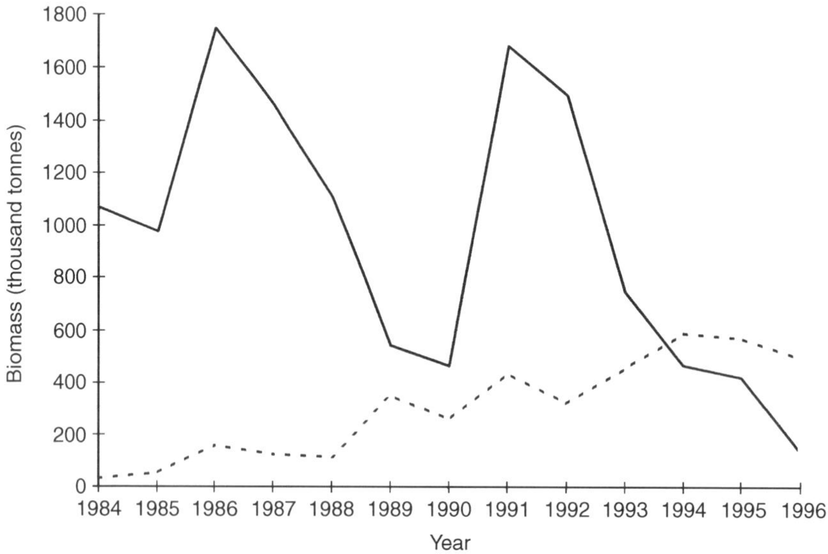

Fig. 5.1 Estimates of anchovy (solid line) and sardine (broken line) spawner biomass in South Africa obtained from hydroacoustic surveys undertaken each year from 1984 to 1996. (Source of data: Cochrane *et al.*, 1997, and Sea Fisheries Research Institute Cape Town, unpublished information.)

ment on a compromise approach. As a result, optimal management of the combined resource has been hindered by a failure to reconcile conflicting objectives, even in a relatively simple fishery with only two conflicting objectives and many parties common to both sides.

Linefish fisheries

The linefish fishery is probably the most heterogeneous fishery discussed in this chapter, and could be considered to be a number of discrete fisheries, including, for example, the small-scale inshore fishery, using open boats and with crews of three or four, a larger-scale operation with vessels able to stay at sea for several days and travel further offshore, the tuna fishery, the squid jig fishery and others. Taken collectively, in 1994 this was, along with the lobster fishery, the third most economically valuable fishery in South Africa and employed more people than the pelagic fishery (Table 5.1). Several of the stocks upon which this fishery depends have been badly depleted and some, such as the red steenbras, *Petrus rupestris*, geelbek, *Atractoscion aequidens*, and seventy-four, *Polysteganus undulosus*, have been depleted to very low levels (Penney *et al.*, 1989). The albacore, *Thunnus alalunga*, is considered to be overexploited (ICCAT, 1996) and the 'chokka' squid is considered to have declined substantially, at least over the last decade, in response to both incidental exploitation by the demersal fishery and directed mortality from the inshore jig fishery, the latter only established in the early 1980s (Roel and Cochrane, 1995).

As would be expected in a diverse fishery exploiting resources frequently available in shallow water close inshore and caught by line, access to the linefish fishery is held by a relatively large number of fishers. Permits are issued for commercial operations (A-permits), semicommercial (B-permits), tuna fishing and jigging for squid. In 1996, a total of 3223 permits was issued (Table 5.3). Of these, 439 were A-permits. Several purse seiners participating in the pelagic fishery and rock lobster fishers also hold B-permits for linefishing and all fishers licensed to catch squid in the jig fishery are also obliged to hold A-permits in the linefish fishery (SFRI, 1995a). Hence there is scope for, and incidence of, considerable movement between the different sectors of the South African fishery.

Rock lobster fisheries

The rock lobster fishery consists of two distinct fisheries, one exploiting *Jasus lalandii* along the West Coast of South Africa and the other harvesting *Palinurus gilchristi* along the South Coast of South Africa. There is also a third fishery developing off the north-east of the country, exploiting *P. delagoae*. This species, and probably stock, has been exploited by a well-developed fishery in Moçambique for a number of years. Collectively, these fisheries employ just under 4000 people (Table 5.1).

Catches of *J. lalandii* peaked at about 18 000 t in the early 1950s, but since then have declined virtually steadily to a TAC of 1500 t in the 1995/96 season (Table 5.2). Overfishing has certainly been largely responsible for the decline and the resource has been mined out, although conservative management approaches began to be implemented with the introduction of a catch quota in 1970 (Crawford *et al.*, 1987). However, in the late 1980s, the generally declining trend was exacerbated by a still-unexplained reduction in somatic growth rates to levels about one-third of those estimated earlier in the decade (SFRI, unpublished data). The phenomenon has persisted to the present and has had the effect of dramatically reducing the stock production although there have been signs of recovery in some areas. The TAC was increased to 1700 t for the 1996/97 season (SFRI, unpublished data).

There were 103 quota holders in the West Coast rock lobster fishery at the time of writing, giving relatively wider access to this fishery than to some of the others. Nevertheless, there is some concentration within this, and five major quota holders held 60% of the TAC for the 1995/96 season (Table 5.3). West coast rock lobster are easily available to the small-scale fisher and, at present, between 15% and 20% of the annual catch is taken by baited hoop nets (Cruywagen *et al.*, 1994), which are used in shallow water and can be set from open dinghies. The remainder of the catches are taken from larger vessels equipped with traps or lobster pots. As a result of its relatively easy availability, there is considerable interest in and pressure to grant access to this resource to fishers along the Cape West Coast, where socio-economic conditions, as in many fishing communities within South Africa, are generally very low.

The *P. gilchristi* stock, upon which the South Coast fishery depends, is apparently in a healthier state. The TAC for this species was set at 475 t tail mass from the mid 1980s. Recent analyses of the catch per unit effort (CPUE) have indicated that this is above the maximum sustainable yield, and the TAC has subsequently been reduced to 427 t tail mass. However, the resource biomass is still estimated to be above the maximum sustainable yield level (Table 5.2). This fishery is also marked by concentration of quota, with four quota holders gaining access to 94% of the TAC in 1995/96 (Table 5.3). The South Coast rock lobster occurs at depths of between 90 and 170 m and is caught using lobster pots set from vessels equipped to stay at sea for weeks at a time, requiring greater capital investment, and presumably operating costs, than for the catches of its West Coast equivalent.

Abalone fishery

The final commercial fishery to be discussed here is that for abalone, *Haliotis midae*. While a relatively minor component of the South African fishery, it is regionally very important, providing substantial social and economic input (Table 5.1) to areas of the south-western Cape (Fig. 5.2). The status of this resource is really not clear. While there is variability from site to site, in general there have been rising catch rates in the abalone fishery for a number of years. However, there is some debate as to whether these catch rates represent real trends in the abundance of the resource, changes in the distribution of the resource, or changes in fishing behaviour by the commercial fleet (Raubenheimer *et al.*, 1998). This uncertainty, coupled with escalating concern about the level of poaching on abalone, substantially hinders attempts to manage the fishery with an optimal yield.

Another deep concern in this fishery within South Africa in recent years has been the developing conflict over access. For the 1994/95 abalone season there were six quota holders for a TAC of 615 t (Table 5.3). One of these quota holders was the Hermanus Community trust which was granted 10 t for the season. Some steps have since been taken to broaden

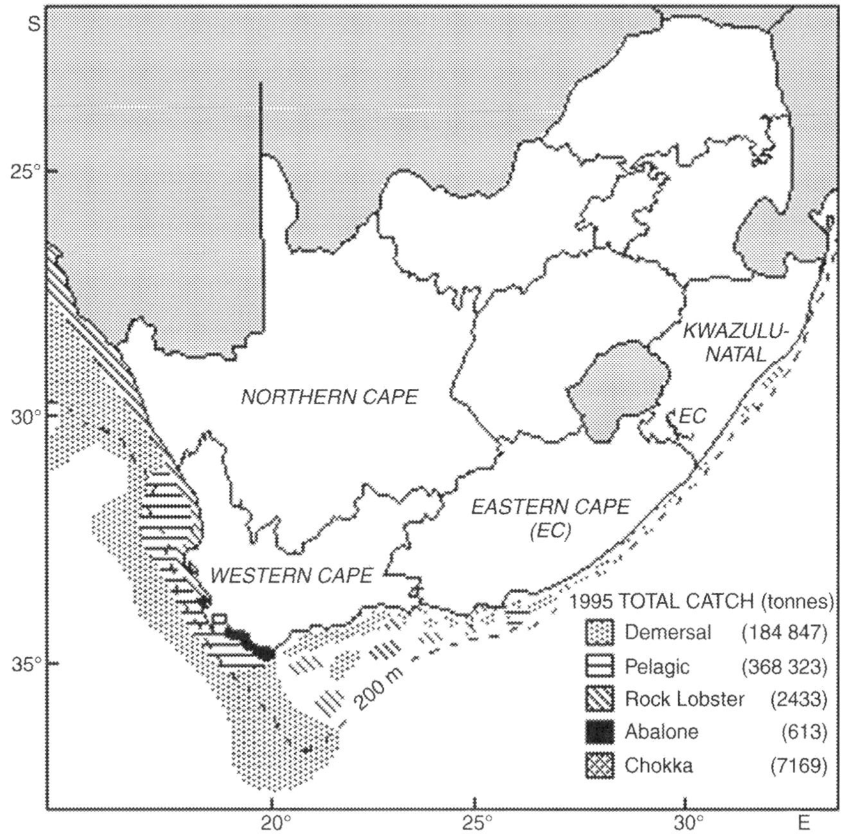

Fig. 5.2 Outline of the coast of South Africa indicating the localities of the coastal provinces and the positions of the major fishing grounds for key fish resources.

access in this fishery and quota was allocated to a total of 16 holders for the 1995/96 season. The original six holders, two of which were from the same parent company, now have access to 93% of the annual TAC (Table 5.3).

Abalone are harvested in this area by divers using hookah equipment (Tarr, 1989), and there are 50 professional divers employed in this way. However, there is enormous pressure to increase access to the resource, and this has led to demonstrations and street blockades in the abalone fishing areas by those demanding access. The problem is exacerbated by extensive poaching, some of which is believed to be linked to drug smuggling, while other poachers claim a right of access, and state that they have been forced into poaching by an unjust system of access (Viljoen,

1995). Violence has broken out and control in some areas is non-existent (Payne and Cochrane, 1995).

Other fisheries

There are other fisheries of lesser social and economic importance in South Africa. These include a midwater trawl fishery for horse mackerel, beach seine and drift net fisheries for mullet and St Josephs shark, a developing fishery for whelks, *Bullia laevissima*, and a crab species, *Ovalipes trimaculata*, and others. These fisheries do have significant roles to play, even if providing employment and income for numbers measured in single digits, rather than in tens, hundreds or thousands. However, they are not explicitly discussed in this chapter, although it is suggested that the overall conclusions will apply in these cases as well.

5.3 RECREATIONAL FISHERIES

There are significant recreational fisheries utilizing South Africa's living marine resources. The most important of these, in terms of the number of participants, is undoubtedly the recreational angling fishery. It has been estimated that 750 000 people in South Africa participate in recreational angling (Branch *et al.*, 1996). While very rough, this figure is generally accepted. It has similarly been estimated that these fishers, on average, fished on 30 days each year and that they caught a total of 17 000 t of fish per annum, with a value of US$29 million, but spent US$200 million in doing so, injecting this latter amount into the economy (C.G. Attwood and S.L. Lamberth, Sea Fisheries Research Institute, Cape Town, pers. comm.; Branch *et al.*, 1996).

There are also important recreational fisheries for abalone and for West Coast rock lobster. More than 54 000 permits were issued for recreational fishing for rock lobsters in the 1994/95 season. It has been estimated that each permit holder caught an average of 20 lobsters during this season (SFRI, 1996). The recreational fishery for abalone generated estimated landings of almost 550 t in 1993/94, approaching the commercial TAC (SFRI, 1995a,b), although they decreased to approximately 375 t in 1994/95 (SFRI, 1996). In 1994/95, more than 34 000 permits were issued for recreational collection of abalone (SFRI, 1996).

Recreational fisheries are also in the centre of the fisheries policy debate, with heated opinions being expressed on the question as to whether recreational fishers should be allowed to sell their catch to defray some of their costs. In general, the commercial fishers are against it and many recreational fishers are in favour of being allowed to sell their catches.

Clearly, this issue is closely linked to the questions of access, and such issues as the role of small-scale and subsistence fishers in the future.

5.4 SUBSISTENCE FISHERIES

Subsistence fisheries in South Africa are not very well developed. Possibly one reason for this has been the existence of racial discrimination, which denied many black South Africans access to much of the coastline and its resources. However, even prior to the establishment of apartheid policies, traditional, subsistence fisheries were not a marked feature of the country and this is perhaps mainly owing to the rugged, frequently dangerous, coastline found around much of South Africa.

One of the longer-standing subsistence fisheries that does exist is for intertidal molluscs and crustaceans, and some finfish, in the south-eastern part of the country. Here some 25 species of mollusc are collected, largely by women and children, with the brown mussel, *Perna perna*, clearly dominating catches. This fishery is essentially unmanaged, apart from a few marine reserves in the area, and the resources have been heavily depleted (Siegfried *et al.*, 1985; Lasiak, 1991). Similar subsistence fisheries for intertidal molluscs occur in Kwazulu Natal (Jean Harris, Natal Parks Board, pers. comm.)

Another subsistence fishery occurs in the north-eastern part of the country, along parts of the coast of Kwazulu Natal (Fig. 5.2). Most of the fishing takes place within the Maputaland Marine Reserve, in the northern parts of the province, and animals such as redbait, mussels and limpets are taken. This fishery is currently reasonably well managed in an experimental co-management programme (Oceanographic Research Institute, Durban, unpublished information; Payne and Cochrane, 1995).

There is certainly also subsistence fishing along the coastline in the form of shore-based handline or rod and line fishing. However, statistics for this activity are not separated from those for recreational angling and the extent of such subsistence utilization is not known, although a nationwide study into the economic value of the South African linefishery is currently underway (M. McGrath *et al.*, unpublished data). This subsistence use could certainly grow in an environment of greater access and continuing poverty.

5.5 CONTROL AND ENFORCEMENT

Effective control is essential for ensuring optimal management and utilization of marine resources. Even where effective co-management is in place, there will still be a need for rigorous control and enforcement of fishing practices and catches to ensure that a uniform and clear set of rules applies to, and is adhered to by, all participants.

In its transitional phase, with applications for access to fishing and uncertainty about access in the future, a strong control arm is particularly important. The Fisheries Policy Development Committee recognized this, and the need for control and enforcement are stressed in the draft Fishing Policy for South Africa (Draft 6, March 1996). That committee also recognized the potential importance of co-management in this regard. Despite the above, funding to support the marine control arm has shrunk in real as well as relative terms and four of the seven small craft used for the purpose are no longer maintained in a seaworthy state (Sea Fisheries, unpublished information). The other three do not go to sea, because of financial constraints, and much of the infrastructure necessary to support a meaningful control service is simply not in place because of fund shortages.

These problems will have to be addressed for effective fisheries management.

5.6 THE SOCIAL AND ECONOMIC CONTEXT IN SOUTH AFRICA

Poverty is an overriding problem in South Africa. This situation is reflected in many social and economic indicators. For example, the total population is about 40 million, of whom 17 million (43%) are estimated to be living below the 'minimum living level' (African National Congress, 1994). There is an estimated shortfall in housing of 3 million units, and 12 million people have no access to clean drinking water.

The brief review of South Africa's fisheries provided above presents a picture of essentially fully developed fisheries and fully exploited to over-exploited resources. However, the emphasis of the new, democratically elected government on management of marine resources 'for the benefit of all South Africans' and on assistance to 'people to have access to the resources' (see Introduction) has raised the expectations of many for a change in the current system of access. In this chapter we make no attempt to discuss the different concepts of access or the relationship between access and ownership. We use access in a broad sense to encompass any right that enables a user to exploit directly one or more marine resources for their own benefit.

The principle of limited access has been accepted as an underlying principle in the policy development process (Draft Fishing Policy for South Africa, Draft 6, March 1996). However, the issue which remains to be resolved at this stage is how to allocate this access. While the process leading to the current position in terms of access is complex (Hirshon, 1995; Usher, 1995), there are strong perceptions that access is currently excessively concentrated and largely in the hands of white holders, to the detriment of other groups with historic claims to fishing rights (African National Congress, 1994; Issel, 1995; Branch *et al.*, 1996; Hutton *et al.*, 1997). One of the key needs within South Africa's fisheries is therefore to devise an access rights system that is accepted by the major interest groups as being fair and equitable.

This new system must be developed in an environment in which the opportunities for expansion of the fisheries are very limited, given the underlying principle of sustainable utilization. Therefore expansion is not the solution to demands by new entrants or to meet social needs. The catch data tend to be good and the methods of assessment of the stocks are generally comparable with the best available elsewhere (Cochrane *et al.*, 1997). Therefore, it is unlikely that the potential productivity of any stock is being grossly underestimated. There are a few stocks which may allow some increases in the total catch and value: the round herring, *Etrumeus whiteheadi*, small mesopelagics and horse mackerel are probably underexploited. However, overall policy makers will have to optimize access to and utilization of what is currently being taken and used.

Redistribution of wealth is one mechanism that has been discussed as a means of overcoming this problem, and the determination of a fair and equitable fisheries policy will be one vehicle that could help to redistribute access to fisheries resources in a manner which also creates the greatest benefits for the people of South Africa in accordance with the objectives outlined in the Introduction. However, redistribution on its own will not solve the country's social and economic problems. The projected mean gross domestic product (GDP) per capita in South Africa for 1996 was US$3370. This compares with a figure of, for example, US$20 160 for Canada (Economist, 1995). Hence, whatever use is made of living marine resources, if it is to contribute to enhancing the living standards of all South Africans, it must be carefully planned to attempt to ensure maximum social and economic benefits on a sustainable basis. The critical role of the new fisheries policy in South Africa is therefore to set out the framework necessary to ensure that the nation's fisheries are managed in the ways necessary to achieve the desired objectives.

5.7 THE POLICY DEVELOPMENT PROCESS

As a part of implementing the RDP, the then Minister of Environmental Affairs and Tourism, in which portfolio fisheries falls, Dr Dawie de Villiers, initiated a policy development process on 27 October 1994. A prominent trade unionist in South Africa, Mr Mandla Gxanyana, was appointed to the chair of the Fisheries Policy Development Committee (FPDC).

The first problem encountered in establishing the FPDC was that of its membership, and this was first thoroughly debated in December 1994. On that occasion, two issues immediately emerged as being of great importance to a large number of those involved in fishing, as opposed to the bureaucrats and the scientists who, arguably, had determined the *de facto* agenda until this process started. These were the dissatisfaction with central control of access to and management of the resources, and the number of people at that time denied access to the resources who felt that they had a legitimate claim (FPDC, 1995a).

Under the new constitution in South Africa, the provinces have been given considerably greater power and responsibility than they had previously enjoyed. At the time of writing, much of South Africa's fisheries activity took place within the Western Cape, with the state fisheries agencies and research institute, most corporate headquarters and most fishing activity occurring there (Fig. 5.2; Payne and Cochrane, 1995). The dissatisfaction of the other coastal provinces, Northern Cape, Eastern Cape and Kwazulu Natal (Fig. 5.2) with this state of affairs was loudly and effectively voiced at the December 1994 meeting, where they successfully negotiated a membership of the FPDC based largely on regional representation (Table 5.4, FPDC, 1995a).

The other issue that came to the fore in this meeting was the aspirations of those fishers who had been deprived of access rights to the fishery in the past, and who now hoped that the political changes would enable them to gain access. This group included fishers involved in poaching abalone and other resources, and who claimed that these activities were forced on them by the unjust existing systems of allocation of access rights. This group successfully negotiated representation on the FPDC as 'informal fishers' (Table 5.4; FPDC, 1995a).

The membership of the FPDC was not finally resolved in 1994, however, and extensive negotiations and discussions on membership continued at a subsequent meeting in March 1995. Here new representation was agreed to, in particular giving the various 'sectors' greater representation (Table 5.4). A significant feature of the debate at this meeting was the strong emphasis on the need for enabling 'grassroots' input to the process (FPDC, 1995a). This has not been a feature of previous fisheries management or policy. As a result of these and subsequent discussions, an expanded FPDC consisting of 61 members was eventually established, with the greater number allowing wider representation (Table 5.4). In addition, a Working Committee (FPDWC, Table 5.4) was established to 'collate and discuss inputs' and 'prepare presentations and recommendations to the FPDC, the Department of Environmental Affairs and Tourism, the Cabinet

Table 5.4 Composition of the Fishery Policy Development Committee (FPDC) and the Fishery Policy Development Working Committee (FPDWC)

Group	Body	Initial proposal (December 1994)	Composition at time of writing	
			FPDC	FPDWC
National government	National govt	1	1	1
Regional government	Northern Cape	1		1
	Western Cape	1		1
	Eastern Cape	1		1
	Kwazulu Natal	1		1
Fishing fora	Northern Cape	4	5	1
	Western Cape	4	5	1
	Southern Cape		5	1
	West Coast			1
	Eastern Cape	4	5	1
	Kwazulu Natal	4	5	1
Sector	Conservation and environmental	1	5	1
	Industrial fishing	1	5	1
	Informal fishing	1	5	1
	Organized labour	1	5	5
	Recreational	1	5	1
	Small business	1	5	1
	Tourism	1	5	1
	Unorganized labour	1		
	Consumers	1		
	Support industries	1		
Total		31	61	22

Committee Meetings' (FPDC, 1995a). The strong representation of labour (a conglomeration of 11 trade unions) on the FPDWC followed their strong appeal for this, on the grounds that they, along with Government and investors, were the major stakeholders in the fishery (FPDC, 1995b).

In addition to these issues highlighted by the policy process itself, the Chairman of the FPDC suggested eight key areas that needed to be addressed by the final fisheries policy (Gxanyana, undated), as well as some principles to govern the policy development process (Table 5.5). These points highlight critical areas and features encompassed by fisheries management. The dominance of human- and societal-orientated issues is striking.

In June 1996, the FPDC completed its work and submitted a draft

Table 5.5 Key issues to be covered in the proposed fisheries policy and the principles to govern the policy development process, as suggested by the FPDC chairman (Gxanyana, undated)

Key issues	Process principles
1. Responsible management and use of marine resources and the environment	1. Consensus seeking
2. Protection of marine resources	2. An open and transparent process
3. Democratic institutional structures	3. Participation at all levels
4. Equitable allocation of access rights	4. Development from the bottom-up
5. Development (presumably social and economic) and social benefits	5. Transfer of information
6. Worker-related issues	6. Accountability to the people
7. Maintenance of an efficient, investor-friendly and competitive industry	
8. Enhancement of South Africa's status as a tourist destination and a fishing nation	

fisheries policy to the Minister on Environmental Affairs and Tourism. However, this draft contained many unresolved issues. Essentially, the question of determining even the broad principles of how to allocate access rights in the future had not been addressed. Consequently, an independent representative panel, made up of two lawyers, a business economist and a social anthropologist, were tasked with investigating options for access rights in South Africa. Their suggestions were incorporated into the final White Paper on Marine Fisheries, which was published in June 1997.

The process has moved on since then. The White Paper has been accepted by the South African Government and a Marine Living Resources Bill (to replace the *Sea Fisheries Act*) has been drafted, tabled and introduced to fishers around the coast. It is scheduled for a hearing in parliament in February 1998 and is intended to become law shortly thereafter. Anticipating sucesss, the lawmakers have also drafted new fisheries regulations. While the process is still incomplete, it has illustrated a number of fundamental principles which apply to fisheries policy and hence to fisheries management.

5.8 IMPLICATIONS OF THE POLICY PROCESS FOR FISHERIES MANAGEMENT

While the fishery issues arising in South Africa have a unique component, arising from unique environmental, cultural and political circumstances and history, they are generically very similar to issues arising around the world. Therefore, it is suggested that many of these conclusions can be extrapolated to other national and multinational fisheries. The major features are grouped here under five subheadings.

Sustainable utilization of resources

Arising from the South African process, it is clear that the users, including the fishers, recognize the need or desirability for sustainable utilization of resources. There has been no widely supported desire expressed to utilize the resources at above their sustainable levels for short-term gains. It must be recognized that there is a large difference between accepting this goal in principle and enacting it in practice when faced with short-term social and economic hardship. However, the acceptance in principle is extremely important and provides the justification within South Africa for maintaining sustainable practices, even when challenged for short-term expediency.

Democratic institutional structures

The need for democratic institutional structures is closely interlinked with all those features that Gxanyana stressed were essential for the policy process itself: consensus seeking; open and transparent process; participation at all levels; development from the bottom up; transfer of information; and accountability to the people. All these are preconditions for, and features of, genuine democratic institutional structures. The call for democratic structures in South Africa is, in part, a justifiable reaction to decades of exclusion from any meaningful democratic participation in government decision making for most of South Africa's population (Hutton *et al.*, 1997). However, the problem in fisheries is not unique to South Africa. Charles (1994) suggested that most fisheries throughout the world have been managed in a centralized, top-down manner, leading to a feeling of competition between the regulators and the regulated. This, Charles suggested, has resulted in a feeling amongst the fishers that ignoring regulations is justified, perhaps even commendable, and there is little peer pressure to encourage compliance.

This view is not new, and there are many examples of governments recognizing the need to involve the stakeholders in management of the resources in which they have interests. In Norway, the Regulatory Council, which advises the Ministry of Fisheries, includes representatives from fishers', seamen's, fish processors' and plant workers' unions or associations (Ratvik, 1994). In the 1980s, the Department of Fisheries and Oceans (DFO) in Canada included systems for widespread consultation in its fisheries management system (Charles, 1995b). In fact, one can go back at least a century and identify systems of co-management, where co-management is a system where a degree of responsibility has been transferred to the users who then participate, with the official management agency, in the fisheries management process (e.g. Hutton *et al.*, 1997). Jentoft (1989) reported on the Lofoten fishery where co-management principles were introduced in the 1890s, and the Japanese system of co-management has been legally operative since 1901 while the practice goes back even further.

The key questions are therefore why is it taking so long to implement more widely what is, at least with hindsight, so obviously necessary, and why are we seeing so few signs of success. The answers to both these questions are undoubtedly varied and complex and this chapter could not hope to set out to answer them. One contributory factor, exemplified in the South African debate, must be the fact that the stakeholders themselves are a diverse and dynamic group. Hence, co-management itself will be hard to implement with any form of consensus, except perhaps for localized, single stocks. Charles (1995b) referred to another example of this problem when he suggested that one of the factors contributing to the collapse of the Canadian Atlantic groundfish fishery, was that DFO saw its 'clients' as being only the fishing industry, excluding other stakeholders, notably the public itself, and in so doing, opted for a sector-based approach to management, rather than a community-based approach. Whether or not Charles' conclusion is valid, it does highlight the fact that co-management, particularly for widespread stocks exploited by diverse groups, will itself be fraught with difficulties.

The importance of effective democratic procedures points to a need for transition in fisheries management. The fact that consultation and co-management is so important to successful management, and involves complex and ever-changing approaches, means that it must be approached in a thoroughly professional manner by fisheries management agencies. Consultation and co-management require that all stakeholders be identified, fully informed and adequately represented wherever decisions are made that affect their interests. Structures therefore need to be created and staffed by appropriately qualified, experienced and trusted professionals to ensure that suitable consultation and information dissemination is taking place. Developments within the fishery need to be continually monitored to maintain appropriate and effective consultative and representative groups. Where full co-management is implemented, mechanisms need to be developed to ensure not just consultation, but joint decision making and shared responsibility and accountability.

In addition to consultation, structures and mechanisms need to be put in place to ensure that decisions are made in a manner that arrives at the optimal solutions under what will inevitably be situations of conflicting goals and needs. Professional application of multiple-criteria decision analysis (Merritt and Criddle, 1993; Pitcher *et al.*, Chapter 3, this volume) should be as much a feature of fisheries management agencies as professional application of the best stock assessment approaches. The use of multiple-criteria decision analysis will not be a guarantee of success, but it could be an invaluable tool in attempting to resolve conflict.

An acceptable system of limited access rights

South Africa has unique problems in terms of access rights, generated by its own peculiar political history. However, in many other parts of the world, including Europe, Canada, the United States of America and many developing countries, the problems of limiting access are being grappled with. Open access is generally accepted as being fatally flawed (FAO, 1997), and in response to this realization, many of the developed countries are moving towards transferable quotas where the market will regulate future ownership (Squires *et al.*, 1995; Munro and Pitcher, 1996). However, a transferable quota approach, or at least an unconstrained transferable quota approach, may not be appropriate for certain fisheries or fishery goals. For example, recreational fisheries, rights of indigenous people (or other groups of fishers requiring special consideration), the incomes of fishers, the promotion of employment, addressing poverty, regional development and other issues are listed by Squires and his colleagues (1995) as requiring explicit consideration when an ITQ (individual transferable quota) programme is being considered.

While many of the decisions on access will be political, the fisheries management agency is the body that is most likely to be responsible for providing the political decision makers with at least some of the advice received on the biological, economic, social and political implications of any given system of access or any selected management approach. Again, the fisheries management agency should have the necessary information and expertise to be able to provide valid advice in these areas, and to evaluate information received from different interest groups. This will require structures, personnel and programmes to monitor and evaluate trends in access, and related indicators of social and economic performance.

Multidimensional nature of fisheries policy

The issues of social and economic development, worker-related issues, maintenance of an efficient industry and enhancement of South Africa's status as a tourist destination (Table 5.5) may all be treated together. They all represent different benefits to the nation or to specific subgroups and together reflect the multidimensional nature of fisheries management and decisions. The reference to tourism raises the question of eco-tourism, introducing ecosystem considerations into this chapter, for example maintaining sustainable populations of sea birds and marine mammals for tourism, as well as for conservation. It is not clear what is intended by enhancement of South Africa's status as a fishing nation (Table 5.5) but it is assumed that good progress in all the other aspects would be the best means of achieving this.

At present, South African fisheries such as the abalone, rock lobster and hake fisheries are probably striving for maximizing efficiency. If different goals are selected for one or more of these fisheries, such as upliftment of coastal communities or maximising employment, it would be highly beneficial to have the background social, economic and ecological information to be able to guide these decisions by estimating relative costs and benefits of different access systems and management approaches. Data to answer similar questions, including, for example, bycatch of sea birds, are now being collected for the hake fishery, but serious and concerted attention also needs to be directed to other fisheries to avoid having to make decisions without vital information.

Again, it would appear that South Africa is not alone in this problem. Merritt and Criddle (1993), in a report on a study of multi-criteria decision making for management of a salmon fishery in Alaska, USA, wrote 'The emphasis placed on political, social and economic concerns that are not commonly in the domain of fishery managers was striking.' In a 1996 meeting on trends in fisheries management, Davis (St Francis Xavier University, Nova Scotia, unpublished information) expressed the view that, with the exception of some resource economists, fisheries social research seems to have received little serious attention in fisheries natural science and management. Felt (Memorial University, Newfoundland, unpublished data), also from Canada, stated that the integration of social and natural science within fisheries management plans has 'proven to be fairly elusive'. Similarly, Jentoft (Institute of Social Science, Toromsø, unpublished data), from Norway, suggested that social scientists are essentially not represented in fisheries management decision making, again apart, perhaps, from economists. He said 'Rarely do sociologists, social anthropologists and political scientists serve in any advisory role to ... government agencies ... in the same way as biologists do.' Addressing these deficiencies will, again, require sections and skilled staff to be able to undertake such studies. The domination of the South African access rights panel by lawyers and human-scientists (section 5.7) suggests that South African decision makers have woken up to the reality that many of the issues dealt with in fisheries management are human-centred. However, the general question still arises as to whether the data, as well as the expertise, will be generally available to address all the important fisheries issues in the short to medium term.

New management structures to implement the new approach

In the above discussion, it has been suggested on several occasions that new structures incorporating specialized personnel will need to be created to perform several key but currently neglected or omitted tasks. Clearly there are few countries where increased funds will be readily available for these tasks. Implicit in what has been discussed is that fisheries management agencies will have to engage in drastic restructuring to maintain adequate stock assessment functions, without which social and economic planning will be impossible, while introducing new consultation, co-management and human science research functions. The extent and nature of the restructuring necessary will obviously vary substantially from organization to organization and from country to country.

5.9 CONCLUSIONS

Fisheries management this century evolved to a point where it was essentially dominated by natural scientists using techniques which, while very imprecise and prone to error, allowed a measure of control over the trajectories of fish stock biomasses with time by regulation of fishing mortality. Fisheries management agencies and fisheries management practice reflected this scientific dominance and in most cases still do (Haggan, Chapter 2, and Williams, Chapter 11, this volume). However, fisheries management has almost inevitably failed, as reflected by declining stocks and, hence, declining yields. It is generally agreed that the primary reason is that fisheries management has neglected the human dimension (Jentoft, Chapter 12, Hart and Pitcher, Chapter 16 and Haggan, Chapter 2, all in this volume). This omission has meant that the scientific recommendations frequently have not been implemented or have not been observed, and hence, with or without good stock assessment advice, overall fishing mortality has been in excess of that sustainable by the stocks. The lesson for fisheries management agencies should be clear. They must drastically alter their structures, their approaches and their goals to undertake the necessary research, consultation and decision making to ensure that fishing strategies and regulations to be implemented, based on valid biological recommendations, will be generally accepted, supported and adhered to (Fig. 5.3) and will come closest to meeting the desired objectives for the stocks, including that of sustainability.

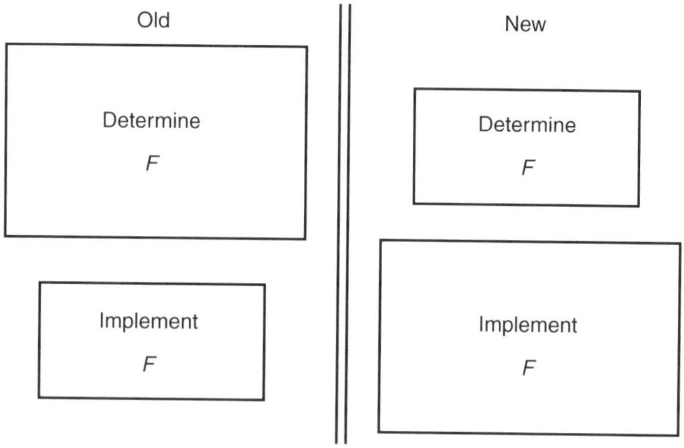

Fig. 5.3 Schematic of the primary conclusions from this analysis indicating, by area of the boxes, the need for a change in allocation of fisheries management effort, to place greater emphasis on realizing a selected fishing mortality (F) in relation to the effort expended in determining an appropriate F.

There is a final conclusion that can be drawn from the South African policy process. The complex and dynamic nature of fisheries, in particular the human component, means that there is unlikely to be a unique correct solution to any given management problem. In fact it is probably more correct to say that there will not be any 'correct' solution. Fisheries management is marked by conflicting objectives. Taking these objectives together, compromises are invariably necessary to meet them. In some cases, satisfactory compromises between all interest groups and interests may be possible, but in many cases one or more interest groups will not be able to be accommodated to their satisfaction. Conflicts generated by this failure will clearly be at their maximum when the demands on the resources exceed their production potential. The statistics provided in the introduction indicate that, all too often, the biggest loser amongst the 'interest groups' has been the resources themselves. The best science, information, consultation and institutional structures available will not provide a solution within a fishery to satisfy demands that exceed supply. The only solution in these cases is the existence of the political will to ensure the sustainability of the resources and to identify and utilize alternative sources, other than overexploitation of the resources, to meet the demands while simultaneously making optimal use of the sustainable production.

ACKNOWLEDGEMENTS

The comments of two anonymous referees on an earlier draft of this chapter arc gratefully acknowledged. Tony van Dalsen and Cathy Boucher of the Sea Fisheries Research Institute assisted with the figures and Lesley Johnstone of the Foundation for Research Development provided information on the Fisheries Policy Development Committee. We thank them for this invaluable assistance.

REFERENCES

African National Congress (1994) *The Reconstruction and Development Programme. A Policy Framework*, Umanyano, Johannesburg, 147 pp.

Branch, G., Baird, D., Butterworth, D., Cochrane, K., Devine, D., Moola, Z., Sowman, M. and Zulu, P. (1996) Review of access rights options for South Africa. Final report of the Access Rights Technical Committee. Unpublished report, Fisheries Policy Development Committee, Cape Town, 71 pp.

Charles, A.T. (1994) Towards sustainability: the fishery experience. *Ecol. Econ.* **11**, 201–211.

Charles, A.T. (1995a) Fishery science: the study of fishery systems. *Aquat. Living Resour.* **8**, 233–239.

Charles, A.T. (1995b) The Atlantic Canadian groundfishery: roots of a collapse. *Dalhousie Law J.* **18**, 65–83.

Cochrane, K.L., Butterworth, D.S. and Payne, A.I.L. (1997) South Africa's offshore living marine resources: the scientific basis for management of the fisheries. *Proc. Roy. Soc. Sth Afr.* **52**, 149–176.

Crawford, R.J.M., Shannon, L.V. and Pollock, D.E. (1987) The Benguela ecosystem. 4. The major fish and invertebrate resources. *Oceanogr. Mar. Biol. Ann. Rev.* **25**, 353–505.

Cruywagen, G.C., Cochrane, K.L. Melville-Smith, R. and Goosen, P. (1994) Estimation of the pristine biomass of the South African west-coast rock lobster. Unpublished Report WG/06/94/WCL1, Sea Fisheries Research Institute, Cape Town, 34 pp.

Economist, The (1995) The World in 1996. *The Economist*, London, 135pp.

FAO (1993) Review of the state of world marine fishery resources. *FAO Fish. Tech. Pap.* **335**, 136 pp.

FAO (1996) *Fishstat PC. An Analytical Tool for Global Catch Time Series*, Release 5094/96. Computer database, FAO, Rome.

FAO (1997) *FAO Technical Guidelines for Responsible Fisheries. 4. Fisheries Management*, FAO, Rome, 82 pp.

FPDC (1995a) Minutes of the Fisheries Policy Development Meeting, 6th March, 1995. Unpublished report: 11 pp.

FPDC (1995b) Minutes of the Fisheries Policy Development Meeting, 5th, 6th and 7th September, 1995. Unpublished report: 21 pp.

Gxanyana, M. Undated. Broad framework for policy development. Unpublished document, Fisheries Policy Development Committee, Cape Town, 6 pp.

Hirshon, G. (1995) Angling for more equity. *Financial Mail*, 17 February 1995, 28–30.

Hutton, T., Cochrane, K.L. and Pitcher, T.J. (1997) Post-apartheid fisheries management policy in South Africa: the need for a change in management philosophy, in *Developing and Sustaining World Fisheries Resources: The State of Science and Management* (eds D.A. Hancock, D.C. Smith, A. Grant and J.P. Beumer) (Proc. 2nd World Fish. Congr., Brisbane, July 1996), CSIRO Publishing, Collingwood, VIC, Australia, pp. 228–232.

ICCAT (1996) Report for biennial period, 1994–1995. Part II (1995) – 2. Int. Comm. Conserv. Atl. Tunas, Madrid. 236 pp.

Issel, J. (1995) Fisherfolk demand government intervention. *S.A. Comml. Mar. Mag.* **4**(2), 4.

Jentoft, S. (1989) Fisheries co-management. Delegating government responsibility to fishermen's organizations. *Mar. Pol.* April: 137–154.

Lasiak, T. (1991) The susceptibility and/or resilience of rocky littoral molluscs to stock depletion by the indigenous coastal people of Transkei, southern Africa. *Biol.Conserv.* **6**, 245–263.

Merritt, M.F. and Criddle, K.R. (1993) Evaluation of the analytic hierarchy process for aiding management decisions in recreational fisheries: a case study of the chinook salmon fishery in the Kenai River, Alaska, in *Proc. Int. Symp. Management Strategies for Exploited Fish Pop.* Univ. Alaska Sea Grant College Programme, Fairbanks, Rep. 93–02: 683–703.

Munro, G. and Pitcher, T.J. (eds) (1996) *Individual Transferable Quotas* (Special Issue), *Rev. Fish Biol. Fish.* **6**, 1–116.

Payne, A.I.L. (1989) Cape hakes, in *Oceans of Life off Southern Africa* (eds A.I.L. Payne and R.J.M. Crawford), Vlaeberg, Cape Town, pp. 136–147.

Payne, A.I.L. and Cochrane, K.L. (1994) Managing fisheries in a changing society with a well-developed science base. Paper presented to ICES, September 1994. *ICES C.M.* 1994/T:26, 15 pp.

Payne, A.I.L. and Cochrane, K.L. (1995) How management and science are trying to develop fisheries policy in South Africa. Paper presented to ICES, September 1995. *ICES C.M.* 1995/S:8, 13 pp.

Penney, A.J., Buxton, C.D., Garratt, P.A. and Smale, M. (1989) The commercial marine linefishery, in *Oceans of Life off Southern Africa* (eds A.I.L. Payne and R.J.M. Crawford), Vlaeberg, Cape Town, pp. 214–229.

Ratvik, I. (1994) Devolved and regional management systems for fisheries. Norwegian Baseline Report. Unpublished report, Norwegian College of Fish. Sci., Univ. Tromso, 43 pp.

Raubenheimer, C.M., Butterworth, D.S. and Cochrane, K.L. (1998) Towards adaptive approaches to management of the South African abalone *Haliotes midae* fishery. *S. Afr. J. Mar. Sci.* (in press).

Roel, B.A. and Cochrane, K.L. (1995) Assessment of the chokka squid. Unpublished report, Sea Fisheries Research Institute, South Africa, WG/95/03/SQ01, 21 pp.

Rosenberg, A.A., Fogerty, M.J., Sissenwine, M.P., Beddington, J.R. and Shepherd, J.G. (1993) Achieving sustainable use of renewable resources. *Science* **262**, 828–829.

SFRI (1995a) South African commercial fisheries review 1993. Sea Fisheries, Dept of Environment Affairs, Cape Town, 51 pp.

SFRI (1995b) Research highlights 1994–1995. Sea Fisheries Research Institute, Dept of Environment Affairs, Cape Town, 41 pp.

SFRI (1996) Research highlights 1995–1996. Sea Fisheries Research Institute, Dept of Environment Affairs, Cape Town, 41 pp.

Siegfried, W.R., Hockey, P.A.R. and Crowe, A.A. (1985) Exploitation and conservation of brown mussel stocks by coastal people of Transkei. *Env. Conserv.* **12**, 303–307.

Squires, D., Kirkley, J. and Tisdell, C.A. (1995) Individual transferable quotas as a fisheries management tool. *Rev. Fish. Sci.* **3**, 141–169.

Stephenson, R.L. and Lane, D.E. (1995) Fisheries management science: a plea for conceptual change. *Can. J. Fish. Aquat. Sci.* **52**, 2051–2056

Stuttaford, M. (ed.) (1996) *Fishing Industry Handbook. South Africa, Namibia and Mocambique*, Marine Information cc, Stellenbosch, 434 pp.

Tarr, R. (1989) Abalone, in *Oceans of Life off Southern Africa.* (eds A.I.L. Payne and R.J.M. Crawford), Vlaeberg, Cape Town, pp. 62–69.

Usher, D. (1995) The sea appropriate start for the new SA. *S.A. Comml. Mar.* **4**(2), 6 and 27.

Viljoen, C. (1995) Perlemoen crisis. *S.A. Comm. Mar.* March, 12–13.

Walters, C. and Maguire, J.-J. (1996) Lessons for stock assessment from the northern cod collapse. *Rev. Fish Biol. Fish.* **6**, 125–137.

Chapter six

Beyond the status quo: rethinking fishery management

Anthony T. Charles

ABSTRACT

This chapter explores several controversial challenges for fishery management of the future. First, emphasis is placed on the need for appropriate involvement of the various stakeholders, including fishers, communities and the general public. In particular, the public, as holders of property rights to fish resources, deserve a greater involvement in fishery decision making. Second, the implications of a precautionary approach to uncertainty are explored, from the perspectives of fishery science (methodologies of stock assessment) and management (notably the transition from rigid to adaptive harvesting plans). The discussion points to a clear need to adjust the 'burden of proof' in decision making so that by default, such decisions will favour conservation. Third, the nature of quota management is discussed, highlighting the inherent lack of resilience and the negative conservation impacts produced by an over-reliance on such controls. Finally, the chapter highlights the importance of shifting from a 'sectoral' view, in which the fishery is treated in isolation, to an integrated 'systems' view, with the fishery recognized as part of the larger coastal and regional economy.

Reinventing Fisheries Management. Edited by Tony J. Pitcher, Paul J.B. Hart and Daniel Pauly. Published in 1998 by Kluwer Academic Publishers, London. ISBN 0 412 83410 3.

6.1 INTRODUCTION

This chapter argues that the key challenge in fishery management lies not so much in developing new techniques or obtaining better data, but rather in rethinking the philosophy of managing for sustainability. Who should be involved in the process? What are the guiding principles? The chapter focuses on three particularly crucial yet controversial challenges:

- enlarging the role of 'stakeholders';
- applying a precautionary approach to fishery management and the 'burden of proof';
- rethinking quota management, in the light of limitations on fishery 'controllability'.

These challenges, widespread throughout the world's fisheries, are considered here within the context of the early-1990s collapse of Canada's Atlantic groundfish fishery, analysed in greater detail in Charles (1995).

6.2 RETHINKING THE ROLE OF STAKEHOLDERS

The fishers

Historically, fishery management has been built around a simplistic view of fishers and managers. Within this perspective, fishers are no more than selfish profit-maximizers, driven to take as much fish as possible, while regulators have sole responsibility for conservation, to be achieved by controlling the fishers. This view of the world, shaped by Hardin's (1968) *Tragedy of the Commons*, led to authoritarian attempts to implement controls over the fishing industry. The response from fishers was a rejection of such management controls in practice. Hardin's flawed perspective (Berkes *et al.*, 1989) became self-fulfilling: when fishers were excluded from management, they indeed had no other role than to catch as much fish as possible, and to thwart regulations whenever possible. Under such circumstances, no level of enforcement, however extensive, has been able to prevent illegal fishing and overharvesting.

To overcome this 'us versus them' attitude, many fishery regulatory bodies shifted to a consultative model, in which government discusses management with the industry prior to implementation. However, consultations did not imply decision-making power, and as a result, in most cases fishers did not 'buy into' government-imposed regulations.

The next step in the evolution of fishery management lies in the recognition that, for pragmatic reasons, effective management must be designed and implemented with the involvement of those being regulated, the fishing industry. Increased stakeholder involvement can be found in the trend toward 'co-management', the creation of suitable institutions in which stakeholders and government work together to develop and enforce regulations jointly (Berkes, 1989; Pinkerton, 1989, Brown, Chapter 13, this volume). The move toward co-management by government and industry is gaining momentum, and clearly has many positive aspects.

The public

Yet those who catch fish are not the only stakeholders. Fish in coastal waters are typically state property, with the citizens of the nation having collective property rights over the fish. While owners of valuable property usually have some say in its management, this is rarely the case in fisheries. Instead, many fishery agencies focus exclusively on 'serving their clients' in the industry. This has meant in the past that an implicit objective of fishery bureaucrats has been to keep the industry happy, sometimes at the expense of conservation efforts. To reinvent fishery management, it is crucial to understand that resource owners are legitimate stakeholders. Management must better involve these resource owners, while at the same time recognizing the pragmatic need for a 'buy in' by other stakeholders, notably the resource users.

Coastal communities

Moreover, there is a third component of stakeholders: coastal communities and their residents. In many, perhaps most instances, such communities have been ignored in fishery decision making, despite their close ties to, and dependence on, the fishery. This unfortunate situation remains in effect in many jurisdictions, despite an expanding body of research that highlights the conservation and socio-economic benefits of local involvement in management. The vehicle of choice for this is 'community-based co-management', the process of joint decision making at the local level, by government, fishers and relevant institutions in coastal communities. This process tends to promote more efficient management, because the involvement of community institutions together with government and industry maximizes the built-in pressure, i.e. the moral suasion, for those in the fishery to comply with regulations in support of conservation.

6.3 RETHINKING THE BURDEN OF PROOF

Given the inherent uncertainty pervasive in every fishery, decisions must always be made by balancing risks – typically a risk of resource collapse through excessive exploitation or environmental damage, versus a risk of foregone economic benefits if harvests are lower than necessary (Walters, Chapter 21, this volume). The question is: in considering this trade-off, where should the burden of proof lie?

Stock and recruitment

Is there a relationship between the number of adult spawners in a population and the resulting recruitment of young fish? Logic would suggest an affirmative answer, but given natural fluctuations in the environment and in the corresponding data available to scientists, there is often no scientific 'proof'. In the Atlantic Canadian groundfishery experience, an absence of such proof meant that spawner–recruit links between fish generations were not incorporated into the assessment process (Hutchings and Myers, 1994), and the fundamental idea of maintaining a critical spawning stock biomass (to preserve the reproductive process) was not put into practice. On the other hand, application of the precautionary approach (Garcia, 1994) suggests that it may be preferable to rethink the burden of proof, assuming a positive connection between spawners and recruitment, unless the evidence is strongly opposed.

What causes a collapse?

Are fishery collapses caused by overfishing or environmental change? When Canada's northern cod, *Gadus morhua*, stock off Newfoundland collapsed in the early 1990s, the government issued an initial government press release that made no mention of human impacts on the resource, nor of problems with the fishery management process. Instead, it was stated that 'the devastating decline in the stock of northern cod' was due 'primarily to ecological factors' (Department of Fisheries and Oceans, 1992). Yet it seems that while ocean conditions might have acted as a trigger to initiate a stock collapse, the principal underlying cause of the collapse was high levels of resource exploitation. Although the dynamics of collapses are complex, the typical sequence of events is often as follows:

1. when ocean and environmental conditions are good or average (from the perspective of the fish), fundamentally unsustainable harvest levels may appear to be sustainable;
2. inevitably, and quite naturally, ocean conditions will change at some point, so that heavily-harvested fish populations become subjected to additional stress – environmental conditions that inhibit growth and reproduction;
3. faced with intense fishing pressure, and a trigger in the form of an adverse environment, the fishery collapses.

This scenario corresponds to considerable experience around the world, from Canada's Pacific coast herring, *Clupea harengus pallasi*, fishery collapse of the 1960s (Hourston, 1978) to the 1972 Peruvian anchovy, *Engraulis ringens*, collapse, triggered by ocean cooling known as El Niño but resulting fundamentally from massive exploitation (contributions in Pauly and Tsukayama, 1987; Hilborn and Walters, 1992). Recent research on the Atlantic Canadian groundfishery also supports this interpretation (Taggart *et al.*, 1994). Indeed, Hutchings and Myers (1994) and Walters and Maguire (1996) indicate that fishing pressure alone is sufficient to explain the collapse of northern cod. Thus, invoking bad ocean conditions as an explanation is simply unnecessary. In any case, a precautionary approach would suggest focusing more on reducing human impacts, rather than on blaming the ocean in the wake of fishery collapses.

Conservation first

In many fisheries, priority has been placed on avoiding disruption to the harvesting process rather than on meeting declared conservation goals. At times, this arose at the political level, such as the Canadian government's decision to set the 1990 quota for northern cod at 197 000 tonnes, instead of 125 000 tonnes as recommended by its scientists (Department of Fisheries and Oceans, 1990).

More often, however, high quotas were produced within the management process, through well-meaning yet counter-productive efforts to reduce the socio-economic effects of downturns in stock abundance. For example, under the so-called '50% rule' (Department of Fisheries and Oceans, 1991, p. 5), when scientific evidence pointed to the need for large reductions in catches of an Atlantic Canadian groundfish stock, the government would reduce the total allowable catch slowly, by 50% of the difference each year, rather than taking immediate action to cut quotas to desired levels (unless scientists could prove a conservation crisis). However, this gradualism did not apply in the opposite direction; if scientific analysis suggested an increase in the allowable harvest, the full increase could be made immediately.

Progress has been made in adopting a more conservationist management philosophy within Atlantic Canada's groundfishery. One example, created by the government following the 1992 collapse of the northern cod stock, is a new conservation institution, the Fisheries Resource Conservation Council (FRCC). A partnership of industry and academic members (including this author), the FRCC provides public advice on conservation needs in Atlantic fisheries, to help the government 'achieve its conservation, economic and social objectives for the fishery' (FRCC, 1995, 1996a). Perhaps the most fundamental of the Council's efforts to date has been its role in bringing a solid 'conservation first' perspective to the fishery. This has been evident, for example, in its recommendations to close many Atlantic Canadian cod fisheries (despite the obvious dislocation this produced), to set total allowable catches well below $F_{0.1}$ levels in the case of depressed stocks, and to supplement catch quotas with effort controls, as a safeguard measure (FRCC, 1995, 1996a).

6.4 RETHINKING QUOTA MANAGEMENT

Quota management is among the most fundamental of fishery management approaches. This often involves (a) estimating the biomass for each stock, (b) determining a total allowable catch (TAC) as a fraction of that biomass, (c) subdividing the TAC by sector (by gear and/or boat size), and (d) subdividing further into individual fisher quotas in some sectors. Quota management has the apparent benefit of setting a firm limit on what can be removed from the ocean. In practice, however, fisheries are not totally controllable – it has been impossible to properly determine quota levels (Walters and Pearse, 1996; Walters, Chapter 21, this volume) or to fully enforce these quotas, given the ease of cheating.

The unknown number

To properly set quotas requires knowledge of the fish biomass, something that can never be known with certainty. The two major sources of assessment information both have problems.

- The fact that research vessel surveys may cover only a fraction of the fishable area (for example, often missing inshore areas of Newfoundland, south-western Nova Scotia and elsewhere) can be a significant problem depending on stock structure and migratory patterns.
- The use of commercial fishery catch rates can produce highly misleading results, owing to the erroneous assumption that success in the fishery, as indicated by high catch rates, is a direct indicator of strong fish stocks. In reality, even as stocks decline, fishers often are able to find and catch the remaining fish; good catch rates create the illusion of a healthy stock.

Compounding these concerns is the impact of misreporting and anti-conservationist behaviour (discussed below) on the catch data crucial to the stock assessment process. Not only is the quality of the data compromised, but additional biases are also introduced, tending to produce faulty assessments of stock status, and overestimates of feasible catch levels.

Anticonservationist incentives

Quota management creates inherent incentives to harvest more fish than is allowed in the established quotas. This overfishing can occur by: (1) exceeding the quota, whether this be a TAC, an allocated share of the TAC for a particular sector of the fleet, or an individual quota; (2) 'high grading' to maximize the value of what is reported as caught, typically by dumping lesser-valued fish overboard; and (3) dumping prohibited fish species (including those for which the quota has been reached) or fish sizes (notably small fish) so as to be able to continue fishing. These problems can be overcome, but only at a significant cost. Addressing the first problem requires reliable and independent catch monitoring, while addressing the second and the third, which take place at sea, can be eliminated only by placing full-time observers on every vessel. (While expensive, this has been implemented in fisheries where the catch per boat is sufficient to warrant the cost.)

Incentives for high grading and dumping increase as quotas are subdivided to gear sectors and then to individuals. Under a simple global quota, high grading or dumping present direct costs to the fisher in the form of lost income, but provide relatively few advantages in return (except if hold capacity is a limiting factor), because the benefits would necessarily be shared by the entire fleet.

On the other hand, under an individual quota system (or trip limits), the fisher's direct cost of high grading or dumping remains the same (the foregone fish), but the benefits of such actions are personalized; the fisher can increase the value of what is landed by high grading, or can engage in dumping so as to continue fishing other species. The relative benefits of anticonservationist actions are especially strong when the fisher must pay for the catch quota (e.g. if the market value of small fish in the catch is less than the cost of quota to account for them), but can be reduced somewhat if quotas are transferable.

The illusion of certainty

A less well-recognized aspect of quota management is the illusion of certainty it creates, as quantities of fish (pieces of what is in fact a pie of very uncertain size) are allocated to the various players. For example, in the groundfish fishery of Atlantic Canada, total allowable catches were adjusted from year to year, but were at least implicitly seen as fixed, firm and sacrosanct *within* the fishing season. Although allowance was made for in-season changes to annual catch quotas 'as a consequence of major changes in the scientific advice' (Department of Fisheries and Oceans, 1991), in reality, such changes were very rare. It was felt that any adjustment to the TAC should be left to next year's fishing plan. Not only were changes inconvenient to management and industry, but they were not thought to be particularly crucial from a conservation perspective. Such an attitude allowed the fishing industry to adopt annual business plans, comparable to those in non-resource sectors.

This in-season rigidity in catch quotas not only contributed to overharvesting, it also led to dislocation when mid-season changes became necessary. For example, in 1993, the government (following FRCC recommendations) closed some fisheries and reduced quotas on other stocks at mid season, in response to mounting evidence of groundfish collapses. Unfortunately, because such changes had not been well anticipated in management plans, considerable difficulties arose. This was particularly apparent in the individual transferable quota (ITQ) system, which had been promoted by the government largely through the claim that an ITQ gave the fisher property rights over shares of the TAC that could be caught at any time desired. This system was built around the illusion of certainty – the idea that the TAC, and thus the ITQ, was a 'sure thing', never to be changed in mid season in response to conservation needs. When such mid-season quota reductions turned out to be required, the system was poorly equipped to produce an equitable sharing of the reduced quota.

While quota management has its advantages, the concerns discussed here suggest that there is scope for a critical assessment of such controls as conservation tools. This is certainly controversial, however, because the view persists in the fishery, with respect to quota management, that 'the system works'. Furthermore, much of the allocation structure in fisheries has developed around the idea of dividing up the pie (i.e. the TAC). One intermediate option lies in supplementing quota management with effort controls, to provide a double check that conservation measures are not thwarted (FRCC, 1996b). For example, if the overall effort limit is reached before the quota is caught, this may indicate that the quota had been set excessively high. Accordingly, the fishery could be closed as a precautionary move.

6.5 CONCLUSIONS

This chapter has focused on three controversial challenges to conventional fishery management.

- First, there is a need to remind ourselves that fish constitute a public resource, and that those with property rights to the resources (the public, and particularly those in coastal communities) are stakeholders who deserve to be involved in fishery decision making.
- Second, fishery scientists and managers must be guided by the precau-

tionary principle in the face of uncertainty. This implies shifting the burden of proof (Walters, Chapter 21, this volume). Rather than requiring proof that increasing harvests will deplete fish populations, that reducing the spawning stock will tend to produce fewer recruits, that dragging fishing gear along the ocean floor can damage the habitat, and so on, these conservation-orientated perspectives should be assumed to be valid unless proven otherwise. In other words, the default view of the world should be one that favours conservation.
- Third, emphasis must be placed on designing resilient management systems, capable of keeping fisheries on the desired conservation and socio-economic paths even if our understanding of the state of the world is incorrect. This requires more than a reliance on quota controls.

In addition to these fishery-specific challenges, it is also important to highlight the need to see fisheries as an integral part of the larger coastal and regional economy. This recognition has several implications for fishery management itself.

- First, it is important to broaden beyond a focus on the fishery's *output* (such as the total allowable catch or 'sustainable yield') to a multidimensional view emphasizing ecological, socio-economic, community and institutional sustainability of the fishery *system* (Charles, 1994).
- Second, because the choices made in fishery management affect those outside the fishery system, and vice versa, an integrated approach to coastal (or watershed) planning is necessary to optimize the overall sustainable benefits to the local residents.
- Third, fishery management should draw upon a 'portfolio' of approaches, to provide multidimensional solutions for the multidimensional problems faced in fishery and coastal systems. We must be on guard against single-minded 'quick fixes' that attempt to deal with problems in the fishery, but that may well have a net negative effect on the coastal economy and beyond.

In conclusion, this chapter has advocated a reassessment of several rather ingrained approaches to fishery management. The key question amounts to whether the system works, or whether fundamental change is needed. If the latter holds, and no change is implemented, history is likely to repeat itself; somewhere in the world, the next fishery collapse will begin to gather momentum.

ACKNOWLEDGEMENTS

I am grateful to Michael Sinclair for helpful comments on an earlier version of this chapter. The chapter benefited as well from reviews by two anonymous referees and the editors. As usual, the views expressed here, and any remaining errors, are the sole responsibility of the author. Financial support is acknowledged from the Natural Sciences and Engineering Research Council of Canada, grant no. A6745.

REFERENCES

Berkes, F. (ed.) (1989) *Common Property Resources: Ecology and Community-Based Sustainable Development*, Bellhaven Press, London, 302 pp.

Berkes, F., Feeny, D., McCay, B.J. and Acheson, J.M. (1989) The benefits of the commons. *Nature* **340**, 91–93.

Charles, A.T. (1994) Towards sustainability: the fishery experience. *Ecol. Econ.* **11**, 201–211.

Charles, A.T. (1995) The Atlantic Canadian groundfishery: roots of a collapse. *Dalhousie Law J.* **18**, 65–83.

Department of Fisheries and Oceans (1990) *Highlights: 1990 Canadian Atlantic Groundfish Management Plan*. Press release no. B-HQ-90-003E, 2 January 1990. DFO, Ottawa.

Department of Fisheries and Oceans (1991) *1992 Atlantic Groundfish Management Plan*. Minister of Supply and Services, Ottawa.

Department of Fisheries and Oceans (1992) *Crosbie Announces First Steps in Northern Cod (2J3KL) Recovery Program*. Press release no. NR-HQ-92-58E, July 2 1992. DFO, Ottawa.

Fisheries Resource Conservation Council (1995) *Conservation: Come Aboard. 1996 Conservation Requirements for Atlantic Groundfish: Report to the Minister of Fisheries and Oceans*. Minister of Supply and Services, Cat. no. FS 1-61/1996E, Ottawa, 109 pp.

Fisheries Resource Conservation Council (1996a) *Building the Bridge. 1997 Conservation Requirements for Atlantic Groundfish: Report to the Minister of Fisheries and Oceans*. Minister of Supply and Services, Cat. no. FS1 61/1997E, Ottawa, 224 pp.

Fisheries Resource Conservation Council (1996b) *Quota Controls and Effort Controls: Conservation Considerations*. Discussion Paper FRCC.96.TD.3. Fisheries Resource Conservation Council, Ottawa, 9 pp.

Garcia, S.M. (1994) The precautionary principle: its implications in capture fisheries management. *Ocean Coastal Manage.* **22**, 99–125.

Hardin, G. (1968) The tragedy of the commons. *Science* **162**, 1243–1247.

Hilborn, R. and Walters, C.J. (1992) *Quantitative Fisheries Stock Assessment: Choice, Dynamics and Uncertainty*. Chapman & Hall, New York, 570 pp.

Hourston, A.S. (1978) The decline and recovery of Canada's Pacific herring stocks. *Fish. Mar. Tech. Rep.* no. 784, Minister of Supply and Services, Cat. no. Fs 97–6/784, Ottawa, 19 pp.

Hutchings, J.A. and Myers, R.A. (1994) What can be learned from the collapse of a renewable resource? Atlantic cod, *Gadus morhua*, of Newfoundland and Labrador. *Can. J. Fish. Aquat. Sci.* **51**, 2126–2146.

Pauly, D. and Tsukayama, L. (eds) (1987) *The Peruvian Anchoveta and its Upwelling Ecosystem: Three Decades of Change* (ICLARM Studies and Reviews No. 15), International Center for Living Aquatic Resources Management, Manila.

Pinkerton, E.W. (ed.) (1989) *Co-operative Management of Local Fisheries*, Univ. British Columbia Press, Vancouver.

Taggart, C.T., Anderson, J., Bishop, C., Colbourne, E., Hutchings, J., Lilly, G., Morgan, J., Murphy, E., Myers, R., Rose, G. and Shelton, P. (1994) Overview of cod stocks, biology, and environment in the Northwest Atlantic region of Newfoundland, with emphasis on northern cod. *ICES Mar. Sci. Symp.* **198**, 140–157.

Walters, C.J. and Maguire, J.J. (1996) Lessons for stock assessment from the Northern cod collapse. *Rev. Fish Biol. Fish.* **6**, 125–137.

Walters, C.J. and Pearse, P.H. (1996) Stock information requirements for quota management systems in commercial fisheries. *Rev. Fish Biol. Fish.* **6**, 21–42.

Chapter seven

Reinventing the formulation of policy in future fisheries

David B. Preikshot

ABSTRACT

This chapter was inspired by the participants in, and the debates following, the presentations of a theme session on 'the role of policy in responsible fishing' at the symposium forming the basis this volume. These suggested that most past fisheries policy failures were not anticipated because the biological indicators used to assess the policy were poorly understood or misreported. These failures may have been averted through a multi-disciplinary approach providing information from sources that are under-utilized at present, such as the social sciences, which back up or reinforce information from biological and ecological sources. It is argued that solutions to the complex issues surrounding the global fisheries crisis may be found by adopting more of an interdisciplinary approach to policy making.

7.1 INTRODUCTION

Three topics dominate the debate over the future of fishing policy formulation. The first is uncertainty, which has consistently bedevilled the making and the implementation of fisheries policies (Peterman *et al.*, Chapter 29,

Reinventing Fisheries Management. Edited by Tony J. Pitcher, Paul J.B. Hart and Daniel Pauly. Published in 1998 by Kluwer Academic Publishers, London. ISBN 0 412 83410 3.

this volume). The second is determining who the legitimate participants in the process of policy making are (Rice, Chapter 14, this volume). The third is the attempt to understand the meaning, and legitimacy, of information from the analytical tools used to talk about fisheries: biology, economics, and social sciences (Jentoft, Chapter 12, this volume). There seems to be an honest recognition by most people involved with fisheries that to better respond to global fisheries crises, all three disciplines and their interactions will have to be examined. Unfortunately, this recognition is often manifested by adherents of particular disciplines, or theories, attempting to point out how fisheries problems could best be identified and resolved if only their views were given more weight. Instead, improvements in making fisheries policy may lie in the greater use of interdisciplinary and multidisciplinary approaches. Such exchanges of knowledge would allow fisheries scientists to help develop the multifaceted policy options which an increasingly complex spectrum of user groups will demand in an increasingly complex future.

7.2 THE CERTAINTY OF UNCERTAINTY

One of the continuing goals of fisheries science has been the quantification of stock sizes. It is with these numbers that policy makers cast management regimes. However, estimating stock size to within one third or one half its actual value is considered a success (Wilson *et al.*, 1994). Such information, when projected upon the unstable population patterns that are typical of even simple fish communities (Wilson *et al.*, 1994), complicates the issue of predicting stock sizes. Even in some of the most intensively studied aquatic ecosystems 'ecological interactions often produce surprising and counter-intuitive results, and simplistic explanations, predictions, or interpretations are seldom correct' (Walters, 1995). Lastly, more of the uncertainty that fisheries scientists deal with arises from missing or inaccurate information. For example, catch rates and catch statistics, traditionally the source of much biological information for policy making, are suspected of being subject to massive global underestimation because of purposeful failure to report discards and illegal catches (Pitcher, 1993).

There are two remedies for this. The first is that 'missing information', especially that pertaining to social science and economics, has often been, in fact, collected by some governmental or non-governmental organization (Harris, 1996). Unfortunately there has been poor or little communication between social scientists and fisheries scientists. In addition, aboriginal and local traditional knowledge about fish and fisheries has been ignored (Kuhn and Duerden, 1996; Haggan, Chapter 2, this volume). This is a result of the need, in science, for *quantitative* information, whereas much of the untapped traditional knowledge is of a *qualitative* nature. The challenge, too often, is not simply finding such information, but accepting its validity and convertibility, thus avoiding the greater trouble of obtaining it afresh. While such information may not be directly applicable to the economic regulation of a large industrial fishery, it may nevertheless be a rich source of untapped biological information collected over many generations by the attentive people who pursue fisheries resources.

The second remedy for uncertainty and complexity is the computer. This is not only a tool with which we can analyse increasingly complex systems, but also one enabling us to access previously untapped knowledge. One spin-off will be access, by consumers, to information on issues related, for instance, to policy making (Sproul, 1996) or to markets (Sutton, Chapter 8, this volume). After all, not only scientists can extract information from the diverse sources that computers have made available to all. A further benefit would thus be the education of many people in the complexities inherent in managing fisheries.

Much interest has also been generated by negotiation processes, often the most difficult and uncertain part of policy making and implementation. The experience of the World Bank is interesting in this respect, as fisheries scientists, like the bank, must deal with uncertainty in the context of the formation and implementation of high-stakes, high-risk policies. The experience of the World Bank is that much work often must be done to determine what makes each negotiation special (van Santen, 1996). Similarly, fisheries scientists are familiar with the conundrum of trying to explain how ecosystems work in general, while accounting for the diversity and nuances in particular of different ecosystems.

Political and bureaucratic policy makers and fisheries scientists must also avoid trying to outguess each other, resulting in situations in which the best decisions become 'no decision'. Bureaucrats and politicians charged with maintaining or rebuilding a fishery know that scientists tend to be conservative and averse to risk; thus they expect they are being 'low-balled' by the scientists' recommendations. Further, bureaucrats and politicians have to work in the arena of public opinion. Lastly, bureaucrats and politicians realize that even if they go beyond the recommendations of fisheries scientists, environmental variations (leading for example to an increase in recruitment) may mask the effect of a bad decision, or make a difficult decision (such as reducing effort) less urgent. Thus, a rational decision may be to do nothing, and not risk public displeasure at cutting back a fishery. Scientists, in turn, are aware of this view and adjust their recommendations to it, completing the cycle (Walters, 1995).

Further, the admitted *uncertainty* of knowledge by fisheries scientists is often reinterpreted as a *lack* of knowledge owing to the public misperception of science as a set of known facts, rather than a process. This mismatch in expectations is exacerbated by the perception that fisheries science ought to be as precise and predictable as, for example, physics and engineering, despite the problems of complexity facing even the most basic aspects of fisheries science. Realizing this complexity may help fisheries scientists and others to accept the legitimacy of complex answers to what appear to be simple questions. Communicating this acceptance to other groups may be facilitated by references to traditional knowledge and to the increasing public familiarity with databases and computers. Garcia (1994) suggests the following goals that fisheries science should pursue to deal with uncertainty: (1) improving the statistical power of methods used to assess fishery model parameters; (2) developing multispecies ecosystem models that account for environmental fluctuation; (3) testing model sensitivity to changes in parameters and how this affects resulting management performance; (4) performing risk assessment via analyses of ranges of options and models; and (5) experimenting with different management systems, and helping to improve fishing gear and practices.

7.3 WHO GETS A SAY: THE POLICY OF POLICY MAKING

The dilemma of who should participate in the formation of policy parallels that of the multi/interdisciplinary debate, discussed below. This dilemma refers to the question of which views should be considered as legitimate when deciding who has a large enough interest, or information deemed relevant, to be included in the policy formation process. As fisheries science becomes more interdisciplinary, new stakeholders are introduced to the policy process. This increase in number of stakeholders could complicate the debate, at a time when the issues to be resolved are more complex than ever. McGoodwin (1990) suggested that fisheries policy makers will have to consider five issues to succeed in the future: '...(1) human needs, values, and social equity; (2) biological conservation and resource productivity; (3) economic productivity and efficiency; (4) administrative feasibility; and (5) political acceptability.' It is therefore not surprising that many fisheries scientists yearn for the days when theirs was a more 'user-friendly' science, which sought to help fishers develop new ways to exploit fish stocks in a world where only government and business needs were addressed (Williams, Chapter 11, this volume).

Such developments elicit much concern among biologically trained fisheries scientists, who wonder whether there will be a decreasing role for them in the future of fisheries policy making (Rice, 1996). One response is that a change in the role of fisheries biology is occurring, not an increase or decrease. Biologists may become freed from simply focusing on stock assessment, to assume an expanded role and address new issues (Cochrane, 1996). This change implies that biologists must now concern themselves with the nuts and bolts of establishing the key parameters of the fisheries they study and must focus on the problem of overcapacity, which must be addressed in and of itself (Cochrane, 1996). We risk, otherwise, the development of well-intentioned policies with plenty of stakeholder input, offering no real solutions. Thus, fisheries scientists will have to spend more of their energy on helping create and implement policies that will ensure biologically sustainable limits. There is much urgency as more than 50% of the world's stocks are either fully or overexploited (Garcia and Newton, 1997). Clearly, without first addressing the biological bottom line, there may soon be no resource to pursue.

Whether or not 'us versus them' attitudes can be reconciled is another aspect of this discussion. Community participation in fisheries management and policy making has great potential for ameliorating both polarization and ossification. However, it will be vital to devise mechanisms to help reconcile the disparate interest groups within and outside most fisheries. While community participation brings information and views from a variety of sources and disciplines, and further complicates fisheries issues, the inclusion of people other than managers, scientists, and industry could also foster conciliation among user groups and valuable exchanges of knowledge.

McGoodwin (1990) points out that a source of antagonism between managers and fishers is the ill effects of overspecialization by both. Often fishers are forced to modernize and specialize, without consultation, while rarely seeing their standard of living improve. Fisheries science and management often is too rigid in its structure and incapable of responding in a timely fashion to dynamic processes. McGoodwin suggests, therefore, that fishers and social science should have a greater say in the formulation of policy, as a mechanism to redress this ossification. By doing so 'economies of managerial effort might thus be realised, as well as a reduction of conflict between fishers and fisheries managers' (McGoodwin, 1990).

As new participants enter the policy process it need not be assumed that they are necessarily ignorant of the issues and/or facts. Of course, this is the 'best case' scenario for the effects of the computer age on information. The real course of events may, however, tragically parallel that taken by television, where it is the trash that dominates the information. Often the public is subjected to great quantities of subtle and persuasive disinformation. In fisheries this imbalance would serve only to hamper policy formation.

There are means of addressing such issues, however, as illustrated by the International Standards Organization (ISO) guidelines on ecolabelling, ISO 14000. This protocol of environmental standards attempts to more objectively assess the environmental impact of the products. It creates a means of judging whether or not current environmental policies, of governments and businesses, are effective (Tibor and Feldman, 1996). In such a way the general public could influence policy by voting with their pocket-books on fishery policy. For any large industrial company that harvests fish, one of the favourable aspects of ISO 14000 would be the ability to consult and 'address environmental issues in a systematic way and thereby improve its environmental performance' (Tibor and Feldman, 1996). These standards are an improvement over past ecolabelling schemes in that they are intended to help the company establish a process to meet environmental goals, and not just to establish goals in a policy vacuum (Sutton, Chapter 8, this volume).

Another potential means of addressing the problems of decision making is within the 'precautionary principle', which can help in the derivation of environmental goals and the means to achieve them. A weakness of past fisheries policy making was that fisheries scientists were asked to clearly demonstrate where a fishery was being harmed before policy could be created or put into action (Garcia, 1994). The precautionary principle turns this process on its head, arguing instead for caution until clear evidence is found to support more intensive fishing (Garcia, 1994). By thus framing the context in which fisheries are discussed, a part of the uncertainty problem could be resolved.

With respect to who participates in the policy process, discussants are often divided into two camps; those who want to keep the status quo, and those who want to increase the participation of new groups. In Canada, the *Fisheries Act* and the *Oceans Act* represent the former point of view. In writing these, the federal government decided who the relevant participants were for each act and then elicited opinions and views accordingly (Charles, 1996). While seemingly inclusive, this model does not outline whether any appeal process is available to those groups and individuals who may feel they should have been included. Pauly (1997) has also shown that, in many developing countries, the phenomenon of Malthusian overfishing, resulting from increasing populations of fishers pursuing resources which at best remain constant, is exacerbated by the marginalization of small-scale fishers' views. Because Malthusian overfishing results, in large part, from the cognitive and political remoteness of fishers from the levers of power, there is usually little room for the often politically weak fisheries departments to take really effective action to deal with policy problems.

However, while it may appear well intentioned to elicit as many views as possible, it may introduce an admixture of superfluity to the debate. For example, while all citizens of a given country have some interest in the fisheries within its waters, there can be little doubt that the direct effect of a fisher on the fisheries, and vice versa, is greater than that of a city dweller who has never seen the sea. Yet the number of city dwellers who vote, and hence help select competing policies, will usually be far greater than the number of fishers. Such are the difficult and contentious moral issues that must be addressed when deciding who participates.

A possible solution to this set of problems is that relatively independent institutions, such as universities, might take on the role of facilitators. The different interest groups could be brought together in a constructive, non-intimidating forum. To do this it would be necessary for academics not to become too closely tied to any particular interest group. Unfortunately, such close association is a temptation to which it is often easy, both financially and organizationally, to surrender.

7.4 BIOLOGY, ECONOMICS AND SOCIOLOGY, BUT NOT NECESSARILY IN THAT ORDER

A recently developed aim of fisheries science is to determine whether its several specialized subdisciplines can, together, contribute to fisheries policy. However, this agreement on ends might not easily be translated into agreement on the means. A case study in such an evolution of fisheries policy has been created by the introduction of democratic rule in South Africa. The overall goal of the new fisheries policy there is the creation of a more equitable and effective system (Cochrane and Payne, Chapter 5, this volume). This goal requires input from social sciences and economics. This aspect is emphasized in the eight 'guiding considerations' of fisheries policy:

1. 'responsible' use and management;
2. 'democratic' institutional structures;
3. 'protection' of the resource;
4. 'equitable' allocation of rights;
5. creation of social development and benefits;
6. consideration of worker-related issues;
7. maintenance of an efficient, 'investor friendly', and competitive industry;
8. enhancement of the tourist fishery.

(Cochrane and Payne, Chapter 5, this volume). It should be noted that of these principles, only (1) and (3) could be said to be a direct function of the traditional, biological domain of fisheries science. The other principles indicate that there will be far greater weight given to the opinions and views of those involved with the social and economic spheres of society. This emphasis could provide valuable lessons for the world as to how these disciplines may synthesize changes in the formulation of fisheries policy. New questions will be asked about the ramifications and meaning of policy to fishers' lifestyles, their families, and their place in society as defined by economics, ethnicity or class. It also seems apparent that to change the effect, in fisheries, of broad societal policies, such as those left over from apartheid in South Africa, it will be necessary to begin focusing on non-biological policies (Davis, 1996).

Some of the surprising ways that matters from social science and economics can affect the formation of fisheries policy have been documented in Canada. An examination of changes in the management of the Bay of Fundy groundfish fishery shows that unemployment insurance is an issue in the management process that must be considered (Sinclair, 1996). For many biologically trained fishery researchers such concerns are surprising, even superfluous. Yet it is repeatedly emphasized that such government policies often do affect the fisheries *milieu* in significant ways. An example of how seemingly unrelated political events can influence the government formation of fisheries policy was provided in the case of the Canadian constitutional debate surrounding the status of Quebec in confederation. Not only has this debate diverted resources, and the public's attention, from other policy issues such as fisheries, but it potentially involves changes to the structure of the government itself and thus changes in whether the federal or provincial government regulates the fishery. As a nuance of the federal–provincial power/policy debate, the possible change of constitutional authority in fishery policy is a side show, yet it could undermine the success of even the most carefully considered fishery policy.

Some may argue that a multidisciplinary approach puts analysts and decision makers at risk of becoming bogged down in a swamp of non-pertinent information and incapable of discerning the data most applicable to the questions at hand. Such a critique is disingenuous as it assumes that fisheries managers, scientists, and decision makers are incapable of dealing with different kinds of information. Yet the chief cause of poor policy decisions in the past has been the *lack* of information resulting from unidimensional, single-discipline approaches that oversimplified complex, dynamic fishery systems. In a world where many fisheries policy makers have direct access to the information storage, processing, and sharing that computers provide, it would be foolish not to examine all of the information available (Schnute and Richards, 1994; Richards and Schnute, Chapter 30, this volume).

The computer will allow the expansion of the discipline to intellectual terrains that were previously inaccessible and unexplored. The construction of fish and fishery databases, incorporating interdisciplinary knowledge, has already begun with the introduction of resources such as FishBase (Froese and Pauly, 1997). This database was developed with the idea that everyone, from the most sophisticated academic to the interested layperson, could access information on all the fish that are exploited by humans. FishBase allows the standardization of qualitative information and the identification of cross-relationships that may not previously have been noted (Froese and Pauly, 1997). Fisheries science must embrace technologies such as databases if it is to successfully address the challenges of future fisheries (Schnute and Richards, 1994). Given that the source of the information is reliable, databases will provide the means for relating new knowledge to that which fisheries scientists traditionally use. An important aspect of this assimilation will be the incorporation of traditional environmental knowledge (TEK) into such databases. TEK usually encompasses systems of environmental knowledge and species classification, over long time periods and environmental fluctuations, in stories and oral histories that can be abstracted and used for conventional resource management (Kuhn and Duerden, 1996; Haggan, Chapter 2, this volume). This qualitative information can help 'tune' national fisheries management policies to the needs of different communities.

Interdisciplinary fisheries studies would also help diagnose the causes of policy failures such as with the Atlantic cod on the East Coast of Canada. There, policy makers became so caught up in the economic rationalization and modernization of the fishery that, despite warning signs from the biological and social spheres, they could not recognize the approaching stock collapse. For example, catches by the nearshore fishery declined throughout most of the modernization, resulting in much social displacement (Sinclair, 1992), long before the cataclysmic displacement created by the ultimate closure of the fishery. Problems were also apparent in the changing patterns of class relations and catch allocations, which made unprofitable small-scale fishing operations that had been at least modestly profitable for generations (Sinclair, 1992). While the biological warnings for the collapse were eventually well understood, they were too late to help. In many other fisheries, especially those in the developing world, such information is even less likely to be timely. The use of social sciences and economics to identify policy priorities in fisheries can therefore become tools to construct useful policy when biological information is lacking or absent (Pitcher *et al.*, Chapter 3, this volume).

7.5 CONCLUSIONS

For too long, decision makers and analysts have acted as paternalistic controllers of the world's fishery resources. In many nations this paternalism has resulted in the alienation of the fishers from the people who are studying the fisheries and from those who make policy recommendations.

This alienation has created the perception of many involved in the process of fisheries policy making that expressing one's views antagonistically is the only appropriate way to express them. These diverse players may, however, have more in common than they realize. Most participants in fisheries policy now accept the position that most of the world's fisheries are overexploited. It will be impossible to act upon this crisis in any coherent way unless there is a systematic effort to establish a policy process by which the different players can work together and talk to each other in meaningful ways. The use of multidisciplinary techniques in the analysis of fisheries provides a tool with which to attain this goal. Such an approach would foster the unification of different players by creating the acceptance, and establishing the legitimacy, of points of view and information other than one's own. By adopting a multidisciplinary point of view, new mixes of ideas may open paths to solutions we may not previously have been able to realize.

ACKNOWLEDGEMENTS

I would like to acknowledge Daniel Pauly, Tony Pitcher and Trevor Hutton for their comments, suggestions and insights.

REFERENCES

Charles, T. (1996) Comment, in general discussion on theme three, in *Reinventing Fisheries Management* (ed. T.J. Pitcher), *Fish. Centre Res. Rep.* **4**(2), 50.

Cochrane, K. (1996) Answer and comment, in discussion of keynote address theme three, in *Reinventing Fisheries Management* (ed. T.J. Pitcher), *Fish. Centre Res. Rep.* **4**(2), 43.

Davis, A. (1996) Comment, in discussion of keynote address theme three, in *Reinventing Fisheries Management* (ed. T.J. Pitcher), *Fish. Centre Res. Rep.* **4**(2), 43.

Froese, R. and Pauly, D. (eds) (1997) *FishBase 97: Concepts, Design and Data Sources* ICLARM, Manila, 256 pp.

Garcia, S.M. (1994) The precautionary principle: its implications in capture fisheries management. *Ocean Coastal Manage.* **22**, 99–125.

Garcia, S.M. and Newton, C. (1997) Current situation, trends and prospects in World capture fisheries, in *Global Trends: Fisheries Management* (Am. Fish. Soc. Symp. **20**) (eds E. Pikitch, D.D. Huppert and M.P. Sissenwine), American Fisheries Society, Bethesda, MD, pp. 3–27.

Harris, C. (1996) Comment, in general discussion on theme three, in *Reinventing Fisheries Management* (ed. T.J. Pitcher), *Fish. Centre Res. Rep.* **4**(2), 50.

Kuhn, R. G. and Duerden, F. (1996) A review of traditional environmental knowledge: an interdisciplinary Canadian perspective. *Culture*, XVI (1), 71–84.

McGoodwin, J.R. (1990) *Crisis in the World's Fisheries: People, Problems, and Policies*, Stanford Univ. Press, Stanford, 235 pp.

Pauly, D. (1997) Small-scale fisheries in the tropics: marginality, marginalization, and some implications for fisheries management, in *Global Trends: Fisheries Management* (Am. Fish. Soc. Symp. **20**) (eds E.K. Pikitch, D.D. Huppert and M.P. Sissenwine), American Fisheries Society, Bethesda, MD, pp. 40–49.

Pitcher, T.J. (1993) Stewardship and sustainability of Pacific fishery resources: the need for critical insight and an encyclopedia of ignorance, in *Our Common Shores and Our Common Challenge: Environmental Protection in the Pacific* (Proc. Fourth Int. Symp. Conf. Asian and Pan-Pacific Univ. Presidents) (ed. D.G. Shaw), Anchorage, Alaska Sea Grant College Program, Univ. of Alaska, pp. 5–28.

Rice, J. (1996) Question, in discussion of keynote address theme three, in *Reinventing Fisheries Management* (ed. T.J. Pitcher), *Fish. Centre Res. Rep.* **4**(2), 43.

van Santen, G. (1996) Answer, in discussion session of 'Politics and fisheries' by Gert van Santen, in *Reinventing Fisheries Management* (ed. T.J. Pitcher), *Fish. Centre Res. Rep.* **4**(2), 47.

Schnute, J.T. and Richards, L.J. (1994) Stock assessment for the 21st century. *Fisheries* **19**(11), 10–15.

Sinclair, M. (1996) Modifications of Scotian Fundy groundfish management for sustainable use, in *Reinventing Fisheries Management* (ed. T.J. Pitcher), *Fish. Centre Res. Rep.* **4**(2), 47.

Sinclair, P.R. (1992) Atlantic Canada's fishing communities: the impact of change, in *Rural Sociology in Canada* (eds D.A. Hay and G.S. Basran), Oxford Univ. Press, Toronto, pp. 84–98.

Sproul, J. (1996) Comment, in general discussion on theme three, in *Reinventing Fisheries Management* (ed. T.J. Pitcher), *Fish. Centre Res. Rep.* **4**(2), 43.

Tibor, T. and Feldman, I. (1996) *ISO 14000: A Guide to the Environmental Management Standards*, Irwin Publishing, Chicago, 237 pp.

Walters, C. (1995) *Fish on the Line: the Future of Pacific Fisheries*. The David Suzuki Foundation, Vancouver, 82 pp.

Wilson, J.A., Acheson, J.M., Metcalfe, M. and Kleban, P. (1994) Chaos, complexity and community management of fisheries. *Mar. Pol.* **18**, 291–305.

Chapter eight

Harnessing market forces and consumer power in favour of sustainable fisheries

Michael Sutton

ABSTRACT

Fisheries are the last major world industry exploiting wild natural resources for food. Yet the history of modern fishery management is replete with spectacular failures. In virtually every case, the *realpolitik* of fisheries has rendered long-term sustainability of catches a futile management goal. All too often, politicians have compelled fishery managers to ignore the implications of the best available science. Management actions that might have prevented the disastrous collapse of fisheries but which carried a price unacceptable to industry have been scrupulously avoided. Society has simply lacked the political will to forestall the fishing industry's tendency to use up its living capital and thereby destroy itself.

Reversing the fisheries crisis will require a major overhaul of contemporary fishery management. Only this, coupled with heightened public interest and powerful economic incentives, has the power to stop chronic overfishing and to shift the focus of fishery management from development and exploitation to conservation and sustainability. This chapter presents a series of necessary reforms and suggests how, by working together, conservationists, scientists, fishers, industry, and governments can help

Reinventing Fisheries Management. Edited by Tony J. Pitcher, Paul J.B. Hart and Daniel Pauly. Published in 1998 by Kluwer Academic Publishers, London. ISBN 0 412 83410 3.

shape the future of world fisheries. WWF and Unilever Corporation have taken a first step by launching the Marine Stewardship Council, an innovative plan to bring market forces and the power of consumer choice to bear in favour of sustainable, well-managed fisheries.

If marine fish populations are to be sustained – both as an important source of food and as a vital component of ocean ecosystems – increased public support and consumer power must be used to create social, economic, and political incentives for fishing that is both sustainable and clean. That will not be easy; fish neither sing like whales nor look like pandas. But the stakes are high: the future of world fisheries, their associated marine ecosystems, and the millions of people that depend on them for food and employment.

8.1 INTRODUCTION

The need for fundamental reform of marine fishery management has become abundantly clear over the past decade (Earle, 1995; FAO, 1995a; Safina, 1995; Weber and Gradwohl, 1995; House of Lords 1996). Fisheries that have sustained coastal communities for generations have suffered catastrophic declines. In some areas, excessive fishing has driven staple species such as Atlantic cod commercially extinct (Collins, 1994). Increasingly volatile fish wars, such as the 1995 dispute between Canada and Spain, have erupted over what remains. Governments pay tens of billions of dollars each year in fisheries subsidies to catch only $70 billion worth of fish (FAO, 1993; WWF, 1997). Gone forever are the historical estimates that marine catches could top 500 million tonnes per year. Many scientists now believe that modern fisheries have exceeded the limit of the seas (Earle, 1995; Myers *et al.*, 1995; Safina 1995; Cook *et al.*, 1997).

The essential question is therefore not whether the past model of marine fishery management has failed, but why? What lessons can be drawn for the future? Throughout modern history, governments have largely managed marine fisheries for the growth and development of their associated commercial fishing industries. Decision makers have paid scant attention to the sustainability of those fisheries, much less the health of their associated ecosystems or the needs of artisanal fishers exploiting the same species. In virtually every case, the short-term social and economic needs of a region's fishing industry have rendered long-term sustainability of catches a futile management goal. In many parts of the world, subsidized fleets have become grossly overcapitalized. Unsustainable fishing has become an industrial addiction.*

*Sustainable use means using renewable natural resources such as marine fisheries at rates within their capacity for renewal (IUCN/UNEP/WWF, 1991).

This predicament cannot be attributed to a lack of scientific information. Fisheries scientists have for years provided more or less accurate models of fish population dynamics and estimates of fishery production. But all too often, fishery managers more concerned with political than scientific realities have been compelled to ignore the implications of the best available science. Politicians, often at the highest levels, have frequently intervened in decisions about specific fisheries. Society has simply lacked the political will to forestall the fishing industry's tendency to use up its capital and thereby destroy itself (Alverson, 1995).

Conservationists have responded to this growing crisis by seeking to strengthen laws and treaties and to bolster government action. However, the fishing industry, dependent on a steady income to sustain boat mortgages and marginal businesses, has steadfastly resisted change. Thanks largely to this political stalemate, the decline of world fisheries has proven virtually impossible to reverse. Governments have typically devised politically expedient solutions and then described them as environmentally necessary. These efforts have mostly been too little, too late. Management actions that might have prevented the disastrous collapse of fisheries but which carried a price unacceptable to industry have been scrupulously avoided.

Turning this situation around will require more than just minor fine-tuning of contemporary fishery management. Two heretofore overlooked influences – public support and market forces – will have to be harnessed to help create new international standards. First, greater public involvement in the fishery management process must be generated. The same worldwide public concern that banned the trade in elephant ivory and outlawed commercial whaling must be brought to bear. Second, market-led economic incentives must be created to promote sustainable fishing. Conservationists, working with responsible, progressive seafood companies and other stakeholders, must develop market reforms that will encourage fish buyers to purchase their products only from sustainable, well-managed fisheries (Sproul, Chapter 9, this volume).

This chapter does not specifically address inland fisheries, as they make up less than 8% of the world commercial catch (excluding aquaculture) and are subject to a different set of problems than are marine fisheries (FAO, 1997). However, many of the reforms suggested in the following section would also undoubtedly benefit inland fisheries.

8.2 REFORMING FISHERY MANAGEMENT

To reverse the fisheries crisis, long-term solutions that are environmentally necessary must be developed and then made politically feasible through the use of public pressure and economic incentives. The following ten reforms will be essential to speed the transition to sustainable, well-managed, and ecologically sound fisheries.

1. **Strengthen national, regional, and international capacity to manage marine fishes.** Governments must allocate sufficient funds to develop the institutional capacity necessary to adequately manage their marine fisheries. This is especially necessary in developing countries, where such capacity is largely underdeveloped or non-existent. Those nations suffering from a fishery management system rife with conflicts of interest must reform their laws (WWF, 1995). International commissions charged with managing shared fisheries and those on the high seas must immediately implement the provisions of the 1995 UN Agreement on Straddling Fish Stocks and Highly Migratory Fish Stocks. Among other things, this will require regional bodies to open their decision-making procedures to public scrutiny. In addition, fishery management at all levels must be relieved from political interference aimed at satisfying the short-term economic needs of industry rather than the long-term requirements of fish populations and the marine ecosystem (but see Hart and Pitcher, Chapter 16, this volume). To help ensure that these reforms are carried out, the United Nations or other international authority should create a high-profile Global Oceans Forum to place fishery conservation and other ocean issues higher on the international political agenda. The Oceans Forum could exercise oversight of regional fishery management bodies and report annually to the UN Secretary General and the public on the state of the oceans and marine fisheries.
2. **Focus management programmes on limiting effort and restricting access to fisheries.** Past efforts at fishery management, particularly in developed countries, have been characterized by ineffective measures such as mesh-size restrictions and trip limits that simply mandate inefficiency (WWF, 1995). These techniques should be abandoned in favour of management schemes that limit fishing effort, especially in fisheries that are overfished or depleted. Limited-access management should reduce effort to levels consistent with sustainable fishing and the recovery of depleted species. Such programmes should prevent new entry into fisheries that are fully subscribed and should form a part of comprehensive management plans for each fishery (compare Scott, Chapter 15, and Hannesson, Chapter 19, both this volume).
3. **Enact and implement recovery plans for depleted species.** Most overfished species, even those that have been severely depleted, are not subject to any kind of recovery plan. Fishery managers should, as a matter of priority, develop and implement effective recovery plans,

including target population sizes and timetables for achieving them. A necessary prerequisite in developing countries may be national and international efforts to build the institutional capacity to develop recovery plans. In any event, the targets and pace of these plans should be driven primarily by the biological requirements of the fish populations involved, not the short-term demands of the local fishing industry. Well-managed fisheries that are allowed to recover from past overfishing would help restore the health of the marine ecosystem and concurrently yield more to fishers. In fact, the FAO estimates that improved management could yield 10–20 million tonnes in additional catch worldwide (FAO 1997).

4. **Reduce and eliminate the subsidies that sustain commercial fisheries.** The billions of dollars in subsidies that are propping up unsustainable fisheries should be eliminated. This includes subsidies for shipbuilding and construction, refitting of fishing vessels, market research and development, industry bail outs, low-cost industry loans, and development of fisheries for underutilized species, so called because they have not yet been overexploited. Where subsidies are provided, they should be part of a comprehensive plan for the future of the fishing industry, including decommissioning of fishing vessels and retraining of fishers, where necessary (WWF, 1997).
5. **Accelerate programmes for decommissioning excess fishing fleet capacity.** Overcapacity owing to uncontrolled growth of the world's fishing fleets is one of the most serious problems facing marine fisheries today (Mace, 1997). The only practicable way to address this problem in the near term seems to be government-funded programmes to decommission or buy back excess fishing vessels (Collins, 1995). At a minimum, funds for such programmes should be doubled as quickly as possible so as to expand these programmes to achieve a rapid reduction in fishing effort. Funding of future decommissioning programmes should be dictated by capacity-reduction targets specific to each fishery. Governments ought to make the appropriation of these funds a top priority in drawing up their annual fisheries budgets.
6. **Expand programmes for retraining fishers displaced by overfishing and effort limitation.** The legacy of past open-access fisheries is a global population of fishers that far exceeds the number required to catch what fish are available. Many of these people have few skills or professional abilities other than fishing or fish processing. Retraining programmes are urgently needed, especially in northern countries, to move displaced fishers into productive employment in other sectors as quickly as possible. Funds for retraining fishers should be a priority and should go hand in hand with those for decommissioning programmes. However, redeploying fishers into other economic activities

will depend on the availability or creation of alternative jobs in other sectors. Where there are no alternative forms of employment available (e.g. in artisanal fishing communities), governments must ensure that fishers are given preferential access, where appropriate, to traditional fishing grounds and resources (FAO, 1995b, Article 6.18).

7. **Develop social and economic incentives for sustainable, well-managed fisheries.** Today's social and economic forces mostly provide an incentive for unsustainable, destructive fishing. Positive incentives for sustainable, well-managed fishing are urgently required. Public concern, market forces and consumer power must be harnessed to create such incentives. Governments should take the lead in developing and enacting economic incentives for sustainable fishing, thus providing a carrot rather than merely the stick of prescriptive regulations. In addition, voluntary, market-led incentives for sustainable fishing, such as the Marine Stewardship Council (MSC) initiative described in this chapter, must be created so as to swing consumer power behind efforts to recover and sustain well-managed fisheries. This will require conservation organizations and progressive seafood companies to work together to educate consumers (both corporate and individual) about the large potential effect of their purchasing decisions.
8. **Reduce the footprint of northern fleets on fisheries in developing countries.** International standards for distant-water fishing should be developed as a matter of priority and enforced by UN mandate. No country should be permitted to purchase fishing rights from a foreign government without a full assessment of the impacts of such fishing on the specific fisheries involved, their associated marine ecosystem, and upon indigenous fishers that have relied on the same resources for generations.
9. **Eliminate destructive fishing practices such as the use of poisons and explosives.** Destructive fishing practices should be phased out immediately in favour of more sustainable, less destructive alternatives. Such practices are usually a sign of Malthusian overfishing (Pauly, 1990). Funding should be provided for programmes that rectify the underlying unsustainable social, environmental and economic situation. Fishers must also be trained to use techniques other than poisons and explosives. Enforcement of laws that proscribe activities such as cyanide fishing, bleach fishing, and the use of explosives should be strengthened. Where practicable, governments and industry should interdict supplies of poisons and explosives intended for fishers. Where specific fisheries have traditionally used poisons, such as cyanide in the live fish trade of East Asia, simple tests should be devised that allow inspectors and customers to determine when poisons have been used and to act accordingly (Johannes and Reipen, 1995).

10. **Reduce and eliminate the bycatch of marine wildlife in commercial fisheries.** Commercial fisheries currently kill and waste 18–40 million tonnes of fishes and other marine wildlife annually (Alverson *et al.*, 1994). To add insult to injury, modern fishing practices have a devastating effect on marine biological diversity and the physical environment of the oceans (Dayton *et al.*, 1995; Lee and Safina, 1995; National Research Council, 1995). This enormous destruction and waste of marine life should be prevented by the use of effective bycatch reduction devices on fishing gear. Mandatory programmes should be imposed immediately to require the use of such devices wherever they are available. Incentives such as bycatch quotas (combined with observer coverage to discourage dumping at sea and misreporting) should be imposed or made available to encourage the use of the least destructive fishing gear and practices. When implementing programmes to reduce waste by allowing the landing of bycatch, governments should use extreme caution to ensure that these programmes do not impede bycatch reduction efforts. The reduction of waste should go hand in hand with the elimination of bycatch.

8.3 BUILDING PUBLIC SUPPORT

The public has long been unaware of the developing crisis in many fisheries. The news media focus mainly on acute threats to the marine environment such as oil spills and have accorded far less coverage to chronic, long-term threats such as overfishing. Nonetheless, there are hopeful signs that this lack of awareness is changing, including a cover story in the November 1995 *National Geographic* (Parfit, 1995) and the lead article in *Scientific American* the same month (Safina, 1995). High-profile fish wars and the collapse of fisheries have also received significant press coverage (Swardson, 1995; Cook *et al.*, 1997). In April 1996, the press noted US Secretary of State Warren Christopher's statement in a major speech at Stanford University on threats to the global environment that overfishing of the oceans is putting thousands of fishers out of work (UPI, 1996).

In addition to increasing press coverage, more and more conservation organizations are beginning to focus their attention on fisheries conservation. For example, the Marine Fish Conservation Network was formed by WWF and four other US non-governmental organizations in 1993. The Network eventually grew into a broad-based coalition of more than 100 national, regional, and local organizations whose primary goal was to reform and strengthen the management of US marine fisheries so as to promote their long-term sustainability. In 1996, little more than three years after its formation, the Network achieved its objective of improving the Magnuson Fishery Conservation and Management Act, the principal US fisheries law.

The Network and similar coalition efforts in the US and other countries are important mechanisms that help raise public awareness and concern about marine fisheries. They also help focus activism by concerned citizens. For example, the Network held a series of workshops for citizen activists around the US during 1994 to muster greater support for reform of the Magnuson Act. Much more of this kind of activity will be needed before marine fishes begin to receive the recognition and public support usually accorded to terrestrial vertebrates.

8.4 HARNESSING MARKET FORCES

Although a vital component of the long-term solution, increased public support alone will not reverse the fisheries crisis. Market forces must also be engaged to counter unsustainable fishing and its powerful proponents. Fortunately, an approach is available that has succeeded in other areas: working in partnership with industry to design and implement market-driven incentives for sustainable, well-managed fishing.

Probably the best example of this strategy was the highly successful (though not uncontroversial) tuna–dolphin campaign of the early 1990s (Joseph, 1994; Marine Mammal Commission, 1996). Hundreds of thousands of dolphins were being killed in purse seine fisheries for tuna in the eastern Pacific. Public outrage and consumer power helped provide the political incentive for the US to embargo imports of tuna caught in a manner resulting in excessive mortality of dolphins. Thanks to the successful marketing of dolphin-safe tuna by major manufacturers, the killing of dolphins in tuna fisheries was quickly reduced. However, the goal of that campaign was dolphin protection, not fisheries conservation. No one has ever brought market forces to bear directly on the fisheries crisis itself.

To succeed in this effort, the conservation community must forge alliances with progressive members of the seafood industry. The tuna–dolphin experience suggests that finding corporate allies and redirecting market forces in favour of conservation can be very powerful. One thing is certain; where public opinion, industry and the market lead, governments will likely follow.

In February 1996, two global organizations announced a partnership to create economic incentives for sustainable fishing by establishing an independent Marine Stewardship Council (MSC) (Maitland, 1996). The World Wide Fund for Nature (WWF), the world's largest private, non-profit conservation group, seeks a new approach to ensure more effective management of marine fisheries. Anglo–Dutch Unilever Corporation, one of the world's largest buyers of frozen fish and manufacturer of the world's best-known frozen fish products under such brands as Iglo, Bird's Eye and Gorton's, is interested in long-term fish stock sustainability to ensure a future for its successful fish business. These organizations have different motivations, but a shared objective – to ensure the long-term viability of global fish populations and the health of the marine ecosystems on which they depend.

The MSC, established in early 1997, is an independent, non-profit, non-governmental body. Following worldwide consultations, the organization is developing a broad set of principles and criteria for sustainable fishing. The second draft of this document, developed in late 1997, deals with biological, environmental, economic, and social issues. Ultimately, only fisheries meeting these standards will be eligible for certification by independent, accredited certifying firms. Seafood companies will be encouraged to make commitments to purchase fish products from certified sources only. Products from fisheries certified to MSC standards may be marked with a logo on the package. This will allow seafood consumers to select fish products that come from a sustainable source.

The MSC was inspired by the successful Forest Stewardship Council (FSC), established by WWF, other conservation organizations, and the timber trade in 1993 to promote a market-led solution to the ongoing destruction of the world's forests. By the end of 1996, FSC-accredited companies had certified more than 20 forests comprising more than two million hectares as conforming to sustainable forestry practices. Perhaps more important, more than 100 timber companies and retailers worldwide had made commitments to purchase only timber certified to FSC standards (Bendall and Sullivan, 1996; Bendall and Murphy, 1997).

The creation of the MSC has the potential to significantly alter fishing practices everywhere in favour of more sustainable, less destructive fisheries. When Unilever and other major seafood companies make commitments to buy their fish products only from well-managed and MSC-certified fisheries, the fishing industry will be compelled to modify its current practices. Governments, laws, and treaties aside, the market itself will begin to determine the means of fish production. The costs of environmental protection and the development of sustainable fisheries will thus be distributed through the market chain, rather than falling disproportionately on the shoulders of one or more sectors.

8.5 CONCLUSION

Fisheries are the last major world industry exploiting wild natural resources for food. Only a series of fundamental reforms of fishery management, coupled with heightened public interest and powerful economic incentives, has the power to stop chronic overfishing and to shift the focus of fishery management from development and exploitation to conservation and sustainability. If marine fishes are to be saved – both as an important source of food and as a vital component of ocean ecosystems – increased public support and market forces must be used to create social, economic, and political incentives for fishing that is both sustainable and clean. That will not be easy; fish neither sing like whales nor look like pandas. But the stakes are high: the future of world fisheries, their associated marine ecosystems, and the millions of people that depend on them for food and employment.

ACKNOWLEDGEMENT

The author is grateful to Ms Indrani Lutchman for presenting this paper at the symposium on Reinventing Fisheries Management in February 1996.

REFERENCES

Alverson, D.L. (1995) Fisheries management: a perspective over time. *Fisheries* **20**(8), 6–7.

Alverson, D.L., Freeberg, M., Murawski, S.A. and Pope, J.G. (1994) A global assessment of fisheries bycatch and discards. *FAO Fish. Tech. Pap.* **339**, vii + 19 pp.

Bendall, J. and Murphy, D. (1997) Stewardship councils: superheroes seeking solutions. *Green Futures* no. 3, 23–26.

Bendall, J. and Sullivan, F. (1996) Sleeping with the enemy? Business–environmentalist partnerships for sustainable development: the case of the WWF 1995 Group, in *Environmentalist and Business Partnerships: a Sustainable Model?* (eds R. Aspinwall and J. Smith, Cambridge Environmental Initiative), White Horse Press, Cambridge, UK, pp. 3–33.

Collins, C. (1994) *Beyond Denial: The Northeast Fisheries Crisis, Causes, Ramifications and Choices for the Future*, Northeast Fisheries and Sustainable Communities Project, Watertown, MA, 75 pp.

Collins, C. (1995) *Beyond Denial, Revisited: A Progress Report on the Northeast Fisheries Crisis*, Northeast Fisheries and Sustainable Communities Project, Watertown, MA, 19 pp.

Cook, R.M., Sinclair, A. and Stefansson, G. (1997) Potential collapse of North Sea cod stocks. *Nature* **385**, 521–522.

Dayton, P.K., Thrush, S.F., Agardy, M.T. and Hofman, R.J. (1995) Viewpoint: environmental effects of marine fishing. *Aquat. Conserv. Mar. Freshwat. Ecosyst.* **5**, 205–232.

Earle, Sylvia A. (1995) *Sea Change: A Message of the Oceans*, G.P. Putnams Sons, New York, 362 pp.

Food and Agriculture Organization of the United Nations (FAO) (1993) Marine fishes and the law of the sea: a decade of change. *FAO Fish. Circ.* **853**.

Food and Agriculture Organization of the United Nations (1995a) *The State of World Fisheries and Aquaculture 1994*, FAO Fisheries Dept, Rome, 50 pp.
Food and Agriculture Organization of the United Nations (1995b) *Code of Conduct for Responsible Fisheries*, FAO Fisheries Dept, Rome, 41 pp.
Food and Agriculture Organization of the United Nations (1997) *The State of World Fisheries and Aquaculture 1996*, FAO Fisheries Dept, Rome, 125 pp.
House of Lords Select Committee on Science and Technology (1996) *Report: Fish Stock Conservation and Management*. HMSO, London, 77 pp.
IUCN/UNEP/WWF (1991) *Caring for the Earth: A Strategy for Sustainable Living*, IUCN–The World Conservation Union, UNEP–United Nations Environment Programme, WWF–World Wide Fund for Nature, Gland, Switzerland, 10.
Johannes, R.E. and Reipen, M. (1995) *Environmental, Economic, and Social Implications of the Live Reef Fish Trade in Asia and the Western Pacific*, The Nature Conservancy, Washington, DC, 82 pp.
Joseph, J. (1994) The tuna–dolphin controversy in the eastern Pacific Ocean: biological, economic, and political impacts. *Ocean Dev. and Int. Law* **25**(1), 1–30.
Lee, M. and Safina, C. (1995) The effects of overfishing on marine biodiversity. *Current: J. Mar. Educ.* **13**(2), 5–9.
Mace, P.M. (1997). Developing and sustaining world fisheries resources: the state of science and management, in *Developing and Sustaining World Fisheries Resources: The State of Science and Management. Second World Fisheries Congress, Brisbane 1996. Vol. 2: Proceedings* (eds D.A. Hancock, D.C. Smith, A. Grant and J.P. Beumer), CSIRO, Collingwood, Victoria, pp. 1–20.
Maitland, A. (1996) Unilever in fight to save global fisheries. *Financial Times*, February 22. p. 3.
Marine Mammal Commission (1996) The tuna–dolphin issue, in *Annual Report to Congress for 1995*, MMC, Washington, DC, pp. 99–109.
Myers, R.A., Barrowman, N.J., Hutchings, J.A. and Rosenberg, A.A. (1995) Population dynamics of exploited fish stocks at low population levels. *Science* **269**, 1106–1109.
National Research Council (1995) *Understanding Marine Biodiversity*, Washington, DC, National Academy Press, 114 pp.
Parfit, M. (1995) Diminishing returns, exploiting the oceans' bounty. *Nat. Geogr.* **188**(5), 2–37.
Pauly, D. (1990) On Malthusian overfishing. *NAGA–The ICLARM Quarterly* **13**(1), 3–4.
Safina, C. (1995) The world's imperiled fish. *Scient. Am.* **273**(5), 30–37.
Swardson, A. (1995) In the end, a victory for the world's fish. *Int. Herald Trib.*, April 19, pp. 1,7.
United Press International (UPI) (1996) Christopher assesses world environment. April 9, 1996.
Weber, M.L. and Gradwohl, J. (1995) *The Wealth of Oceans*, W. Norton, New York, 256 pp.
World Wide Fund for Nature (1995) *Managing US Marine Fisheries: Public Interest or Conflict of Interest?* WWF, Washington, DC, 57 pp.
World Wide Fund for Nature (1997) *Subsidies and Depletion of World Fisheries*. Endangered Seas Campaign, WWF, Washington, DC, 136 pp.

Chapter nine

Green fisheries: certification as a management tool

John T. Sproul

ABSTRACT

To protect future global food security and fisheries health, an urgent need exists to initiate long-term market-based changes that link the consumer's choices of seafood with the impact that consumption has on marine ecosystems. These market adjustments will significantly contribute toward countering the environmentally destructive economic forces that have motivated counterproductive fishing behaviour. Introducing environmental information as a value-added component of marine products can form constructive linkages between purchases of seafood and their correspondingly different environmental consequences. This chapter explores the concept of applying internationally defensible certification of sustainable fisheries and recommends putting the idea into practice by integrating existing protocols for environmental management systems with sustainable fisheries standards.

9.1 INTRODUCTION

Historically, a product's reasonable quality, price and physical usefulness to the consumer were sufficient to dictate its demand in the market-place.

Reinventing Fisheries Management. Edited by Tony J. Pitcher, Paul J.B. Hart and Daniel Pauly. Published in 1998 by Kluwer Academic Publishers, London. ISBN 0 412 83410 3.

This was especially true when a diversity of similar products were not available to the average consumer. Generally, other factors such as where, how, why and by whom the product was made were aspects that did not influence the individual's decision to buy. However, as economic prosperity and commodity availability have increased, society has increasingly become concerned with issues of process. For example, the diversity of world-views regarding humanity's role in or apart from our surrounding ecosystem has begun translating into economic behaviour. Consumer boycotts initiated by cultural movements of environmentalism and animal rights, are direct consequences of these trends in society.

Through common daily economic activity and behaviour, both at the individual and corporate level, many current commercial affiliations exist among those holding contradictory philosophies of resource use. This results from both the vast complexity of the economy and a lack of widespread information on ecological process available to consumers. For the most part, we simply are unaware of these inconsistencies because the information/knowledge has not accompanied the products we choose. In the world's seafood sector, harvesters, processors, distributors, and consumers hold varying values concerning the issue of marine ecosystem stewardship. Some views are compatible, others potentially mutually exclusive. The presence of credible product information would make it possible for seafood consumers to understand with whom and with what they are associating themselves when purchasing a particular seafood commodity or service. On occasion, budgetary constraints or other factors may overrule philosophical merit. However, the trade-off between that factor and a specific environmental impact will have become clear and personalized. This process will allow acts of consumption to link directly with their ecological consequences at the level of individual choice.

Insight into the fisheries management and extraction methods that lie behind the seafood we buy can provide society with an appreciation of our environmental, economic, and social linkages. This will lead to an education of the consumer about the trade-offs associated with our support of alternative types of harvesting practices and/or fisheries management strategies. This approach exposes consumers, retailers, distributors and processors to a deeper level of understanding and responsibility with respect to their individual opportunity to support sustainable fisheries practices via alternative commercial relationships.

The absence of sustainable fisheries bolstering market forces has contributed to collapse in many local fisheries. The effects have devastated their dependent communities, both socially and economically. The broader negative implications reflect larger issues including food security, biological diversity and global ecosystem health (FAO, 1995a,b; Fraser Board, 1995; Freese, 1994; Sproul and Queirolo, 1994; Rapport, 1995). Mostly, our current seafood markets fail to incorporate information on social and environmental practice associated with processes necessary in bringing commodities from their points of extraction to consumption. In the context of presenting defensible ecological information between harvesters, producers and consumers, there is clear market failure in the international trade of marine products.

The market's failure to incorporate social and environmental externalities in the price mechanism is well recognized (Munro and Scott, 1985; Pearse, 1982; Clark, 1990; van Kooten, 1993). Seafood prices observed by consumers partially reflect a fishery's operational costs associated with a given harvesting method; species availability, distribution cost, taxes or subsidies, and profit mark-up potential also contribute significantly to retail price. Tragically absent in the pricing mechanism is a fishery's 'impact cost' to the ecosystem. Fishers harvesting in an ecologically harmful manner can expect to receive the same price as their counterparts pursuing the same resource in a more ecologically beneficial manner. Also, unsustainable means of exploitation and production have been considered less expensive for industry to operate than environmentally sustainable practices. Thus, seafood consumers, either unaware of, or indifferent to, the negative environmental and social consequences that they may be endorsing when buying the 'cheapest' fish, can reinforce ecologically harmful harvesting. For seafood, as with nearly all goods and services, in-tandem presentation of defensible and concise socio-environmental product information does not yet adequately exist for consumers.

In the 21st century, consumer demand can be the means of creating economic market incentives that endorse and promote sustainable fisheries harvesting and management practices. What is needed to facilitate this demand is the creation of a heretofore non-existent market mechanism that rewards fishery sustainability, socio-environmental education, and consumer responsibility. Through daily transaction activity, this process could educate all interested parties regarding the documented environmental practices and social initiatives of all participants in the seafood industry. Consumer-orientated product information must be fostered. Environmental and social information must be transformed into a primary value-added component of seafood products and must be credibly presented in the market-place just as notably as price.

A more symbiotic relationship between the environment and our economy will arise if knowledge on long-term productive capacity, natural capital stewardship (Daly and Cobb, 1994), optimal use of harvested material, and societal health considerations are presented together with commodity price. Our challenge is to coordinate the technological tools, human resources and institutional networks that will realize these informative linkages between consumption and industrial practice and process.

9.2 IMPLEMENTATION

Defensible certification procedures would be required to document social, economic and environmental practices. Locally appropriate and internationally recognized criteria should be used by a third party eco-certification programme to evaluate a fishery and its operators. A more holistic approach would also extend to life-cycle evaluation of the product and processes associated with an item of seafood from its point of extraction to consumption and to the disposal of its packaging (Cohen, 1996). Evaluation of energy efficiency and use of materials associated along the 'green chain' of the product's life could extend to processors, distributors, retailers, consumers, and waste managers. In effect, this would allow clients to distinguish between 'clean' and 'dirty' suppliers and improve international market competitiveness in the arena of a product's ecological impact (Sakurai, 1995). An international protocol must be created to standardize fishery certification and solidify its successful acceptance and implementation worldwide.

Currently at least two international initiatives are poised to move forward on potentially complementary aspects of industrial certification that can affect all economic sectors related to natural resource extraction. First, the existing International Organization of Standardization's environmental management system (ISO 14000) is capable of addressing much of the necessary conditions for 'green chain' certification from production to disposal. Second, the Marine Stewardship Council (Sutton, Chapter 8, this volume) has potential for its ability to complement ISO 14000 by certifying sustainable harvesting practices prior to value-added processing.

9.3 ISO 14000: ENVIRONMENTAL MANAGEMENT SYSTEM

An international standard framework must be used for developing broad principles of fishery certification within which specifications particular to fisheries could be devised. Several standards for fisheries principles exist including the UN Code of Conduct for Responsible Fisheries (FAO, 1995a), and the UN Conference on Straddling Fish Stocks and Highly Migratory Fish Stocks (FAO, 1995b). To facilitate global acceptance and implementation of sustainable fisheries (and aquaculture) principles, they should be framed within a broad arena of environmental management standards.

ISO 14000 is a rapidly developing methodology for companies to incorporate into their activities internationally defensible environmental management measures. ISO is a Geneva-based harmonizer of world standards in industrial practice. Its members are governmental standardization organizations of 111 countries. Motivated to foster world trade after World

War II, it was created in 1946 to address electronic, communication, trade, and manufacturing standards. Generally, ISO standards are initiated on a voluntary basis, yet countries regularly adopt them, thereby making them mandatory (Tibor and Feldman, 1996).

In the early 1980s, ISO branched out to develop quality standards for management. The initiative resulted in the globally implemented ISO 9000 series on industrial quality control. Aspects of ISO 9000 have subsequently been incorporated into international policy to facilitate trade and remove barriers (Cascio, 1994). In 1992, the ISO Technical Board authorized establishment of Technical Committee number 207 (TC 207). This body, whose secretariat is in Ottawa at the Canadian Standards Association, has the responsibility to oversee and coordinate the diversity of activity associated with international environmental management systems development. ISO has an extensive intergovernmental and ministerial network at the national policy-development level and long experience of international standardization. Therefore, the ISO 14000 protocol should be pursued as a principal international forum for nurturing global debate on the mechanisms required to implement and monitor appropriate 'resource-friendly' commercial activity in the major sectors of the global fishing industry.

As might be expected, attempting to generate an internationally acceptable environmental management protocol has yielded numerous specializations. To visualize this, Fig. 9.1 presents the various aspects of the ever expanding ISO 14000 agenda. The forum encompasses a wide variety of aspects including environmental auditing, eco-labelling, certification, and life-cycle analysis. Years of work by numerous teams of international experts have produced guideline specifications in the ISO 14000 EMS (environmental management systems) series. Several were approved in Oslo in 1995 and finally ratified in Rio de Janeiro, in 1996 (ISO 14001, 1995a; ISO 14004, 1995b; ISO 14010, 1995c; ISO 14011/1, 1995d; ISO 14012, 1995e).

The fishing industry is well suited to implement various aspects of an international EMS, whether at the level of a vessel, a firm or the complete fishery. The ISO 14000 approach focuses on an individual firm developing an appropriate EMS with guidance from registered environmental systems managers. As an example, fishery cooperative associations could make these EMS managers available to their membership at a reasonable cost. The EMS would be specific to a firm/industry/fishery and follow appropriate ISO guidelines that would provide the basis for its development and implementation. Before ISO certification occurred, the system and its implementation procedures would undergo an environmental audit by an independent third party. (To avoid conflict of interest, the EMS developer and final auditor must be different parties.) If this audit was satisfactory

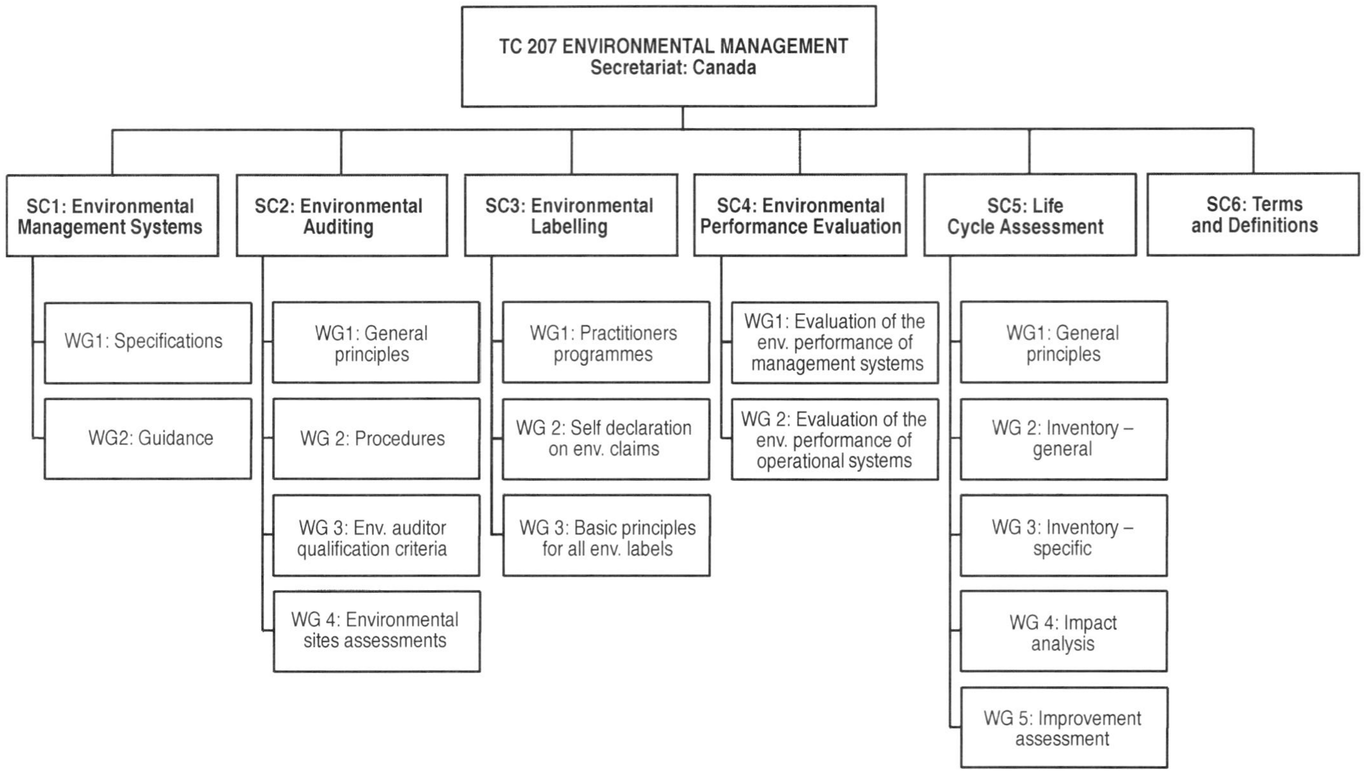

Fig. 9.1 The ISO 14000 TC 207 listing of subcommittees (SC) and working groups (WG).

('pass'), periodic internal performance evaluations would occur as well as follow-up third party environmental audits. Independence between these monitoring and enforcement reviews ensures continuation of compliance with certification and encourages management to adapt and improve practices in the context of EMS and the actual fishery impact the firm is making.

Fisheries auditing will be an important field of employment for nationals in both developed and developing countries. Primary consideration must be given to the training of national auditors. This is especially important for developing nations with substantial fisheries resources but which have limited training facilities and capabilities. If necessary, a consortium of developed nations should provide trainers and workbook materials. Governments of participants from developing countries should in turn provide contributions in kind toward the training, such as translation and workshop coordination services. Local environmental auditors in fisheries (and agriculture, forestry, etc.) will significantly complement the human resource pools in national economies that are resource-based and focused on international trade. Certification of a country's products via locally provided human resource teams of independent 'eco-auditors' would then be articulated in a consistent way to both domestic and international clientele by way of an international 'eco-label' protocol. It is easy to see why cross-boundary standardization of the procedure is crucial, not only to optimize defensible EMS documentation, but also to avoid non-tariff trade barriers that could materialize in the chaos associated with a plethora of potential certification schemes.

In general terms, the ISO eco-label simply confirms that the product is what it says it is. Also certified is the documentation describing the process by which the product came about. However, if it were specifically applied to fisheries, and FAO criteria were employed for example, the process could certify and label the sustainability of a fishery and institutional ecological practice. The ISO 14000 series on environmental management includes both organizational evaluation and product/process evaluation practices (Fig. 9.2). These continue to be refined by the ISO TC 207 (Wolfe, 1995). The international eco-labelling protocol is an important aspect of ISO product evaluation that is actively being addressed.

Technical Committees exist within the ISO 14000 development activity as working groups where specifically applicable guidelines and criteria are debated. There is currently an extensive network of international professionals working in different Technical Committees to describe guidelines appropriate to their specialization. These professionals include academics, industry, and government representatives. Currently, no ISO Technical Committee on fisheries exists. At the earliest possible time, fisheries specialists should become involved in establishing a Technical Committee

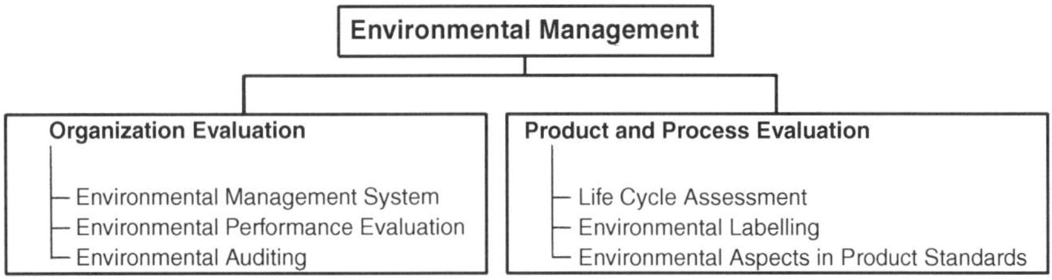

Fig. 9.2 The International Organization for Standardization's development structure for environmental management standards.

on this critical component of the global environment and economy. The Marine Stewardship Council's 'blue ribbon panel' of fisheries experts could contribute toward this process*. In this way, the inevitable development of fishery evaluation, certification, and labelling practices would be consistent with general international methodologies backing the protocol dealing with environmental management systems. With this approach, inter-industry cooperation can move forward more rapidly toward articulating and implementing coordination of ecological stewardship in all aspects of commercial human activity, whether fisheries-orientated or otherwise.

A coordinated approach would ensure a more rapid worldwide acceptance of this vitally important initiative to incorporate environmental practice within the fishing industry's production cycle. In addition, it would provide political and economic incentives for conveying that information in a consistent, concise and defensible way to all consumers. In the absence of a unified advance, several potentially conflicting fishery certification initiatives will emerge and confuse the public and dilute or destroy the credibility of the practice for all. The stage is potentially set for such a confrontation in fisheries certification. The existing ISO 14000 and Marine Stewardship Council (Sutton, Chapter 8, this volume) must coordinate their efforts in this field. In fisheries certification, multiple standardization schemes such as those occurring in forestry can prove counter-productive, and could be avoided. In forestry, for example, separate certification initiatives range from private corporate activity (Jack, 1994) to the ISO 14000's Forestry Certification programme (Campbell *et al.*, 1995), and work of non-governmental organizations such as the World Wide Fund for Natures's Forest Stewardship Council (FSC, 1994). The ISO Forestry and FSC participants are now beginning to coordinate

* This group's first gathering was in September 1996 in London, and the second in New York in December 1997.

their efforts. From the outset in fisheries certification, division of common purpose should be avoided to maintain public integrity of the cause and economy of effort and resources. Anything short of early cooperation will be viewed as mere posturing for influence, because coordination is inevitable, given the interwoven characteristics of standardization protocols and international law and trade policy.

9.4 CONCLUSIONS

It is now essential for all of humanity to write its own legacy on sustainable development. We have within our grasp the opportunity to initiate and institutionalize free-market, democratic instruments of choice. Current and future generations could utilize and build upon it to foster resource responsibility and guardianship for our surrounding natural endowments. With determination, this system could also integrate mechanisms furthering human dignity and healthy creativity.

As natural resource specialists we should seriously apply ourselves collectively to this new initiative. Sustainable fishery certification will incorporate into everyday consumer decisions the process of learning how to minimize their environmental impacts while financing socially constructive organization. As we utilize our surrounding physical materials with these broader influences in mind, through market forces a fundamental shift in thinking will take place. The economy will begin to incorporate ways for people to respond in the market-place to the realization that constructive and sustainable processes are at least as important as their end-products. Linking the product with its process is the key.

A respectable international sustainable fisheries certification system will result in more opportunity for mutually beneficial relationships among consumers and industry, and will increase the underlying productivity of our global marine ecosystems. As educators, managers, and students, we must believe in and seek diverse means of increasing knowledge among humanity, worldwide. In one form or another, the market-place reaches every corner of the globe, at every level. Certification has a real potential for contributing toward transforming consumer behaviour into a genuine learning experience in ecosystem management. To accomplish this, educational information can and must become a highlighted, value-added aspect for future goods and services. Realizing this vision will facilitate increased symbiosis among our human economies and natural environments. It will provide a constructive contribution toward a more market-based system of resource conflict resolution for our successors in the next century. In short, tomorrow's marine ecosystems, and their related human communities, will together be more sustainable than today's. Their sustainability will be judged not just by levels measured by abstract industrial historians who will record our attempts at reinventing fisheries, but by individuals during personal and corporate choices of informed consumer behaviour. An informed consumer is a reformed abuser.

In conclusion, we should now begin to build tomorrow's demand-based information markets that will link specific alternative purchases with social, ecological and economic trade-offs, while also informing consumers of their contribution toward probable alternative futures.

ACKNOWLEDGEMNETS

I wish to express my appreciation to the faculty, staff, students, and professional associates who participated in the University of British Columbia 'Fraser Basin Ecosystem Study: 1993–1996' as sponsored by the Canadian federal Green Plan and administered by the Tri-Council Secretariat of the federal Granting Council. My extended participation in this interdisciplinary research project fostered the creative forum enabling my pursuit of the concepts shared here.

REFERENCES

Campbell, C., Curie, D. and McMahon, J. (1995) ISO 14000 and forestry, in *ISO 14000 Conference Proceedings – Environmental Management for the 21st Century* (ed. K.C. Ayers), Irwin Professional, Bellevue, WA, pp. 146–154.

Cascio, J. (1994) International environmental management standards: ISO 9000s less tractable siblings, *ASTM Standardization News*, April 1994.

Clark, C. (1990) *Mathematical Bioeconomics*, 2nd edn, Wiley-Interscience, New York.

Cohen, D. (1996) Integrated life-cycle management, in *Design for the Environment: Creating Eco-Efficient Products and Processes* (ed. J. Fiksel), McGraw-Hill Press, New York, pp. 127–151.

Daly, H. and Cobb, J.B. Jr (1994) *For the Common Good*. Beacon Press, Boston, MA.

FAO (1995a) *Code of Conduct for Responsible Fisheries*. Twenty-eighth Session Food and Agriculture Organization of the United Nations, 1995. C 95/REV. 1.

FAO (1995b) *United Nations Conference on Straddling Fish Stocks and Highly Migratory Fish Stocks*. UN General Assembly Sixth Session. Distr. General, Annex 3 A/ Conf. 164/37. New York, USA.

Fraser River Sockeye Public Review Board (1995) *Fraser River Sockeye 1994: Problems and Discrepancies*, Canada Communications Group, Ottawa.

Freese, S. (1994) *West Coast Salmon Fisheries: Disaster Relief*. National Oceanic and Atmospheric Administration (NOAA). National Marine Fisheries Service (NMFS). US Commerce Dept. Federal Register, Vol. 59, No. 106, 3 June 1994.

FSC (1994) *Forest Stewardship Council: Principles and Criteria for Natural Forest Management*, Board Approved Version, June 1994. Oaxaca, Mexico.

ISO (1995a) ISO/DIS 14001, *Environmental Management Systems – Specifications with Guidance for Use*. ISO/TC 207 SC1 N82. International Organization of Standardization, Tech. Com. 207.

ISO (1995b) ISO/DIS 14004, *Environmental Management Systems – General Guidance on Principles, Systems and Supporting Techniques*. ISO/TC 207 SC1 WG2 N92. International Organization of Standardization, Tech. Com. 207.

ISO (1995c) ISO/DIS 14010, *Guidelines for Environmental Auditing – General Principles of Environmental Auditing*. ISO/TC 207 SC2 N63. International Organization of Standardization, Tech. Com. 207.

ISO (1995d) ISO/DIS 14011/1, *Guidelines for Environmental Auditing – Audit Procedures – Part 1: Auditing of Environmental Management Systems*. ISO/TC 207 SC2 N64. International Organization of Standardization, Tech. Com. 207.

ISO (1995e) ISO/DIS 14012, *Guidelines for Environmental Auditing – Qualification Criteria for Environmental Auditors*. ISO/TC 207 SC2 N65. International Organization of Standardization, Tech. Com. 207.

Jack, R. (1994) The Trillium Corporations Rio-Condor project: sustainable tropical forestry. *Seattle Times*, 10 April 1994.

van Kooten, G.C. (1993) *Land Resources Economics and Sustainable Development*, UBC Press, Vancouver.

Munro, G. and Scott, A.D. (1985) The economics of fisheries management, in *Handbook of Natural Resource and Energy Economics*, Vol. 2 (eds A.V. Knees and J.L. Sweeny), North Holland, Amsterdam, NY, pp. 623–676.

Pearse, P.H. (1982) *Turning the Tide: The Commission on Pacific Fisheries Policy*. Department of Fisheries and Oceans, Pacific Region, Vancouver.

Rapport, D.J. (1995) Ecosystem health: what's in a name? *Ecosystem Health* **1**(1), 6–13.

Sakurai, K. (1995) Integrated summary, in *Cleaner Production for Green Productivity: Asian Perspectives*. A Report of the APO Basic Research VII on Productivity and Environment (RES-93/94) (ed. K. Sakurai), Asian Productivity Organization, Tokyo, pp. 1–10.

Sproul, J.T. and Queirolo, L. (1994) Trade and management: exclusive economic zones and the changing Japanese surimi market. *Mar. Fish. Rev.* **56**(1), 31–39.

Tibor, T. and Feldman, I. (1996) *ISO 14000: A Guide to the New Environmental Management Standards*, Irwin Publishing, Chicago, 237 pp.

Wolfe, J. (1995) Keynote presentation by TC 207 Secretary: Canadian Standards Association and the ISO 14000, in *ISO 14000 Conference Proceedings – Environmental Management for the 21st Century* (ed. K.C. Ayers), Irwin Professional, Bellevue, WA, pp. 13–47.

Chapter ten

Aboriginal fisheries and a sustainable future: a case study from an agreement with the Nisga'a nation in British Columbia

Michael R. Link and Karl K. English

ABSTRACT

By settling the land claim of the aboriginal Nisga'a people of north-western British Columbia, Canada, governments have an opportunity to put in place an institution that will promote sustainable fisheries management of River Nass fish stocks. We describe five features of the recent Agreement-in-Principle between the Nisga'a, the Canadian federal government and the British Columbia provincial government that are intended to address common problems encountered in fisheries management. While the Nisga'a Fisheries Agreement takes advantage of unique features of the Nass River area, some elements of the Agreement would be applicable to other areas. By studying fisheries management under the land claim settlement with the Nisga'a, fisheries managers may gain insight into the common fisheries management issues of community-based management, property-rights-based fisheries on highly mobile fish stocks, and mechanisms for sustainable funding of research and management.

Reinventing Fisheries Management. Edited by Tony J. Pitcher, Paul J.B. Hart and Daniel Pauly. Published in 1998 by Kluwer Academic Publishers, London. ISBN 0 412 83410 3.

10.1 INTRODUCTION

Most fisheries managers can imagine a better fisheries management world than we find ourselves in today. This better world might be one where:

1. fish stocks are productive and their ecosystems healthy;
2. resource user groups and managers trust and respect each other;
3. bycatch is avoided by using harvesting techniques that are selective and catch only the target species;
4. stock assessments are accurate and timely;
5. the commercial harvesters of the resource make average or above-average incomes compared with other members of society; and
6. resource management decisions reflect societal values and not those of special-interest groups.

We think that a reinvented fisheries management should result in many of the features listed above. The fact that it is easy to come up with long list of better conditions is a sign that fisheries management needs to be reinvented, or more simply, needs to be changed. However, when we attempt to list actions that will ensure that we get to this better fisheries management world, our task becomes extremely difficult.

The paucity of solutions to the problems facing fisheries management attests to the fact that it is very difficult to get to better fisheries management from where we are today (Larkin, 1979). We argue that the path from here to there is, and may always be, seriously constrained by the issues and problems facing fisheries managers.

In this chapter we present one method that we believe will get us from here to there in a salmon fishery on the north-western coast of British Columbia. We describe the essential features of the preliminary Fisheries Agreement recently reached between the aboriginal Nisga'a (Nis·ga') people of northern British Columbia, and the governments of Canada and British Columbia (BC), and describe how they will promote a sustainable fishery and management system. We explain how treaties with First Nations elsewhere may be unique opportunities for changing the way many fisheries are managed. Finally, we describe how fisheries managers from around the world may gain insight into some practical solutions to common problems by studying the implementation of the Nisga'a settlement.

10.2 STUDY AREA

The Nisga'a Traditional Territory is composed of 20 000 km^2 of land and sea in north-western British Columbia (Fig. 10.1). K'alii Aksim Lisims (K'älee ax'·em L' is·emz) (the Nass River) flows through the heart of the

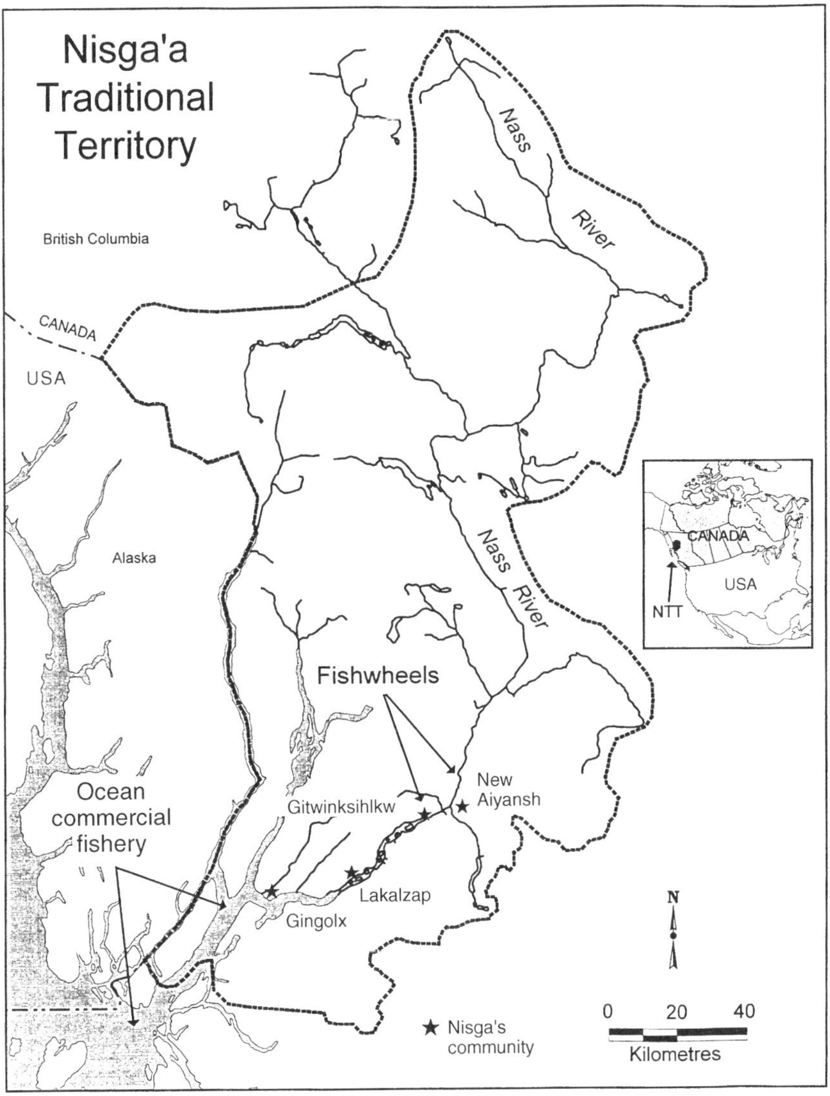

Fig. 10.1 The Nisga'a Traditional Territory and its position relative to BC and Canada.

Territory and supports a rich salmon resource that has allowed the Nisga'a and their culture to flourish. The average annual returns of all species of Pacific salmon (*Oncorhynchus* spp.) to the Nass is in the range of 1.5 to 2.5 million fish (total wet weight, 3000 to 5000 t $year^{-1}$). These stocks support a commercial ocean fishery that harvests about 1 million fish annually (2000 t year^{-1}) with a landed valued of CAN$5–7 million (1 Canadian dollar is about US$0.70 (in December 1997) – *ed.*).

There are 2500 Nisga'a who live in four communities in the Nass Valley and 3000 Nisga'a who live outside the valley. In addition to actively participating in the general commercial ocean gillnet and seine fisheries, the Nisga'a annually harvest about 50 000 sockeye salmon (*O. nerka*), 7000 chinook salmon (*O. tshawytscha*) and smaller numbers of the other species of salmon in a river gillnet fishery (200 t year^{-1}). Salmon caught in the Nass River have been used traditionally for food, and for social and ceremonial purposes.

10.3 LAND CLAIM NEGOTIATIONS

The long history of land claim negotiations in Canada dates back to 1763 when, under direction from the King of the British Empire, the colonial government of Canada was told that no land was to be occupied by European people without first settling land claims by First Nations people. This instruction was largely carried out as European settlers moved west across Canada. However, for 75% of BC and the entire Yukon Territory, the regional governments of the day essentially ignored land claims by First Nations people.

The Nisga'a sent their first land claims delegation to Victoria, the provincial capital of BC, in 1881. After this and several other attempts were turned aside, they appealed to the highest court in the land by presenting a petition to Privy Council in London, UK, in 1913. In the late 1980s, under public pressure to resolve the uncertainty over land tenure and rights to natural resources, the BC provincial government entered into its first land claim negotiations, joining the Nisga'a and federal Canadian governments to begin a tripartite treaty negotiation process. The negotiations have been long, comprehensive and have included many aspects other than fisheries (i.e. lands, forestry, wildlife, Nisga'a government, justice, taxation). However, fisheries has always been a major component. In 1992, the Nisga'a Tribal Council undertook a large-scale fisheries research programme to better understand and quantify the fisheries resources of the Nass and to develop new and innovative harvesting and management techniques. After extensive negotiation, and using fisheries information gathered from 1992 to 1995, an Agreement-in-Principle (AIP) was reached between the three parties on 15 February 1996. The AIP is a precursor to a final treaty expected by 1998.

The land claim negotiation process presents an opportunity to change the way fisheries are managed. The importance of fish to First Nations peoples ensures that fisheries issues will always have high priority during treaty negotiations. By seizing the opportunity offered by the treaty process, fisheries managers can change local fisheries management techniques and demonstrate to other fisheries managers around the world ways of doing things differently. Specifically, the Nisga'a settlement offers insight into the issues of community-based management, property-rights-based fishing, integrated harvesting and management, and sustainable funding for management and research.

In Canada, treaties will be constitutionally protected and therefore treaty-designed fisheries management offers security to the entire process that will eliminate many of the problems associated with the short-term political horizon that has plagued resource management decisions (Walters, 1995). However, by the nature of their permanence, it is critical that treaties be adaptable; fisheries aspects of treaties must not be set in stone with little room for learning and adapting to new knowledge. Ecosystems and social systems are dynamic and it is impossible to foresee the direction that either will take 10, 50 or 100 years from now. As described below, the Nisga'a Fisheries Agreement acknowledges the dynamic nature of fish stocks, ecosystems and Nisga'a culture.

10.4 THE FISHERIES AGREEMENT

The negotiating team had to address many concerns to arrive at an agreement. In short, the Nisga'a wanted:

1. a secure share of the fishery resource;
2. improved fisheries management; and
3. the means to rebuild and sustain their valuable resource in the long term.

The federal and provincial governments' primary goals were:

1. to define the previously undefined Nisga'a aboriginal right to fish;
2. to maintain final authority over the fishery; and
3. to minimize impacts on non-Nisga'a fishers.

A common concern among all three parties was the need to ensure that the agreement was manageable and deliverable.

The Fisheries Agreement described in the AIP was intentionally limited to definition of resource shares, financial commitments and basic management structure for Nisga'a fisheries and a general outline of each party's responsibilities. The details required to understand how this agreement can and will be implemented were discussed, analysed and documented but were not included in the AIP. The parties believed that the details of fisheries management must be permitted to evolve and their inclusion in the AIP and subsequent Treaty would create a barrier to this evolution. A separate document, which provides details on current fisheries management goals, the operation of the JFMC (see p. 155), stock assessment procedures, biological concerns, annual Nisga'a fisheries management plans and harvest accounting protocols, is currently being completed. The stock assessment and in-season fisheries management procedures described in this document are not theoretical ideals but procedures actually implemented through the Nisga'a Fisheries Program (NFP) over the past 3–5 years. The fisheries component of the Nisga'a AIP has been crafted to provide a framework within which the harvesting, fisheries management and stock assessment procedures developed through the NFP can continue to thrive and develop.

The tripartite fisheries negotiating team took an interdisciplinary approach to building a sustainable fishery and fishery management system for the Nass. The negotiating team drew on a firm understanding of historical problems (and touted solutions) of salmon fisheries and fisheries management systems and all available information on Nass Area fish stocks. The five features of the Fisheries Agreement (Table 10.1) reflect this interdisciplinary approach and were, in part, intended to avoid problems created by traditionally narrow approaches to fisheries management (discussion in Crutchfield and Pontecorvo, 1969; Pearse, 1981).

Table 10.1 Five features of the Nisga'a Fisheries Agreement and their anticipated results

Feature	Anticipated results
1. Communally held entitlement to a share of the salmon returns	Ecologically and economically sustainable harvesting
2. Joint Fisheries Management Committee (JFMC)	Meaningful participation in management Community-based management Adaptable management
3. Integrated research, management and harvesting	Substantially lower overall costs Resource rent captured Ecologically sustainable species-selective harvesting
4. Final authority with the Minister of Fisheries	A check and balance to ensure that guiding principles are adhered to
5. Lisims Fisheries Conservation Trust Fund	Sustainable funding Sustainable management system

Communally held entitlement to a portion of the annual returns of salmon

A share of the fishery resource was to be allocated to the Nisga'a. This could have been accomplished in several ways. The Agreement proposes to allocate a fixed portion of the annual returns of salmon (above conservation levels) to the Nisga'a Central Government (NCG) and that this share be held communally (approximately 26% of the average annual Canadian harvest of Nass Area salmon; AIP, 1996). This arrangement is intended to promote the development of an ecologically and economically sustainable fishery. By making the entitlement communally held and a function of the abundance of returns, the NCG will face powerful incentives to harvest fish with the most ecologically and economically sustainable methods available. Ecologically sustainable methods will be encouraged because:

1. the magnitude of future catches for the Nisga'a is directly correlated to the health of the Nass ecosystem and its salmon stocks;
2. harvests in mixed-stock ocean fisheries will be reduced to make fish available to Nisga'a in-river fisheries;
3. the requirement to deliver the Nisga'a entitlement will result in more conservative management of ocean mixed-stock fisheries; and
4. returns below a defined minimum escapement level will result in the complete closure of all fisheries for these stocks.

Healthy stocks will benefit the Nisga'a and non-Nisga'a users of the Nass fish resource. By holding a significant share of the salmon harvest communally, the NCG will be in a position to take advantage of ecologically sustainable species-selective harvesting methods (see below). Economically sustainable harvesting will be promoted because there is considerable resource rent (revenue above a normal return on harvesting costs) available from the Nass salmon stocks if a single entity (i.e. NCG) can manage the fish harvesting. Although they would have only a portion of the resource, the NCG will act similar to a 'sole owner' (Scott, 1955) of the resource. The NCG will be in a position to maximize the resource rent rather than be encouraged to dissipate it, as is currently done in common-property, open-access fisheries (Clark, 1973).

Joint fisheries management committee

Fisheries management under a Fisheries Agreement has been designed to include the Nisga'a and to be adaptable. This will be achieved by a Joint Fisheries Management Committee (JFMC) made up of two representatives from each of the three governments (NCG, BC and Canada) that will oversee all research, management and harvesting plans. The JFMC will annually review the scientific understanding of the fish stocks and the performance of management, and then recommend management plans to the Minister of Fisheries. In reviewing these plans, the Minister must take into consideration the conservation requirements of the fisheries resource, ensure the wise use of the resource to benefit all Canadians, make sure that accepted scientific procedures are used for managing the fisheries resources, and ensure that harvesting and management activities are efficient and integrated. The Minister may modify the annual plans to coincide with these guiding principles. As understanding of the fisheries resource changes, the JFMC can incorporate this new information into how the resource is managed and harvested.

Integrated research, management and harvesting

The Fisheries Agreement will promote integration of research, management and harvesting activities by delegating a large share of these activities to a single organization, the NCG. To make research, management and harvesting in the Nass River more effective and efficient, a system has been designed that integrates all three activities. This has largely been possible by the introduction of live-capture fishwheels (Meehan, 1961; Donaldson and Cramer, 1971, Link and English, 1996). Fishwheels are currently used on the Nass to: (1) collect basic biological information from migrating fish of all species; (2) assess the abundance of sockeye salmon on a daily basis during the season to manage the ocean commercial fisheries; (3) obtain annual, escapement and biological information on all species of salmon that enter the Nass; and (4) selectively harvest the abundant sockeye salmon amidst less productive co-migrating species of salmon (Link and English, 1998). In recent years, the NCG has clearly demonstrated a genuine interest in ecologically sound and economically efficient salmon harvesting by developing and operating a species-selective fishwheel-based fishery to harvest surplus sockeye salmon.

By integrating these management and harvesting activities, the NCG is in a position to capture nearly all the potential resource rent from their salmon harvest. The resource rent may be captured because: (1) the cost of the fishwheel-based harvesting is low relative to most net harvesting technologies used today; (2) the cost of harvesting in general is very low because there are substantial economies of scale and savings by integrating these activities; and (3) the quality (and price) of the fish harvested using the fishwheels is higher than the fish harvested in the nearby ocean gillnet fishery (e.g. no net marks, held live until ready to process). The resource rent captured through the integrated management and harvesting will be available to contribute to funding for future research and management activities (see below).

Final authority remains with the Minister of Fisheries

To safeguard against harvesting and fishery practices that maximize short-term profits at the expense of long-term ecological sustainability, annual fishing and management plans of the NCG must be approved in advance by the Minister of Fisheries of the Government of Canada (see above). Third parties to the JFMC will have an opportunity to review annual fishing and management plans and make depositions to the Minister. This final check will help to ensure that fishing plans and management activities are consistent with Canada-wide principles of resource conservation.

Lisims Fisheries Conservation Trust Fund

To make the proposed management system financially secure, a sustainable funding arrangement will be created through the establishment of a trust fund. The proceeds from the fund will finance the management programme in perpetuity. The Lisims Fisheries Conservation Trust Fund will be started with a combined donation of CAN$13 million from the Nisga'a and Canadian governments. This fund will be further built up by depositing resource rent derived from efficient fish harvesting, and by philanthropic contributions from corporations, organizations and individuals. We expect this fund to eventually provide from $650 000 to $1 million (Canadian dollars) annually and in perpetuity. The trust will buffer the wide fluctuations in funding levels that would result from ecological variability if funding were entirely dependent on revenues from fish harvesting.

10.5 COMPARATIVE FEATURES OF THE AGREEMENT

Unique aspects of the Nass

The Nisga'a are a unique people and the Nass Valley is a unique place compared with many other First Nations and their territories in BC, and other Aboriginal people around the world. First, the Nisga'a are the only First Nation people that live year-round in the Nass Valley. This contrasts with most other large river systems in BC where several First Nations live within single watersheds. Second, the Nass Area fisheries resource is valuable compared with the fisheries resources in many other areas. Lastly, the topography of the Lower Nass is suitable for the deployment of fishwheels, a technology that has made possible the integrated fisheries management and harvesting programmes. As a result of these unique features, the Nisga'a solution will not be suitable for, or achievable in, all areas. Rather, other negotiated fisheries agreements will almost certainly be different from the Nisga'a Fisheries Agreement. The success of these other agreements will likely depend on negotiators' ability to capitalize on the local conditions and opportunities.

Features in common with other areas

Despite the unique features of the Nisga'a and the Nass, there are several common features between them and other First Nations and territories. A process to negotiate and settle treaties with First Nations has been put in place by the federal and provincial governments. Settlement of these treaties will bring First Nations in BC, a key resource user that has been traditionally excluded, into meaningful management and stewardship of the fisheries resources. Fisheries issues will always be important in land claim negotiations because fish have played a crucial role in the culture and survival of most First Nations people in BC. The process of negotiation should ensure that the unique features of each area and people are taken into account in constructing fisheries agreements. In addition, the negotiation process takes considerable time (many years) and like the Nisga'a, each First Nation should be given the opportunity to demonstrate their capability to manage and steward the resource before a final treaty is signed. The results from this learning and demonstration process can then be incorporated or considered in the preparation of a final agreement. Lastly, there will be excellent opportunities to put in place sustainable funding for management and research activities because monetary settlements for land claims will be large enough to establish trust funds similar to the Lisims Fisheries Conservation Trust.

These common features of all upcoming land claim negotiations in BC and what we have described about the Nisga'a Fisheries Agreement are strong reasons for optimism that better fisheries management is possible with resolution of native land claims. The land claims negotiations process may not be the only mechanism for achieving substantial improvements to the current fisheries management system, but it is an opportunity with a great deal of potential.

10.6 CONCLUSIONS

To reinvent fisheries management, institutions are needed that will actually change the way fisheries are managed. One does not have to look far to find beautifully articulated expositions of the problems facing fisheries management in the 21st century; there are numerous examples dating back to at least the 1950s (Gordon, 1954; Hardin, 1968; Clark, 1973; Copes, 1981; Pearse, 1981; Walters, 1986, 1995). The persistence of the problems facing fisheries management supports the conclusion that institutions that have the ability to effect change are rare (Pinkerton, 1989). We know much of what is wrong; what we need to reinvent in fisheries management is research on, and development of, institutions that can bring about change.

Fisheries managers from around the world can gain insights into several commonly held solutions to fisheries management problems by studying fisheries management under a land claim settlement with the Nisga'a. First, by integrating management, research, and harvesting at a local level, a Nisga'a treaty may provide evidence for or against the effectiveness of community-based management. Second, the allocation of a share of the resource to the NCG will provide insights into the effectiveness of property-rights-based fisheries for highly mobile, semelparous fishes such as Pacific salmon. Third, funding annual research and management programmes from a trust fund composed of resource rent and government seed money offers a potential solution to highly variable and shrinking budgets faced by most fisheries management agencies.

Settling land claims with First Nations may help foster the right kinds of institutions for sustainable fisheries. Some have mistakenly interpreted the negotiations of treaties in BC as a process which will include the wholesale transfer of resource management responsibilities to First Nations. Nothing could be further from the truth. Treaties will, for the first time, clarify the roles of each party and establish rules for conservation, resource sharing and resource management. Through treaty-designed fisheries management systems, managers can explore new ways of doing things within a secure legal framework and a permanent funding arrangement. Constitutionally protected treaties with First Nations may be seen as irreversible or permanent by many, but as we have described, treaty-designed fisheries management does not have to be permanent but instead can be adaptable over time. In negotiating modern treaties, fisheries managers have a unique opportunity to change the way things are done. The treaty negotiations process has provided Nisga'a and non-Nisga'a fisheries managers and scientists with a unique opportunity to develop and test improved procedures for harvesting and assessing salmon resources in the Nass Area. The extension of the Nisga'a experience to other species and other areas where treaties are being negotiated could substantially improve our understanding of local fisheries resources and management options.

The accumulated wisdom of our predecessors must be used to design and implement more rational fisheries management systems. To counter Larkin's (1979) pessimistic outlook on fisheries management, maybe we *can* get there from here, if we seize rare opportunities such as treaty negotiations to gain understanding and effect real change.

ACKNOWLEDGEMENTS

The Nisga'a Agreement-in-Principle and its fisheries component would never have been possible without the dedicated efforts of numerous people over many years. Recently, Nisga'a leadership in fisheries has been provided by President Joseph Gosnell, Harry Nyce, Edmond Wright, Herbert Morven and Ben Stewart. The Nisga'a were assisted in fisheries negotiations by Jim Aldridge, Karl English and Ellen Gunn. Fisheries negotiations on behalf of Canadian government were led by David Osborne with the assistance of Alan Gould, Mike Shepard and David Peacock; while the BC government was represented by Jack Ebbles and Trevor Proverbs with assistance from Marlyn Simpson and Nancy Besharah. The Nisga'a Fisheries Program has been directed by Harry Nyce, Ben Stewart, Karl English, David Peacock, and Kim Hyatt, coordinated by Bob Bocking, Michael Link and Bryan Nass, and delivered by the dedicated efforts of more than 35 Nisga'a fisheries technicians. Funding for the Nisga'a Fisheries Program has been provided through the Aboriginal Fisheries Strategy of Fisheries and Oceans Canada. Steve Johnson, Bill Griffiths and two anonymous reviewers provided many helpful suggestions that improved the manuscript of this chapter.

REFERENCES

AIP (1996) *Nisga'a Treaty Negotiations, Agreement-in-Principle*, available from Ministry of Aboriginal Affairs, Communications Branch, 908 Pandora Avenue, Victoria, BC, Canada, V8V 1X4. 127 pp. + appendices.

Clark, C.W. (1973) The economics of overexploitation. *Science* **181**, 630–634.

Copes, P. (1981) Rational resource management and institutional constraints: the case of the fishery, in *Economics and Resource Policy* (ed. J.A. Butlin), Longman, London, pp. 113–128.

Crutchfield, J.A. and Pontecorvo, G. (1969) *The Pacific Salmon Fisheries: A Study of Irrational Conservation*. Johns Hopkins Press, Baltimore.

Donaldson, I.J. and Cramer, F.K. (1971) *Fishwheels of the Columbia*, Binfords and Mort, Publishers, Portland, OR, 124 pp.

Gordon, H.C. (1954) The economic theory of a common property resource: the fishery. *J. Polit. Econ.* **62**, 124–142.

Hardin, G. (1968) The tragedy of the commons. *Science* **162**, 1243–1248.

Larkin, P.A. (1979) Maybe you can't get there from here: a foreshortened history of research in relation to management of Pacific salmon. *J. Fish. Res. Bd Can.* **36**, 98–106.

Link, M.R. and English, K.K. (1996) The 1993 fishwheel project on the Nass River and an evaluation of fishwheels as an inseason management and stock assessment tool for the Nass River. *Can. Tech. Rep. Fish. Aquat. Sci.* no. 2372: 82 pp.

Link, M.R. and English, K.K. (1998) Working toward sustainable fisheries by integrating management, research and harvesting, in *Proceedings of the Sustainable*

Fisheries Conference (*Am. Fish. Soc. Spec. Publ.* **00**), Am. Fisheries Society, Bethesda, MD (in press).

Meehan, W.R. (1961) Use of a fishwheel in salmon research and management. *Trans. Amer. Fish. Soc.* **90**, 490–494.

Pearse, P.H. (1981) *Conflict and Opportunity: Toward a New Policy for Canada's Pacific Fisheries*, The Commission on Pacific Fisheries Policy, Vancouver.

Pinkerton, E. (1989) *Cooperative Management of Local Fisheries: New Directions for Improved Management and Community Development*. Univ. BC Press, Vancouver.

Scott, A.D. (1955) The fishery: the objectives of sole ownership. *J. Polit. Econ.* **63**, 116–124.

Walters, C.J. (1986) *Adaptive Management of Renewable Resources*, Macmillan, New York, 386 pp.

Walters, C.J. (1995) *Fish on the Line: The Future of Pacific Fisheries*, available from David Suzuki Foundation, 219–2211 West 4th Ave., Vancouver, BC, Canada V6K 4S2.

Chapter eleven

Aquatic resources education for the development of world needs

Meryl Williams

ABSTRACT

Some definitions of 'Development' are reviewed, with emphasis on natural resources and environmental issues. Earlier theories of development, based on industrialization and building of physical infrastructures, with the state and the industrial sectors as key actors, are now being replaced by a focus on education, and empowerment of new actors, notably local communities, women and previously marginalized ethnic groups. This focus is particularly important as it also defines a new era in aquatic resource uses, wherein a narrow resource base imposes much tighter interactions between research (the various forms of which are explored) and management. The implication for tertiary education (including curriculum development) is that it will have to encompass a wider range of skills, including the ability to interact with the clients of the research. Such broad-based education can best be achieved through cooperative arrangements, for example between international centres such as ICLARM and national institutions. Some of the forms these arrangements may take (joint comparative studies, mentoring, etc.) are listed and evaluated.

Reinventing Fisheries Management. Edited by Tony J. Pitcher, Paul J.B. Hart and Daniel Pauly. Published in 1998 by Kluwer Academic Publishers, London. ISBN 0 412 83410 3.

11.1 INTRODUCTION

Peoples' abilities to know and to act must be built if we are to reinvent and sustain progress in natural resource management, including aquatic resource and fisheries management. Education and training are key tools for building capacity. This chapter specifically addresses education needs for aquatic resource management. Developing-world needs are primarily considered here because many developing countries now face the possible consequences of fisheries resource scarcity, adding to already stressed economic, food and employment conditions.

The experience that drives this analysis arises from the work of the International Centre for Living Aquatic Resources Management (ICLARM), which carries out strategic research and related activities for the benefit of low-income people in the developing world. ICLARM's work is aimed at finding ways to improve the productivity, management and conservation of aquatic resources. Working from a research base, ICLARM also has a major objective to strengthen national capacities for research. Education is one mode for achieving this.

Education and training work best when based on the needs of the participants and their sector and therefore this chapter begins by examining the changes in the context of development and the global-scale transition now occurring in the contribution of fish to food security. It then suggests ways in which education can meet the needs of aquatic resource management as now perceived.

11.2 THE CONTEXT

In debating new approaches to managing fisheries, scientists are reacting to the strong perception of the 1990s that change is essential because fisheries management and its servant/creator, fisheries science, have a flawed track record. Such a perception is a recent phenomenon at the global level, although unease had been growing for some time, starting in regions such as the North Atlantic. The global phenomenon crystallized when a 1992 assessment of the state of the world's fish stocks by the Food and Agriculture Organization of the United Nations (FAO, 1992) became headline news in early 1994 (Williams, 1996). The state of fisheries in developing countries, as well as of those in the developed world, is of concern. Conflict, even international conflict, over access to resources and their management is more frequent and can be severe.

Although the fisheries sector*, including research, education and training, has tended to operate independently of other economic sectors, changes in the sector have not occurred in isolation from other developments. Indeed, many changes have been driven by external factors. For example, changes have been influenced by economic and technical drives to develop rural food-producing and livelihood-creating activities, and more lately by environment and resource sustainability concerns.

Development: what does it mean in the fisheries sector and more generally?

Fisheries development is usually narrowly interpreted as the initiation of exploitation of new fisheries resources, or improvements in fishing gear, or improvements in fish product. More generally, the concepts of development have been undergoing change (Hettne, 1995). Prior to the 1970s, 'developing countries' and industry sectors within them were seen as needing to undergo steady modernization. The earlier modernization concepts are now considered Eurocentric. They also involved a bias toward the urban and industrial sectors. Rural food production, such as agriculture and fisheries and especially their subsistence forms, were little regarded (Hettne, 1995). However, they clearly were, and still are, influenced by modernization concepts. A popular modernization pattern of the 1950s and 1960s was attributed to Rostow (1960; see Hettne, 1995). According to this model, societies developed through several sequential stages, namely:

- the traditional society;
- the pre-take-off society;
- take-off;
- the road to maturity;
- the mass consumption society.

Underdevelopment was analysed in terms of 'barriers to modernization' and 'resistance to change'.

In developing countries, fisheries development began in earnest in this period with injections of modern gear, vessels and surveys of potential fishing grounds. Traditional fishing gear, practices, and management were studied only by anthropologists and historians. Fisheries scientists usually did not access these sources of knowledge.

*The terms 'fisheries' and 'fisheries sector' are used in this chapter to refer to both fisheries and aquaculture sectors. 'Aquatic resources' are living resources, including fisheries resources, extracted or otherwise used by humans from seas, rivers, lakes, ponds and other water bodies.

By the 1970s, the 'cultural imperialism' of the outsider's approach to development was being questioned, but a real stimulus to reinventing development concepts came in 1974 from a declaration at the joint United Nations Environment Program and United Nations Council on Trade and Development conference on resources and development (Hettne, 1995). This led to alternative concepts of development which have become progressively elaborated. These more elaborated concepts have not deviated from the concept of the linear path of human development, but rather have increased the number of parameters describing the path, because they embody not just economics but also the processes of development. Community participation and natural resource management are also considered.

The modern concept of development recognizes the need to achieve sustainability, and to consider the needs of disadvantaged groups such as women and minority ethnic groups. In this sense development in the past has been imposed on minority groups by the dominant section of society (Hettne, 1995). The new perspective led Hettne (1995) to a revised concept of development which can be expressed in three points, as follows.

1. Territorialism to replace functionalism (and sectoralism): this implies a geographically bounded community controlling a finite set of resources and united through their cultural values or needs, as well as devolved authority. This is in contrast to functionalism, which defines resources and skills by their function and assumes they are highly mobile. For this model, control has to be highly centralized such as in the nation state.
2. Cultural pluralism: this recognizes the right of cultural groups to continue to reproduce their culture within a society supporting other cultures.
3. Ecological sustainability: development should be adapted to the realities of the resource base – with the rights of future generations in mind.

The first of these seems the most problematic and not fully supported by empirical evidence. Hannesson (Chapter 19, this volume), in his analysis of the management of four cod, *Gadus morhua*, stocks has shown that territorialism in fisheries management may not lead to sustainable management of stocks. The premise that community-based management is a universal remedy for all natural resource management problems may be insupportable. A recent review of community-based coastal resource management projects in the Philippines showed that large investments of time, training and external resources are required to make these work and that perceptions of their success differ widely among groups (Pomeroy *et al.*, 1996). The new conception of development has had an influence on the thinking of all international development agencies.

In summary, fisheries developed during the last four decades along lines which followed a standard modernization model from development theory.

Resource limits halted this approach. Later than in other sectors, the problem of resource sustainability drove development in the fisheries sectors, and many developed countries began to adopt more comprehensive approaches to development. These new approaches incorporate principles of territorialism, cultural pluralism and environmental sustainability. They provide for the greater involvement of communities and civil society in resource management, for the greater integration of interests between fisheries and other sectors and for greater fisheries interactions among nations.

11.3 DEFINING THE NEEDS FOR EDUCATION IN THE NEW ERA OF AQUATIC RESOURCE MANAGEMENT

Education is one tool to help create and reinvent capacities for aquatic resource management. The context described in the previous section leads to the conclusion that education needs are different now from what they were in the past. They differ because:

1. the target actors are different;
2. the actors' roles are different;
3. the suppliers are likely to be different; and therefore
4. the curricula/content will be different; and
5. the institutional context will be different.

Each of these factors will be discussed briefly.

Target actors

The target actors will be not only young people leaving school and seeking higher education for a career in fisheries management, science or industry. They could be anyone involved with the resource. They will need a basic professional education and ongoing in-service training. Women and minority ethnic groups, politicians and the public will need to know more about fisheries management and how to access knowledge. Not just actors in national government, but many in local government and community groups, need basic lessons in resource management. These new actors require many different levels of education.

Roles

In the current climate of resource management, more roles are played by the actors. Many will fill more than one role. Fisheries managers, fishers, scientists and public administrators will act as conflict resolvers and planners. Fisheries management planning has become a new pursuit and many actors now play planning roles or are involved in the consultation process.

Research can fill at least four different roles: it produces knowledge and understanding, identifies issues needing attention, helps resolve conflicts and suggests solutions or new options (Williams, unpublished data).

The pressures on scientists to become conflict resolvers and problem solvers are greater when resources are scarce and stressed. Fisheries managers, scientists and other professionals are called more to account in such times. When the account is found lacking, the approaches of researchers and of advocates in fisheries appear to fall into one of two main types: the iconoclasts who are 'outside the tent', and the orthodox or apparatchiks who are 'inside the tent'.

The iconoclasts are found mostly in universities in developed countries and NGOs (non-governmental organizations) in all countries. They tend to favour radical new approaches to fisheries management, often unmindful at first of how these could be implemented. They are essential if we are to develop better practice. The existence of educational institutions provides a suitable intellectual environment in which new ideas can originate.

The orthodox are found mainly in government fisheries agencies. They accept the reality of implementation. They must frequently defend past management decisions and they tend to favour an expanded or incremental version of the current management practice. The orthodox are usually strongly rooted in single disciplines such as fisheries biology and economics. Both types espouse the importance of multidisciplinary and interdisciplinary studies but neither can readily work across the boundaries of their own discipline.

Another new form of experimental development is 'action-research', a term often used by fisheries development officers and NGOs who are seeking and creating new ways of empowering people in natural resource management. Their projects are documented by in-house publications and often involve educators from local and other universities. Through empowering people and encouraging creativity, new approaches are being developed for more varied local situations, rather than applying old formulae.

Suppliers

Supply and delivery of education and training are changing. In some developing countries, national suppliers are strengthening. Nationally delivered postgraduate courses are available in many Asian, Latin American and some African countries. At the same time, many developed-country institutes are aggressively marketing their courses in developing countries. Schools, universities and colleges have traditional but broadening roles. The private sector is also more relevant in some countries, especially through provision of technology and use of the media. Distance and remote education is having some impact. NGOs, advocacy groups and government agencies are all trying to get their messages out, often through collaborative efforts in education and training.

A recent example of collaborative training is that of a pilot project in the Philippines applying science to community-based coastal resource management. The project, funded by the Rockefeller Brother Fund, is implemented by the Haribon Foundation (a Philippine NGO), the Philippine Department of Environment and Natural Resources, the Bureau of Fisheries and Aquatic Resources, the Philippine Council of Aquatic and Marine Research and Development, ICLARM and the International Institute of Rural Reconstruction. The project illustrates a typically large and diverse number of actors. The courses are being conducted in the provinces and cover not just fisheries management but the management of other coastal resources such as mangroves and coral reef ecosystems. While tailored to the local Philippine situation for resource management, the basic pedagogical method was that developed internationally by the United Nations Development Program, the TRAIN-X method, adapted for coastal management under the TRAIN-SEA-COAST project.

Curriculum

The content of curricula is being challenged in its breadth and depth, and its degree of specialization. The content is being driven by needs that change as countries develop. For example, at the University of Zimbabwe, the Centre for Applied Social Studies (CASS) started its early postcolonial existence with courses and research into self-determination, followed some years later by women's and other interest groups' studies. CASS now focuses on natural resource management work, especially using community-based approaches. Another example is the above-mentioned Philippine community-based coastal resource management training project. Here, fisheries, coastal ecology, resource management planning and community-based development topics are all central to the curriculum.

Most professionals will need knowledge in several disciplines throughout their careers. For example, biologists will need to know something about other disciplines such as sociology and they will also need good social skills. Fisheries managers will require a grounding in biology, planning methods and conflict resolution.

Education is a continuing process, even for specialists. Short courses and in-service training will be vital. For example, a fisheries biologist who trained more than a decade ago will likely need to attend courses in new methods in fisheries resource assessment and quantitative and molecular genetics.

Institutional context

As the points made above show, the institutional context of education for aquatic resource management is changing. The institutional context defines who benefits, who pays, how education is organized and where it happens. Traditional educational institutions still have key roles but they must cope with changing needs. In developing countries, the institutional contexts are varied. In some, mass education and strong university development occurs (e.g. Philippines, India). In some, NGOs run strong programmes to compensate the excluded, e.g. women in Bangladesh, or to reach the less empowered communities (e.g. rural fishing communities in the Philippines), or to promote issues such as conservation. As for most developed countries, fisheries in developing countries are rarely important enough to warrant institutions of their own. Work on fisheries is usually conducted by institutes also catering for aquaculture, agriculture, and/or natural resource management. The University of British Columbia Fisheries Centre, and others like it, have a vital role to fill, not just because they deliver quality education but also because they fill a gap in fisheries schools. Similarly, vocational universities such as the Australian Maritime College in Launceston, Australia, fill an international role as well as a national role in fisheries education.

A key question for educators is whether it can be feasible to put together specialist, targeted courses for the many different actors in fisheries management so that they can capture the benefits of training together. At this conference, Tony Davis highlighted familiarization and acquaintanceship as two such benefits.

In summary, we now have a richer, more diverse landscape of education needs, actors, roles, suppliers and institutions to support aquatic resource management in the world. The needs will be partly met from internal sources in any country but international linkages are important to supply new and alternative courses and/or when national demands are too small to warrant specialist education services in fisheries. Educational institutions must be mindful of the changes occurring.

11.4 MEETING DEVELOPING COUNTRIES' FISHERIES EDUCATION NEEDS THROUGH PARTNERSHIPS

This section suggests how fisheries education needs may be met through relevant partnerships. That the educator is one trustee for the development of human capacity is paramount in many suggestions. Some of the following suggestions are formed with the capacities of an institute in mind such as the Fisheries Centre of the University of British Columbia, where these ideas were first presented.

Understand the context

Educators should develop a good understanding of the context of development and the state of fisheries resources management, worldwide, as the starting point for an analysis of needs and curriculum development. Educators need to be in touch with many sectors of society and need to be experienced in diverse fisheries and related fields such as integrated coastal management, aquaculture and conservation.

Link education to the latest knowledge

Course content should reflect the latest new knowledge. Students would be short-changed if this were not the case. Conversely, researchers and educators must be in tune with the problems of the sector and of students, as these often provide researchable issues.

Study developing-country fisheries

Much good research on developing-country fisheries can be done by students undertaking projects during their degrees. Developed-country institutions should insist that their developing-country students study their own fisheries rather than contributing further to research on developed-country fisheries, such as salmon (*Salmo salar, Oncorhynchus* spp).

Foster comparative studies

Realizing that mutual benefits exist, educators and researchers in universities should be encouraged to teach courses and study problems of a comparative nature in fisheries. These should include developing country cases where possible but should avoid prescribing developed-country solutions.

Promote mentoring

Mentoring or being mentored by partners can provide motivation and impart knowledge. A helpful mentor at the right time is usually critical to a person's career development.

One lesson learned from studies of women in science was that mentoring relationships can be more powerful motivators than role models. This is good news because mentors are people who give the right sort of encouragement at the right time. They do not need to be of the same sex and race as the student. Young scientists in developing countries can use mentors from their own countries and/or from their chosen professions. For the latter, these relationships can be developed at university, on courses or at conferences.

Work for mutual benefit

All actors could capitalize on the fact that aid agencies have two aims. They want to help developing countries but want to involve their own domestic agencies as implementors. Universities in developed countries, e.g. the University of British Columbia, and international research centres such as ICLARM are frequently used by development agencies to conduct studies which will inform future development decisions or solve current problems. Many developed countries, e.g. Canada, Australia, Denmark, Norway, Great Britain and the Netherlands, have explicit policies of mutual benefit for overseas development assistance. Sometimes special funds are available for linkages; otherwise the policy is more pervasive.

Linkages with ICLARM programmes of research and education may be particularly useful to aid agencies. We offer institutional support in several sites in the developing world. Within the bounds of available resources, we can take students and help supervise doctoral and master's studies on issues relevant to a developing country. This increases the quantum of research done on developing-country problems. It offers the benefits of a challenging research topic, e.g. multispecies tropical fisheries.

In the other direction, ICLARM research associates and assistants, and people in our partner institutes in developing countries, are frequently seeking higher education opportunities or cross-experience. Sabbaticals (in both directions) and cross appointments are other options.

Support professional societies

Professional societies such as the Asian Fisheries Society are important elements in the continuing development of fisheries professionals. They play a most important educational role through conferences, scientific publications and professional networking. Educational institutes can support these societies through sponsoring staff attendance at conferences, submitting and helping students to submit papers to journals, assisting in editing papers and through simply encouraging membership.

Provide editorial assistance

Publishing research is still one of the most important contributions to continuing education. The professional will often face the additional challenge of having to write in a second (non-native) language to make results available internationally. Many developing-country fisheries scientists appreciate help with writing articles in a second language. As a special form of mentoring, some journal and newsletter editors have made great efforts to encourage young writers through special editing assistance.

Learn from the real world

On-the-job education for the fisheries professional will include being prepared to learn from fishers, government officials, NGOs and others. Education does not stop with a formal degree from a university or college.

Look at other disciplines

In formulating new directions in fisheries education, what can we learn from other natural resource disciplines? Taking note of subject differences, we can see that fisheries science, like most sectors, has been too insular in its approach. It can no longer afford this insularity as sectoral impacts and interests overlap more and more.

Help build institutions, too

Development relies on good institutions, including private and public-sector organizations, education systems, legal and enforcement systems, communities and property rights.

Education can help build capacity within all of these institutions but it has a tendency to focus only on the development of the individual. Organizational development is not given priority within curricula. Because most individuals spend their entire careers working within formal organizations, the quality and strength of the organization is critical to what well-educated individuals can achieve. To build and strengthen organizations, managers need skills in planning, resource mobilization and management, the setting of research priorities, managing scientists, influencing and responding to the needs of the sector and representation. Educational institutes, often in faculties outside their fisheries centres, can help provide education in many of these topics.

11.5 CONCLUSION

Today, we face difficult challenges in aquatic resource management. To meet these challenges we have few sure answers. Success will call for a wide understanding of the context, and that is why this chapter began with a description of development and the state of fisheries. Managing fisheries in the face of these challenges will require lifelong education and training to build for human capacity. Education and training must be reorientated to cope.

Universities and other traditional educational institutions have a role to play in education and research and in fostering the linkages between the two. I would encourage both the iconoclast and the orthodox views among fisheries professionals and in education, especially until a new paradigm emerges to guide the next phase.

REFERENCES

FAO (1992) Review of the state of world fishery resources. Part 1. The marine resources. *FAO Fish. Circ.* no. 710 (Rev. 8), 114 pp.

Hettne, B. (1995) *Development Theory and The Three Worlds: Towards an International Political Economy of Development*, 2nd edn, Longman Scientific & Technical, Essex, 319 pp.

Pomeroy, R.S., Pollnac, R.B., Predo, C.D. and Katon, B.M. (1996) *Impact Evaluation of Community-based Coastal Resource Management Projects in the Philippines.* (*Fish. Co-manage. Proj. Res. Rep.* **3**), ICLARM, Manila, 282 pp.

Rostow, W.W. (1960) *The Stages of Economic Growth.* Cambridge, Cambridge Univ. Press.

Williams, M.J. (1996) The transition in the contribution of living aquatic resources to food security. Food, Agric. Env. Discussion Paper no. 13, Int. Food Policy Res. Inst., Washington, DC, 41 pp.

Part Three

The Role of the Social Sciences in a Reinvented Fisheries Management

Chapter twelve

Social science in fisheries management: a risk assessment

Svein Jentoft

ABSTRACT

Today, social scientists are effectively absent from the fisheries management decision-making process – perhaps with the exception of the economic profession. Rarely do sociologists, social anthropologists and political scientists serve in advisory roles *vis-à-vis* government agencies in the same way as biologists do. If social scientists were to become involved as responsible co-managers, which they often claim that they should, what could possibly be their contribution? I argue that there are at least two areas where social science is of relevance: (1) in the design of management institutions; and (2) as providers of critical feedback to the management process. However, fisheries management could also benefit from the purely intellectual role of social science. Making insights count in fisheries management requires the social scientist to be committed and imaginative. It does not necessarily require direct participation in the decision-making process. There is also a need for the critical, sceptical, independent and unfettered mind. But can the social sciences have it both ways?

Reinventing Fisheries Management. Edited by Tony J. Pitcher, Paul J.B. Hart and Daniel Pauly. Published in 1998 by Kluwer Academic Publishers, London. ISBN 0 412 83410 3.

12.1 WHY SOCIAL SCIENCE?

In this chapter I will reflect upon the role of social science in fisheries management as it is today and how it could possibly be within a reinvented fisheries management system. What is the relevance of social science to fisheries management? What could its contribution be? To answer these questions, I must start out by recalling what fisheries management is.

First of all, fisheries management is not essentially about the management of fish. If it were, we would refer to it by another, more precise name such as resource conservation. Fisheries is an industry and fishing is a human activity, and it is through regulatory measures of fishing behaviour that we attempt to secure the viability of fish stocks. Therefore, the social scientist would argue the obvious: to manage well, you need to know not only fish, but also fishers and fishing. Secondly, fisheries management is science based, but not science governed. In most countries it is a political battlefield of conflicting interests, and management goals have the character of delicate compromises. Consequently, one also needs to understand the political process of fisheries management. Again, this is basically a social science issue. Thirdly, a fisheries management system is an institutional set-up, and effective management is a question of finding the appropriate organizational mechanisms, i.e. the rules, procedures and incentives, that will help fulfil management goals. If there is one area in which social science can claim to have some expertise relevant to fisheries management, it is in the area of institutional formation, function and consequences.

If all this is true, social scientists should be much in demand by management agencies. The fact of the matter, however, is that we are not. Social scientists are conspicuously absent from the fisheries management decision-making process. Rarely do sociologists, social anthropologists, political scientists and geographers serve in advisory roles *vis-à-vis* management agencies in the same way as biologists do (Hanna, 1995). Even economists are barely represented, even though they have more confidence than the rest of us about having something important to offer the regulatory process. Perhaps they also have more reasons to be disappointed about this process than we do. But, rather than wailing over this rather discouraging situation (and I have been to a great number of conferences where we have all been united in this), we should ask why this is so: why is there little demand for social science in fisheries management?

12.2 SOCIAL SCIENTISTS AS OUTSIDERS

Some of the difficulties that social scientists have with being heard stem from the fundamental difference that exists between natural and social sciences. You simply do not study fishers as you study fish. The reason is that the social researcher and the object of study belong to the same species. Fishers can talk, and we as social scientists can talk with them. This is how social scientists get their data. If fish could speak, think, learn and organize as fishers can, the job of the biologist in fisheries management would be quite different from what it is today. I suppose that the position would be very similar to that of the average social scientist. Rather than being up front, on the stage, the biologists would have had to take a seat with the rest of us in the audience.

While the fish might need biologists to speak on their behalf, it is not obvious at all that fishers need social scientists to speak for them. If fishers were in desperate need of our expertise, we would have been up on stage with the biologists long ago. A strong demand for expert social knowledge coming from the management process, while possibly 'nice' for the social scientists, could in fact be taken as a warning signal. Would not the need for social scientists to speak for the fishers be an indication of a democracy that fails, when those that are most affected by the management system are excluded from the decision-making process?

To argue that this is generally the case, however, is an overstatement. In my own country, Norway, fishers are well represented in the management decision-making process, and their organization is present and heard at all levels of fisheries management. This is part of the established routine, to the extent that the fishers' role as co-managers is taken for granted, by themselves, by the government and by the public. Their involvement is highly institutionalized. Norwegian fishers do not need lawyers to present their case, as I understand is frequently so in the regional management councils set up in the USA. The relative absence of social science and social scientists directly within the regulatory councils and processes should therefore not come as a surprise. If management agencies want information on the fishery, they consult with the fishers' interest organizations, not with the odd fisheries social researcher. And if fishers want to speak to management agencies, they do it themselves, through their organizations (Hoel *et al.*, 1996).

I should add here that Norway is not a special case. In most countries of Western Europe and North America that I know of, arrangements have been developed and implemented to facilitate interaction between government and industry and to provide direct input from fishers' organizations in the regulatory process (Jentoft and McCay, 1995). Many countries have institutions more or less similar to regional management councils in the USA, and these have existed for a very long time. The unilateral, entirely top-down approach to fisheries management, I would argue, is the exception, not the rule. For this reason, in Norway, at least, I have doubts regarding what the character and substance of social scientists' contribution would be, were we inside the regulatory decision-making system along with industry representatives, managers and biologists. However, I do believe that social scientists can make a useful contribution in studying how these systems function and in helping to answer concerns about why they do not work well in many countries.

12.3 THE SOCIAL SCIENCE PERSPECTIVE

Whether or not social science is called for is largely a consequence of the goals of fisheries management, and goals are basically a political issue. Until now, stock conservation and enhancement have been ostensibly the overriding concern of fisheries management, and for good reasons. No wonder, therefore, that biology is the principal among the sciences. The perspective that fish are also an economic resource that should be exploited at cost-optimum, i.e. at a level where the rent is maximized, is a fairly recent consideration in fisheries management. The overcapacity that prevails in the world's fisheries indicates that there have been problems in getting this economic message across to governments. Nevertheless, there can be no doubt that the economics profession is a rising star in fisheries management. When economists talk ITQs (individual transferable quotas), governments are listening. Their hearing is less acute when sociologists and social anthropologists speak of coastal communities and culture.

One could of course argue, as an ITQ enthusiast did in the 'Fishfolk' electronic discussion forum that: 'In the long run, sound economic policy in fisheries management is the best social policy.' If this were true, sociologists and anthropologists would have no task in fisheries management. I do not know if there are absolute, objective criteria of what constitutes 'sound economic policy in fisheries management', but even if there were, and we agreed with the statement that this would also imply 'the best social policy in the long run', few would argue that this would not entail great social and human costs. The opposition to ITQs, particularly among small-scale fishers, testifies to this. They fear that ITQs lead to unemployment and that their communities will suffer. As has been demonstrated in the case of Iceland (Palsson and Helgason, 1995), this threat is real. Neither government, nor the users themselves, can be blamed for being attentive to the social costs that will occur in both the short and long term. For this reason, the Norwegian government has abstained from introducing ITQs, and, in doing so, seems to be running against current fashion.

The social science of fisheries management is not only the social science of markets, efficiency and quotas. It is also, among other things, the social science of households, gender, communities, power, equity, democracy, and knowledge. Sociologists and social anthropologists assume that fishers are not driven by profit motives alone (Hart and Pitcher, Chapter 16, this volume). We work from the assumption that fishing is an activity deeply embedded in, and conditioned by, social networks, institutions and culture. Thus, we anticipate that fishers are also driven by norms, obligations and responsibilities as members of families, communities and social groups, and that such considerations are operative even when at sea. For this reason we tend to have faith in organizations as coordinating devices. This is also why we often have reservations as to the explanatory value of approaches such as game theory (Hart, Chapter 17, this volume). Individual behaviour is neither always strategic nor nakedly self-interested. It is in the potential of human beings to work for the common good, even if it means personal sacrifice and loss.

Just as our analysis is complex and perhaps even true to empirical reality, our message unfortunately is often similarly complex and obscure. This explains another aspect of the peripheral role of sociologists and social anthropologists in fisheries management. We have not had a single fix to offer to fisheries managers as the economists have today with ITQs. But now things are about to change, or at least so it seems. As a sociologist, I am encouraged by the fact that the co-management concept (Brown, Chapter 13, this volume), a construct of social scientists, is receiving widespread attention, even within the bureaucracy of the European Commission (Jentoft, 1989; Pinkerton, 1989). If we could find a good Norwegian translation of the word it could even have the chance of catching the attention of the Norwegian fishing industry and Fisheries Ministry. At least it would help legitimize the institutions of user participation that exist, and which some critics argue are redundant.

There is no consensus as to what constitutes relevant knowledge and information in fisheries management. Neither is there any widespread agreement on goals or means. If one tries to make sense of the management discourse, it often seems that the root problem is that we do not share each other's concerns, maybe not even the same basic values. When there is general agreement on certain issues, such as the usefulness of co-management, it may be because concepts are sometimes too broad, allowing for interpretations that conceal opposing views and for supporting preconceived positions.

It is for the same reasons that we rarely agree on management policies. For example, to many sociologists and social anthropologists, co-management is in accordance with democratic theory, a way of empowering users and local communities in the management process. For the economist it is basically a transaction cost issue, a way of reducing the expenditures of surveillance. For government agencies, co-management seems to be a means of relieving the political pressure of user-groups: 'If you don't like what we do, why don't you handle it yourself?' It is my impression that the European Commission hopes that co-management will contribute to solving the overcapacity problem. The expectation is that when users are permitted to hold a hand on the wheel they will be more supportive of down-scaling. This is not exactly what most advocates of co-management have in mind.

12.4 THE RISKS OF INVOLVEMENT

Whether we as social scientists really want to become more directly involved in the decision-making process as experts equivalent to the biologists, is, I think, a pertinent question. How would we respond if government invited us inside and asked for our advice? I think many of us would feel rather uneasy. Most of us prefer to be at arm's length from government agencies. There is a widespread fear of co-optation, i.e. becoming swamped in the highly politicized process of fisheries management, or even worse, becoming seduced by the gratifications and rewards almost inherent to positions of decision-making, of power. Additionally, there is the question of what is the most preferable position to be in if you want to make an impact: 'outside the tent pissing in, or inside the tent pissing out.' Most of us, I think, appreciate the independence of the former position, while it is most likely that governments would prefer us in the latter role, if for no other reason than that it is easier to contain the mess we often make for them.

Clearly, fisheries management could benefit from the purely intellectual, outsider perspective of social science, which is basically how we contribute today. Making insights count in fisheries management requires the social scientist to be committed and creative. It does not necessarily require direct participation in the decision-making process. There is also a need for the critical, sceptical, independent and unfettered mind. But can the social scientist have it both ways? Can we maintain the critical attitude that is nourished by the independence of the outsider's role while balancing it with the participatory, responsible and constructive role that is obligatory for the insider?

I think the management discourse would benefit if social scientists could occupy several positions in the management system, if some of us could be within it and while others remain on the outside. It would make communication easier and reduce antagonism, especially if we could shift positions from time to time. Today the management process is hampered because academics and managers speak different languages. Likewise, the discourse suffers from the fact that we are seldom allowed to try out different roles.

Fisheries management is still an ongoing learning process, and there seems to be lots of false learning going on because of poor data, including data on social impacts. Effective learning requires critical feedback. The greater the complexity of the system and the uncertainty of the outcome, the more important it is to allow such inputs (Dryzek, 1990, p. 32). I believe that social scientists have an important role to play in this process, regardless of whether we work inside or outside the decision-making system. We are in fishing communities and among fishers all over the world with our questionnaires and tape recorders collecting data on demographics, fishing techniques, catch-sharing systems, crew recruitment, occupational histories, gender roles, fishing practices, management systems, and the like. Management agencies have much to gain by listening to what we learn and can tell them. The literature is there.

12.5 FAUST AS ROLE MODEL?

The social sciences will not necessarily be co-opted by becoming more involved and constructive. Indeed, the critical sociological analyst also has an obligation to be constructive (Løchen, 1994; McGuire, Chapter 18, this volume). We make it too easy for ourselves if we define our role as merely 'deconstructive', as a postmodernist would put it. Social scientists should not be like Goethe's Faust, who did not want to follow in his father's footsteps as a doctor because he thought the medical profession did more harm than good. Faust could not bear the thought of blood-stained hands. Therefore, he retreated to the ivory tower and became a famous scholar. After some time, however, he started to feel restless. He wanted to come out, to get involved, to become what he believed to be a more authentic human being. But when Faust did leave science, it led to catastrophe – mind you, not so much for Faust himself as for the others he met on the road.

There is a lesson in the Faust myth: Faust was too preoccupied with realizing his own projects. He should have listened more to what the people with whom he became involved had to say. He should have been more sensitive to the needs of his fellow human beings. Then, the outcome (with Goethe's permission) would have been less disastrous. Philemon and Baucis, the old couple who stood in his way because they did not want to move and resettle despite the offer of a generous cash settlement, would not have perished, and Faust himself would not have been burdened with the traumas of bad conscience.

Here is perhaps also the way out of our dilemma. As social scientists we should be careful not to impose our own concepts and models on the industry. We should listen more to fishers and fish-plant workers. Their concerns are legitimate. They are not only worried about the conservation of fish stocks. They are also deeply concerned with the sustenance of their communities, the livelihood of their families and the future of their children. Who are we to ignore such concerns or to explain them away?

To conclude, I do not think that Faust should be the role model for social scientists in a reinvented fisheries management system. Under no circumstances should we take on the Mephisto role, that is the role and character of the Prince of Darkness, who helped Faust deny his guilt.

ACKNOWLEDGEMENTS

I thank Petter Holm and Peter Sinclair for useful discussions. Anthony Davis made constructive comments on the original version of this manuscript, and two anonymous reviewers helped me clarify my argument.

REFERENCES

Dryzek, J.S. (1990) *Discursive Democracy: Politics, Policy and Political Science*, Cambridge Univ. Press, Cambridge, UK.

Hanna, S.S. (1995) User participation and fishery management performance within the pacific management council. *Ocean Coastal Manage.* **26**, pp. 23–44.

Hoel, A.H., Jentoft, S. and Mikalsen, K.H. (1996) User group participation in Norwegian fisheries management, in *Fisheries Utilization and Policy, Theme 2. Proceedings of the World Fisheries Congress* (eds R.M. Meyer, C. Zhang, M.L. Windsor, B.J. McCay, L.J. Hushak and R.M. Muth), Oxford and IBH Publishing Co. Pvt. Inc., New Delhi, pp. 287–302.

Jentoft, S. (1989) Fisheries co-management: delegating government responsibility to fishermen's organizations. *Mar. Pol.*, April, 137–154.

Jentoft, S. and McCay, B. (1995) User participation in fisheries management: lessons drawn from international experiences. *Mar. Pol.* **19**, 227–246.

Løchen, Y. (1994) Commitment and imagination in the social sciences: a concern for the future state of sociology. *Innovation* **7**, 405–412.

Palsson, G. and Helgason, A. (1995) Figuring fish and measuring men: the individual transferable quota system in the Icelandic cod fishery. *Ocean Coastal Manage.* **28**, 117–146.

Pinkerton, E. (1989) *Co-operative Management of Local Fisheries: New Directions for Improved Management and Community Development*, Univ. British Columbia Press, Vancouver.

Chapter thirteen

Community-based cooperative management: renewed interest in an old paradigm

Robert C. Brown

ABSTRACT

Cooperative management may be defined in a number of ways, indicating that it is best considered as a broad concept or goal and is not in itself a precise prescription for better management. Community-based cooperative management is a more tightly defined concept, wherein authority and responsibility over local resources is shared between government and local resource users and/or their communities. A renewed interest in community-based cooperative management has developed worldwide owing to the failure of centrally directed management regimes, based on biological (MSY) and/or economic (MEY) goals, to either protect fish stocks or satisfy the needs of fishing peoples and communities. The dramatic growth in the catching capacity of industrial fleets, plus the willingness of governments to transfer property rights over fish stocks to them, has generated frustration and even desperation in many fishing communities.

Community-based cooperative management takes a variety of forms, largely dictated by local custom and geography. It works best when geographic boundaries are easily defined, where there are traditional fisheries,

Reinventing Fisheries Management. Edited by Tony J. Pitcher, Paul J.B. Hart and Daniel Pauly. Published in 1998 by Kluwer Academic Publishers, London. ISBN 0 412 83410 3.

where fishers have a historic role in stewardship, and where traditional authority remains strong. If those conditions exist, cooperative management may lead to better, more informed decision making, may reduce conflict and uncertainty over resource use, and can contribute toward building a sense of community. Several national governments, recognizing the growing alienation of their coastal peoples and looking for new solutions to the management dilemma, have already adopted the cooperative management paradigm. Others have accepted it for the management of specific fisheries in specific regions. Community-based cooperative management will not work in all cases or for all fisheries, but where it does work, it brings us closer to that elusive goal, optimum social yield (OSY).

13.1 INTRODUCTION

There seems to be a growing consensus on the desirability of cooperative management of fisheries. But, like sustainability, cooperative management has become a term that almost everyone likes, but no one is sure what it means. The power of the concept is related to it being sufficiently vague that almost any interested group can buy into it. Governments, for example, may define cooperative management simply as a partnership between government and industry. Community-based fishers may interpret it as meaning a wide range of negotiated agreements between communities of fishers and various levels of government responsible for fisheries management (Pinkerton, 1989). While some might argue that the vagueness of the concept allows a broad consensus, others might see it as a breeding ground for confusion and disagreement.

Cooperative management is often used interchangeably with other terms, such as joint management, collaborative management and community-based management. In reality each is different, but they share common characteristics. In fact there is a hierarchy of cooperative management arrangements, from those in which fishers are consulted by government before regulations are introduced, to those in which fishers actually design, implement and enforce laws and regulations with the advice and consent of government. Thus cooperative management may be a simple partnership arrangement between government and industry, or it may be a much more complex relationship wherein government actually transfers authority over the management of the resource to industry, community or regional users. The key feature common to the various definitions of cooperative management is that there is some arrangement where responsibility for management is shared between government and user groups (Sen and Nielsen, 1996). The amount of responsibility and/or authority transferred will depend on country and site-specific conditions. Cooperative management, therefore, is for the most part a broad concept or goal, not itself a precise prescription for better management.

Community-based cooperative management

From amongst these various interpretations of cooperative management, one which is gaining increasing importance in fisheries is community-based cooperative management. This is a system wherein authority and responsibility over local resources is shared between government and local resource users and/or their communities. It is based on the notion that proprietary rights to the common fishery resource should be allocated to those communities most dependent upon that resource.

Community-based cooperative management, of course, is not a new concept. Many traditional, pre-industrial fishery-based communities embraced cooperative management of some kind. Pacific North-west American Indian systems (Morrell, 1989), the Eastern Indonesian 'Sasi' system (Bailey and Zerner, 1991), and the inshore northern cod fisheries of Newfoundland (Chapter 14 in Pinkerton and Weinstein, 1995) are some examples. While some commercial exchange was associated with these systems, they were for the most part artisanal, subsistence fisheries which were either ignored or legislated away by governments during the emergence of industrial fisheries (Newell, 1993). There also are numerous more contemporary examples of successful community-based cooperative management regimes within the exchange economy which have been described in the literature (Pinkerton and Weinstein, 1995). The rest of this chapter is focused on the background, promise and problems of community-based cooperative management.

13.2 BACKGROUND: WHY COMMUNITY-BASED CO-MANAGEMENT?

A renewed interest in community-based cooperative management has arisen because of the failure of contemporary, mostly centralized management regimes to either protect stocks or satisfy the needs of fishers. Government policies prior to 1977 either directly or indirectly resulted in reductions in access to stocks to many groups of inshore fishers. Then with the Law of the Sea agreement of 1977 (UNCLOS, 1983), governments assumed ownership of the resource out to 200 nautical miles offshore and indirectly, endorsed such practices. These events, plus development of technology that allowed fishers to seek out fish wherever they were to be found, in any kind of weather, brought with them changes which profoundly reshaped fisheries worldwide. Domestic offshore industrial fishing capacity grew enormously in several countries, spurred on by the support of governments interested in creating employment and raising gross national product. For the most part, this took place at the expense of traditional domestic inshore fisheries.

Prior to 1977, many fishing communities in less-developed countries, often employing traditional technologies which limited the distance they could range offshore, for the most part controlled access to the resource. Territorial use rights (TURFS) were common to many societies, often involving elaborate social customs and local-level regulations aimed at effort control and authority to limit harvests (Christy, 1982; Ruddle and Johannes, 1985; Rettig *et al.*, 1989). Under such systems, control over access to inshore fishery resources was mostly in the hands of those most directly dependent upon them. Self-regulatory measures were usually employed which sustained the resource and resource-based communities across generations of users. But these tradition-based access systems changed dramatically following 1977 when, generally speaking, government policies in most countries – whether intentionally or not – resulted in a reallocation of access to stocks from locally based, small-boat inshore fishers to large offshore industrial fleets. Access rights (quotas) sold at modest prices by governments to original owners were resold openly on the market, often pricing small-boat operators out of the market and concentrating (stacking) quota onto large, corporate-owned factory trawlers (Copes, 1995).

There are many examples in the literature which show how this process of transfer of access rights from small-boat inshore fishers to offshore industrial fishers has harmed coastal communities (Parfit, 1995; Blackwood, 1996). For example in Canada, where the federal government indicated publicly that the inshore northern cod fishery would have priority access to stocks, the reality was that the majority of the resource was allocated to other users, such as the domestic offshore fleet, and to foreign, distant water fleets which operated in Canadian waters under bilateral agreements. This process was reinforced and its impact exacerbated by the fact that the offshore fishers got to the stocks first, before the cod began their annual inshore migrations. Thus the offshore fishers were able to fill their quotas, but not enough fish were left to meet traditional inshore fishers' needs. The result was that the combination of overly optimistic stock predictions, an overcapitalized offshore fleet, and, possibly, environmental change, ultimately led to the collapse of a fishery which had sustained coastal Newfoundland communities for over 400 years. The fishery collapse devastated local economies and resulted in massive increases in government welfare costs in the coastal regions.

In many less-developed nations, governments hungry for foreign exchange capital have entered into joint venture agreements with distant-water fishing companies who have moved their operations inshore and forced community-based fishers from their traditional grounds, or reduced stocks and obliged them to have to fish harder and farther offshore to meet their domestic needs (Parfit, 1995). Under such conditions the emergence of conflict was inevitable. Conflict and political activism have grown and brought with them a realization amongst many, including some governments, that an approach to fisheries management must be developed that would re-establish and enshrine the right access to the resources to those who are most dependent upon them. For example in 1991 the government of the Philippines implemented the Local Government Code (LGC) providing policy structures necessary to decentralize the management of coastal resources to local government units, non-government organizations and people's organizations. This has resulted in the establishment of a new approach to fisheries management known in the Philippines as community-based coastal resources management or CBCRM (Pomeroy and Pido, 1995; Carlos and Pomeroy, 1996). CBCRM is a process by which residents of coastal communities are provided the opportunity and responsibility to manage their own resources, to define their needs, goals and aspirations, and to make decisions affecting their own well-being. Other Asian countries, especially Japan, Taiwan and Vietnam, have evolved their own community-based management strategies (Pinkerton and Weinstein, 1995; Verlaan, 1995).

Community-based cooperative management, then, has arisen out of crises caused by stock depletion, a growing perception that central governments are incapable of managing stocks centrally, and political pressure stemming from an increasing sense of alienation amongst coastal communities toward their governments.

13.3 COOPERATIVE MANAGEMENT: CONDITIONS FOR SUCCESS

Community-based cooperative management may not be appropriate for all fisheries. Canada's Department of Fisheries and Oceans (DFO, 1997) argues that it works best when the following factors are in place:

- parties have clear motives compatible with public interest;
- parties agree on the problem/situation/context that has to be addressed;
- respective agendas are compatible, well defined, and openly acknowledged;
- participants in a fishery are organized, ready to participate and speak in a united voice;
- appropriate representative organizations and necessary infrastructures are in place.

Mitchell (1996) adds to these factors by recognizing characteristics common to successful self-governing institutions she has studied. The major common elements are:

- the community of eligible users is clearly defined;
- there are clear geographic or other boundaries to the resource system over which the users have control and the community of users is able (informally or formally) to exclude outsiders;
- the communities involved are highly dependent on the resource and are vulnerable to non-sustainable use;
- the resource users are relatively immobile: if the resource is overused or the resource system is damaged, the users cannot easily move to another location or another livelihood;
- users are able to enforce management rules both against each other and against outsiders;
- although users may not be homogeneous in a cultural sense, they share relatively homogeneous interests in the resource;
- users invest their own resources in activities such as enhancement and enforcement: the costs of management and mismanagement are borne by those who benefit directly from the resource.

From the above, one can conclude that with respect to fisheries, community-based cooperative management initiatives tend to be successful when geographical boundaries are defined; where there are traditional fisheries; where fishers and/or communities have a historic role in stewardship; and where traditional authority remains strong, thus increasing compliance and the effectiveness of enforcement. Generally, they work well for management of sedentary stocks, and less well for management of migratory stocks.

13.4 BENEFITS OF COMMUNITY-BASED COOPERATIVE MANAGEMENT

Pinkerton (1989, p. 26) and her co-authors review a number of '...more and less successful co-management efforts...' from which a number of conclusions may be derived about positive benefits. Her work focuses on co-management in general, but it applies equally well to community-based cooperative management. In her summary, successful co-management:

1. creates co-operation among individual fishers and local fishing groups in planning the improvement or conservation of local fish stocks;
2. creates the commitment among local fishers to share both the costs and benefits of their efforts toward enhancement and conservation;

3. of allocation decisions, creates an appropriate vehicle of conflict resolution among fishers; it increases motivation to negotiate sharing of access which is perceived as equitable;
4. enhances the position of fishers so that a more equal negotiating relationship exists between local fishers and other water resource users;
5. creates a higher degree of organization and mutual commitment among fishers so that they have a more equal bargaining relationship with fish buyers;
6. creates a willingness among both fishers and government to share data about the resource, and therefore to reach collectively a more complete understanding of the resource;
7. creates a willingness among both fishers and government to explore options for regulation which reduce inefficiencies for fishers;
8. of regulations, creates greater trust between fishers and government, and a greater sense of control on the part of fishers, so that motivation to invest in competitive gear for first capture is reduced;
9. creates a higher degree of trust between fishers and government and improved ability to develop and successfully implement enforcement regimes that fishers perceive as appropriate and legitimate;
10. creates a higher degree of trust between fishers and government and greater willingness on the part of government to allow a range of self-management responsibilities to be assumed by fishers.

Other authors have concluded that successful cooperative management can lead to better, more informed decisions, reduces conflict and uncertainty over resource use, and increases public awareness and understanding (Vodden *et al.*, 1997). Moreover, they claim that it builds a sense of community which can empower and build capacity among local citizens and organizations to work toward community economic development in general. In other words, as McGoodwin (1990, p. 107) argues, it moves us beyond the traditional management goals of maximizing the biological productivity (MSY) and economic yield (MEY) of our fishery domains and '...brings us closer to the more elusive goal: management for optimum social yield, or OSY, with "optimum" meaning the optimum sustainable well-being for a fishery's human participants.'

Critics of OSY (Waugh, 1984) may argue that it is an elusive term which is open to a variety of interpretations, but it also may be argued that OSY encompasses both MSY and MEY (Roedel, 1975), and goes further to include the human element in the management of fisheries, including the value society places on the yield from fisheries and the costs to society of taking its yield. It includes what it means to be fishing peoples, and to live in fishing communities. It takes account of the human element in the management of fisheries. It is at the heart of what community-based cooperative management seeks to achieve. McGoodwin (1990, p. 158) concludes, rightly I think, that supporters of cooperative management echo E.F. Schumacher (1973) in wanting to see 'economics as if people mattered'. However, so far OSY remains little more than an ideal, so it is easily assailed by its critics who, while they may concede that it is a worthy goal, point to its greatest weakness: the difficulty of making it work in the absence of ideal conditions.

13.5 CONCLUSIONS

Cooperative management is a broad concept or goal, not in itself a precise prescription for better management. Community-based cooperative management refers to a somewhat more tightly defined concept, wherein authority and responsibility over local resources are shared between government and local resource users and/or their communities.

This chapter has advanced the notion that a renewed interest in community-based cooperative management has developed worldwide owing to the failure of centrally directed management regimes based on biological (MSY) and/or economic (MEY) goals either to protect fish stocks or satisfy the needs of fishing peoples and communities. The dramatic growth in the catch capacity of the industrial fleets, plus the willingness of governments to transfer property rights over fish stocks to them, has generated frustration, and even desperation in many fishing communities.

Several national governments, recognizing the growing alienation of their coastal peoples, are looking for new solutions to the management dilemma. Some, such as Japan and the Philippines, have already adopted the cooperative management paradigm. Others, such as the United States and Canada, have accepted the approach for the management of specific fisheries in specific regions.

Community-based cooperative management is seen as a means by which central governments can share authority and responsibility for management of fisheries with those who have the greatest stake in them, the fishers and their communities. It is a broad prescription for management which, as is well documented in the literature, takes a variety of forms, largely dictated by local custom and geography. It works best when geographic boundaries are easily defined, where there are traditional fisheries, where fishers have a historic role in stewardship, and where traditional authority remains strong. If those conditions exist, cooperative management may lead to better, more informed decision-making, reduce conflict and uncertainty over resource use, and contribute toward building a sense of community. It will not work in all cases or for all fisheries, but where it does work it brings us closer to that elusive goal, optimum social yield (OSY).

REFERENCES

Bailey, C. and Zerner, C. (1991) Role of traditional fisheries resource management system for sustainable resource utilization. Paper presented at the Fisheries Forum of Centre for Research Institute of Fisheries, Sukabumi, West Java.

Blackwood, G. (1996) Past and future goals and objectives in the allocation of the northern cod resource, MA thesis, Memorial University, St Johns, Newfoundland, 99 pp.

Carlos, M. and Pomeroy, R. (1996) A Review and Evaluation of Community-based Coastal Resources Management Projects in the Philippines, 1984–1994. Research Report no. 6, Fisheries Co-Management Project. ICLARM, Manila, 77 pp.

Christy, F.T.J. (1982) Territorial use rights in marine fisheries: definitions and conditions. *FAO Fish. Tech. Pap.* **227**, 48 pp.

Copes, P. (1995) *Problems with ITQs in Fisheries Management with Tentative Comments on Relevance for Faroe Island Fisheries (95–1)*, Institute for Fisheries Analysis, Simon Fraser University, Burnaby, BC, 22 pp.

DFO (Dept of Fisheries and Oceans Canada) (1997) Framework and guidelines for implementing the co-management approval. DFO, Ottawa, 24 pp.

McGoodwin, J.R. (1990) *Crisis in World Fisheries: People, Problems, Politics*, Stanford Univ. Press, Stanford, CA, 235 pp.

Mitchell, D. (1996) Cooperative management: can we get the incentives right? in *Politics, Management and Conflict in Canadian Fisheries: Proceedings of a Public Forum Sponsored by the Maritime Awards Society of Canada* (eds A.P. Dobell and J. Longo), Univ. Victoria Press, Victoria, BC, pp. 103–107.

Morrell, M. (1989) The struggle to integrate traditional Indian systems and state management in the salmon fisheries of the Skeena River, in *Co-operative Management of Local Fisheries: New Directions for Improved Management and Community Development* (ed. E. Pinkerton), Univ. British Columbia Press, Vancouver, pp. 231–248.

Newell, D. (1993) *Tangled Webs of History: Indians and the Law in Canada's Pacific Coast Fisheries*, Univ. Toronto Press, Toronto, 306 pp.

Parfit, M. (1995) Diminishing returns: exploiting the ocean's bounty. *Nat. Geogr.* **188**, 5 Nov. 1–37.

Pinkerton, E. (ed.) (1989) *Cooperative Management of Local Fisheries: New Directions for Improved Management and Community Development*. Univ. British Columbia Press, Vancouver, 299 pp.

Pinkerton, E. and Weinstein, M. (1995) *Fisheries that Work: Sustainability Through Community-Based Management*. David Suzuki Foundation, Vancouver, 199 pp.

Pomeroy, R. and Pido, M. (1995) Initiatives towards fisheries co-management in the Philippines: the care of San Miguel Bay. *Mar. Policy* **19**, 213–226.

Rettig, R.B., Berkes, F. and Pinkerton, E. (1989) The future of fisheries co-management: a multi-disciplinary assessment, in *Co-operative Management of Local Fisheries: New Directions for Improved Management and Community Development* (ed. E. Pinkerton), Univ. British Columbia Press, Vancouver, pp. 273–286.

Roedel, P.M. (1975) A summary and critique of the symposium on optimum sustainable yield, in *Optimum Sustainable Yield as a Concept in Fisheries Manage-*

ment: Proceedings of a Symposium Held during the 104th Annual Meeting of the American Fisheries Society, Honolulu, Hawaii, September 9, 1974 (ed. P.M. Roedel), Am. Fish. Soc., Washington, DC, pp. 79––89.

Ruddle, K. and Johannes, R.E. (eds) (1985) *The Traditional Knowledge and Management of Coastal Systems in Asia and the Pacific*. UNESCO, Jakarta, 326 pp.

Schumacher, E. (1973) *Small Is Beautiful: Economics as if People Mattered*, Harper and Row, New York, 290 pp.

Sen, S. and Nielsen, J. (1996) Fisheries co-management: a comparative analysis. *Mar. Policy* **20**, 405–418.

UNCLOS, United Nations (1983) *The Law of the Sea. Official Text of the United Nations Convention on the Law of the Sea with Annexes and Index. Paper presented at the Final Act of the Third United Nations Conference on Law of the Sea, New York.*

Verlaan, V. (1995) *Co-Management of Living Coastal Resources in ASEAN: Theory, Practice, and Implications for Vietnam. Report from Training Course no. 1: Principles and Practice of Co-management*, Minister of Fisheries, Ha Noi, Vietnam, 86 pp.

Vodden, K., Gallaugher, P. and Panek, P. (eds) (1997) *Exploring Cooperative Management in Fisheries: Briefing Book*, The Pacific Fisheries Thinktank, Vancouver, 219 pp.

Waugh, G. (1984) *Fisheries Management: Theoretical Developments and Contemporary Applications*, Westview Press, Boulder, CO, 247 pp.

Chapter fourteen

Fostering sustainable development and research by encouraging the right kind of institutions

Jake Rice

ABSTRACT

From the perspective of institutions, sustainability is a set of values for how ecosystems are used. From this perspective, for institutions to foster sustainable development and research, the institutions must reflect the proper set of values. If the current values of institutions are not focused sufficiently on sustainability, it becomes necessary to know how institutions change their values.

One initiative which has been intended to change the values of institutions is role-exchanges between scientists and fishers. These initiatives have great value because they foster better communication and respect between groups with different traditions. However, it would be a retrograde step to use these role-exchanges as a vehicle to create wholly egalitarian roles for fishers and scientists. The improved communication and respect for different traditions of knowledge should be the basis for a new synthesis of science and the unrecorded knowledge fishers have of ecology. It has not been easy for institutions, and the people who constitute them, to gain respect and understanding for other ways to knowledge, nor

Reinventing Fisheries Management. Edited by Tony J. Pitcher, Paul J.B. Hart and Daniel Pauly. Published in 1998 by Kluwer Academic Publishers, London. ISBN 0 412 83410 3.

has it been easy for them to change their values as individuals and institutions. However, in the past few decades we have seen both the civil rights movement and the women's movement make great strides on this path. We should view these as precedents, study them, and learn from them how to change the values of our institutions to reflect values of sustainability, and foster excellence in all the people and groups – scientists, managers, users – associated with the institutions either as members or as clients.

14.1 INSTITUTIONAL CHANGE

The title of this chapter was the title of a contemporary meeting announcement. I was both pleased and appalled to see this objective receiving explicit attention within our discipline. It is pleasing to see that the close linkage between institutions and partnerships on the one hand and progress in fisheries (or any applied science) on the other hand is being recognized as important. In introductory courses, biologists all learn about the close association between structure and function. The concept has some application to processes as well as organisms.

It was appalling as well, because most of us already perceive there to be a surfeit of institutions. If we are not fostering sustainability with the level of development and research and the number of institutions and partnerships we have now, how can we argue that more institutions and partnerships will do better? The implication, of course, is not that we have too few institutions and partnerships; rather, the ones we have are the wrong kind; they are not devoted to sustainability. If we believe that, what kind are the right kind?

Since the term 'sustainability' came into fashion, we have seen the word used in many ways. A common thread is that sustainability is a set of values for how we use ecosystems. Although it is hard to itemize those values without resorting to platitudes and generalities, we can address the generic problem of how institutions and partnerships change their values. If we understand how to change institutional values, then as we gain knowledge of the values which comprise sustainability we can steer our institutions and partnerships towards desirable targets. This approach shifts attention from biological science to the social, political and organizational sciences, where I lack disciplinary training, but have substantial first-hand experience.

Judging from my experience and from the literature on change, at least within the Department of Fisheries and Oceans (DFO) (Anon., 1994; Anon., 1995; and lectures such as J.T. Thomas' 'Managing change', a presentation to the Canadian Centre for Management Development, 9 February 1996), it is not the institutions themselves that have values. Although it is becoming common practice for institutions to have explicit mission statements, institutions are really inanimate, collective constructs. The values they reflect are the values of those who choose, or are allowed, to participate in them, weighted by the distribution of power among the participants. So changing the values of institutions is more than changing the values of current participants. It also involves changing who participates and who has power. This chapter reviews two initiatives to which DFO is committed, which are intended to change both those factors.

14.2 RECENT ATTEMPTS AT CHANGE

One conspicuous initiative is the recent emphasis on replacing government-centred institutions dealing with many aspects of fisheries research and management by partnerships (King *et al.*, 1994; Anon., 1995; BC Groundfish Research and Management Society, 1995; DFO Science, 1996). Although these partnerships do not guarantee different participants, the emphasis on replacement certainly suggests that there will be a different distribution of power among participants. Moreover, looking at the institutions moving towards partnerships, for example watershed management boards, coastal zone management bodies, and international fisheries commissions, we do see more frequent representation by communities and by interest groups other than harvesters, processors, and government agencies. The Fisheries Resource Conservation Council in Atlantic Canada, for example, is much more widely representative than either of its predecessors, the Canadian Atlantic Fisheries Scientific Advisory Committee or the Atlantic Groundfish Advisory Council (FRCC, 1994).

The other conspicuous initiative is the trend to have traditional participants exchange roles temporarily. This trend is manifest by fisheries scientists and managers going on commercial fishing trips or industry secondments, and by fishers participating in diverse research capacities, including keeping special logbooks, instrumenting their gear, and providing platforms for research. Such programmes are widespread in Canada (for example the sentinel fisheries programmes) as well as the US, Europe and other national and international jurisdictions (partial review in Charles, 1995).

Will these initiatives succeed in changing the values which underlie fisheries decisions? It is informative to consider the consequences of the role of exchange programmes. One consequence is that each group gains understanding and, one hopes, respect, for the knowledge of the other group. That is an unqualified good. True partnerships are possible only among those who view each other as equals. Therefore, if the role exchanges lead to greater respect and trust, then the partnership initiatives also have a greater chance of genuine enfranchisement of partners and a concomitant change in institutional values. Fortunately, these consequences seem likely to be realized, although any rigorous studies of how successful these programmes are at changing the attitudes of participants towards other potential partners would be reported in the literature of social sciences, of which I have spotty and superficial knowledge.

Another consequence of role exchanges is talked about less openly, but nonetheless is hoped for in many quarters. It has been argued that scientists discount fishers' knowledge as being subjective rather than highly rational (Finlayson, 1994). If this is true, scientists may hope that through role exchanges, fishers would become more scientific in their views and opinions, more rational and less subjective. This egalitarian objective needs careful scrutiny. Consider its mirror image: is it desirable for scientists to become more intuitive and less scientific through their association with fishers? Should scientists defer their judgements on scientific issues to the judgement of a fisher just because the fisher has participated as a research partner? These questions can only be answered with anecdotal observations at this time, but the weight of anecdotes suggests the answer to both questions is 'No!'.

14.3 LOOKING AHEAD

I propose that we should not even want totally egalitarian roles for all partners in our reinvented fisheries management system. Rather, we want all partners to be excellent in their role, and to respect the excellence of the other partners. It may be possible to increase respect for partners by role exchanges – making scientists into pseudo-fishers, fishers into pseudo-scientists, and putting messianic environmentalists in the corporate boardrooms of the fishing industry. However, the excellence has to come from encouraging each partner to work at perfecting what each does best, not from dabbling in other roles.

This second consideration reveals some of the features we need in our new institutions and partnerships. First, we should preserve those parts of existing institutions and partnerships that foster excellence. Second, we should develop a set of meta-institutions and metaprocesses, the role of which would be to integrate the excellence of the diverse participants. The role would not be a court of advocacy and adjudication, as, for example, Greenpeace is arguing that the FRCC should become. Rather, it would be inclusive of the knowledge gained in different ways by all partners, producing an integrated whole from the component pieces.

To appreciate the distinction it might help to return to the task identified at the beginning of this chapter. The task is to change values of institutions and processes – in this case from whatever values are the basis of past unsustainable fisheries management decisions to values that reflect sustainable use of living aquatic resources. The generation that grew up in the 1960s has spent a lot of time dealing with the difficult task of changing institutional values. The success record is chequered at best. I refer in particular to the civil rights movement and the women's liberation movement. Both shared characteristics with the present issue. There was a desire to change the participants in decision making and the implementation of decisions, and to change how power was distributed among participants. In both cases, to the extent that progress has been made, it has not been achieved by making everyone the same in gender or colour.

It is certainly true that some progress followed from individuals and institutions simply showing greater respect for diversity and a more equal sharing of power. However, from my readings, an essential component of progress has been the development of processes and institutions that have recognized and fostered excellence within the pieces of the whole. Values changed in positive ways not because women's literature or Afro-American literature became more like men's literature or European literature. Values changed because critics and scholars acknowledged the legitimacy of diverse modes of literature, diverse ways to arrive at and express the truth. Hand in hand with the acknowledgement of legitimacy was the encouragement of institutions to foster excellence in each area of endeavour.

We should explore these parallels more deeply, because I believe they are not just convenient analogies, they are true precedents. We need to acknowledge that there are many legitimate ways to understand our fisheries and aquatic ecosystems. We need to have institutions and processes that foster excellence in all parties to our reinvented fisheries management system. Then we need institutions and processes that understand that their task is to integrate the individual ways to insight and truth into a whole, rather than barter off some views of one group against some of the others, or serve as a debating ground to see whose values survive. If we learn from the precedents recent history has given us, I feel there is hope of success in inventing fisheries management systems that reflect sustainable values.

REFERENCES

Anon. (1994) DFO National Program Review. 9 Committees; 14 volumes of reports. Internal report.

Anon. (1995) Roles and Responsibilities Document. DFO Pacific Region Stock Assessment Reorganisation Team. Internal Report, 26 pp.

BC Groundfish Research and Management Society (1995) Articles of incorporation as non-profit association for industry–government partnership.

Charles, A.T. (1995) Fisheries science: the study of fishery systems. *Aquat. Living Resour.* **8**, 233–239.

DFO Science (1996) Overview of the status of Canadian Groundfish Stocks in the Atlantic Zone. SSR 96/40, 16 pp.

Finlayson, C. (1994) *Fishing for Truth: A Sociological Analysis of Northern Cod Stock Assessment from 1977 to 1990*, Memorial University of Newfoundland Institute of Social and Economic Research, Publ. no. 52, 176 pp.

Fisheries Resource Conservation Council (1994) Conservation: Stay the Course – Report to the Minister of Fisheries and Oceans on 1995 Conservation Requirements for Atlantic Groundfish. FRCC 94. Minister of Supply and Services, Ottawa, 138 pp.

King, P.A., Elsworth, S.G. and Baker, R.F. (1994) Partnerships – the route to better communication, in *Coastal Zone Canada 94: Conference Proceedings* (eds P.G. Wells and P.J. Ricketts), Coastal Zone, Canada Association, Dartmouth, NS, pp. 596–611.

Chapter fifteen

Cooperation and quotas

Anthony Scott

ABSTRACT

Fisheries cooperatives can manage their own fish stocks by self-regulation, also undertaking data collection and research, enhancing the habitat, and liaising with other fishing groups over the management of migratory stocks. However, government will not impose fisheries cooperation on fishers. The drive has to come by voluntary action.

This chapter lists benefits a cooperative can bring to fishers, and asks why they may nevertheless resist it. There are many theories. Some fishers like the racing fishery the way it is. Some fear that other members will defect and 'free ride' or that other members will take over and exploit the rest. Economists and others also theorize that fisher heterogeneity prevents the trust needed for cooperation from emerging. Trying to regroup members to bring more homogeneity would not help. What members need is the assurance that comes from fixed percentage shares in the catch and in other benefits and in costs. Government is needed to set up this sharing. ITQs automatically provide the fixed-share building blocks that are needed.

15.1 INTRODUCTION

One element in a world of reinvented fisheries management would be a cooperative organization. Marketing cooperatives are already familiar. But this chapter is about a type that could take on stock management and other functions, even on the high seas. It would provide an alternative to management by government or by a private sole-ownership corporation.

Reinventing Fisheries Management. Edited by Tony J. Pitcher, Paul J.B. Hart and Daniel Pauly. Published in 1998 by Kluwer Academic Publishers, London. ISBN 0 412 83410 3.

The members would be active fishers. Their permits or rights would be pooled, forming the basis of a legal person, a holder of freehold or leasehold rights over the local TAC. These rights would give the cooperative the three traditional powers of ownership: to manage, to receive revenues and rents, and to dispose of the assets. These would include specific powers to manage catching activity, and to provide or to finance collective goods and services. Hannesson's Chapter 19 in this volume, like other papers in the literature, shows how ITQs have been introduced and what they have accomplished. My viewpoint goes farther: it shows that ITQs are an invaluable foundation for fishers' self-government through cooperatives.

Fishers who try to organize themselves have three tasks. First, they must learn and bring about agreement that, in their particular case, voluntary collective action (a cooperative) would make them better off than staying under government regulation. Second, they must overcome internal divisive forces. Then, for it all to work, they must have constructive legislation from government. I discuss these matters in order, giving most attention to the second.

15.2 THE ATTRACTIONS OF FISHERS' SELF-REGULATION

Benefits

Fishers will resist being drawn into a cooperative unless they believe it will bring benefits, that these will be as great as they now get from the fishery, and that these benefits will exceed the costs.

Here I provide a sort of check-list of benefits: activities under a cooperative that would serve them better, or at lower cost, than when fishery governance is provided some other way (Scott, 1995). As there is no way of predicting how great the benefits might be for a particular group, my list just indicates what functions a fishers' organization might perform.

Regulation

The central role for a cooperative group would be self-regulation: gathering and obtaining information, making and enforcing rules.

Reinvention

Widespread experimentation with new approaches can also be expected. Offered new products by commercial suppliers of vessels, gear and hardware and software, the organizations will look for new ways to protect and expand their fish stock.

Coordination between organizations

Like governments today, a cooperative organization can contract with its neighbours concerning their combined management and enhancement of migrating and straddling stocks.

Joint fishing

A cooperative can provide a framework for multivessel joint fishing among partners and groups, their vessels divided among searching, catching, transporting and being left thriftily 'in fallow'. The whole cooperative could become one catching unit.

Ground improvement and enhancement

An organization can go in for providing and bettering ports, reefs, nursery areas, navigation aids, water quality, hatcheries, 'marine ranching' and genetic engineering.

Entry and membership

A cooperative can't do much about entry except to lobby and liaise with government. However, see Acheson (1993).

Costs?

This listing has suggested what a cooperative could do in addition to self-regulation. There might even be significant cost advantages. In many places much stock and fishing information could be obtained cheaply by cooperative members' own voluntary and part-time activity. And compliance, enforcement and working with government could be better accomplished at lower cost because regulation would be tailored to local conditions, less intrusive and better understood to be in aid of husbanding the members' 'own' fish stock. Joint fishing could reap economies of scale.

Weaknesses and possible failures

But there are reasons why in some places a cooperative could not do these things as well or as economically as they are done now, under government stock management.

Unmanageable

Some cooperatives may try to manage stocks so unpredictable and fluctuating that collective net benefits (benefits minus costs) may simply be unattainable.

Shirking

Studies of cooperative enterprise have uncovered ways in which cooperatives can be inefficient, especially by shielding inefficient behaviour by their officers or members.

Going for a cooperative organization?

These are just lists of the opportunities and costs of cooperative management organization. A more balanced treatment would need comparisons with these activities under alternatives: government, a monopolistic firm or a trade union. In what follows, the discussion applies only to those fisheries where a cooperative's benefits *would* be greater than its costs.

15.3 INTERNAL DIVISIVE FORCES

Some say fishers can voluntarily manage stocks without a formal organization

Is a formal cooperative strictly necessary? Some enthusiasts say just restricting entry into a fishery can be enough encouragement for scattered local groups to work voluntarily to prevent the fragmentation of commercial fisheries (Hart, Chapter 17, this volume).

For instance, limited entry alone has been enough to mould some fishers into a group that starts court action. Recent class actions against polluters have been settled with compensation to the temporarily united fishers (Brubaker, 1995.) Other informal groups cooperate in lobbying, especially about regulation and other subjects. Forming a union helps in finding a spokesperson. But in Canada at least these are only pressure groups; they don't aspire to manage a particular fish stock.

In the rare cases where all fishers are concerned with one fish stock, a discontented group can attempt management. Johnson and Libecap (1982) provide examples of groups that attempted regulatory management of Gulf of Mexico commercial shrimp fishery. Jentoft (Chapter 12, this volume) has provided evidence of self-regulation and inspection in the

Norwegian cod fishery, in force for 90 years. A third example was once provided by Bay of Fundy herring fishermen (Scott and Neher, 1981). A fourth was a Canadian prawn fishery where a few local vessels took it on themselves to stop fishing until the stocks were replenished: it might have worked but for 'use it or lose it' rules (Cruikshank, 1991.) The best-publicized example has been the system of informal exclusive rights in the Maine lobster fishery (Acheson, 1993). Many recent studies show self-government efforts being made in other limited-access fisheries (Leal, 1995; Hart, Chapter 17, this volume).

Studies by anthropologists have found numerous examples of home-made self-regulation without formal cooperation in developing countries. Schlager (1990) assembled studies of 42 voluntary 'groups' scattered over 30 fisheries in 16 countries. Nearly all worked inshore or in coastal fisheries with fixed/stationary traps, nets and longlines. Others have been well reported in an extensive literature (Johannes, 1978; Christy, 1982; Berkes, 1986; Pinkerton, 1991). These show that some unified self-regulation does indeed exist in less-developed economies.

The existence of these cooperating fisheries had little more support than difficult access. Usually it was physical isolation, restriction to family or farm connections, low fishery yields or arduous work conditions. There being no in-migration or new entry, the fishers in time have regarded themselves as clubs, quasi-proprietors, and so observed their own rules – impressive examples of voluntary organization. However, the rules they provide and enforce are not ambitious, perhaps the size or sex of fish landed. They rarely amount to management, otherwise being confined to allocating or rotating fishing sites, or closing the fishery during the off season or on Sundays. In brief, (a) they can rarely limit access by their own actions, and (b) they almost never limit the total catch.

The above cases of voluntary self-regulation have been written up many times. They are mostly inshore fisheries (however, the offshore exception, whalers' self-regulation, also limited neither access nor catch; Ellickson, 1991, p. 205). From here on I deal with possible cooperation in offshore or 'ocean' fisheries, with many vessels and fishers and scattered home ports. Though government management systems are in place, their methods are necessarily producing such high fishing costs and are having such limited success with stock depletion that fisher self-regulation and enhancement is sometimes considered.

Some fishers are not clamouring for self-regulation

In examining willingness to take part in a new type of management organization, it is that of today's fishers, not tomorrow's. It is today's fishers who must both make the hard choices and bear whatever pain these choices bring. They may think a better managed and harvested fishery will be too delayed for them to benefit.

In addition, many of today's fishers may not regard self-management by a cooperative as a good thing, even if it is successful. The reason is that the preferences of these particular fishers are such that they like things the way they are. Some like the unpredictability and risk of an unmanaged stock, the joyful racing for the first catches, and the skippering of huge vessels and elaborate gear. Some just have no taste for cooperating with anyone. They favour today's regulations that free them from day-to-day interference. Their anti-self-government attitudes are all too observable to anyone who watches current events in a managed fishery.

I suggest a reason for their attitudes. In the evolution of modern fishery regulation, these individuals have been 'selected' for their non-cooperative qualities. People who are averse to risk and who do not enjoy holding huge blocks of capital drift away from the fishery. Outsiders are attracted who value highly a mixture of risk and independence. If I am correct, the current regulations determine the makeup of the current fishery labour force. The regulations, and the adapted fleet, are the main determinants of the local bioeconomic equilibrium and of the stocks associated with them. In turn, these have been influential in shaping the revised regulations, the TAC and the openings. Consequently, it is not to be expected that many of today's selected fishers would welcome taking on the husbandry-type functions of a cooperative. If a cooperative is imposed on them, they might decide to retain short openings and other exciting racing features of their past working lifestyles. But it is more likely that the economies and profits of self-regulation and coordinated harvesting would change the attitudes of some of them, and drive others to sell out to a different kind of fisher. In the meantime, it is the old kind of racing fisher who is objecting to fixed quotas and peaceful self-government.

All fishers can see the difficulties

Even if there is limited entry, and even if they have a positive taste for cooperation and self-government, fishers may hold back from trying to manage their own fishery. This is because they believe the organization itself would not be up to it. I have compressed many possible reasons into two. First, its members would not have enough information. Second, it would be subject to divisive strains and threats to defect.

'We would not have enough information to run our own fishery.'

Fishers may become 'owners' of the stock, but would they have the information to look after it?

Monitoring the catch

For example, they might be able to measure total effort (trips, or days' fishing) but could they reliably record dimensions of the catch, day after day? In regulation they would run up against the many incentives for members to underreport, and their unwillingness to attribute correct catches to others. Hence even catch information would remain difficult (or costly) to obtain.

Biological information

Closely related are difficulties of keeping up with biological information about the stocks they are fishing. Fishers acquire only a little of the shared home-spun knowledge that animal breeders have of farm and pet livestock genetics or behaviour. That is why governments and scientists conduct their own explorations, and are then relied on to set TACs. Unsurprisingly, fishers now believe that only government can get and use the information on which management is based.

Producing their own information?

For me, however, the reliance of fishers on government for information is inconclusive. Does it mean that a lack of information must prevent self-regulation, or that obtaining information must be one of the functions of a self-regulating cooperative? Obviously, a cooperative could buy research and information.

To start, a cooperative can use its own members' gear, trips, time, insights, and experience to collect data, and then to share and even to analyse it. In the process they could decide which costly information was worth having, and which was worth paying for (as in New Zealand as described by MacClurg, 1996). Clearly the information problem could be scaring off some of today's fishers. But they will learn that in a cooperative they would have their own powers to buy, or/and to provide, data and analysis.

'We are too suspicious of each other to work together.'

This is a second reason why fishers may not try to form a regulatory cooperative. They believe fishers can't cooperate. Decades of fierce argument about closures and other regulations have taught them not to trust even their friends. To repeat the thought of Schlager (1990), fishers are capable of seeing what they might do together. Her sample shows that most existing 'cooperative' groups don't try anything very ambitious. The reason, in my submission here, is that they knew their groups would fall apart if they did try. One by one, members would defect. For a while they would free ride on whatever the surviving group could accomplish.

Defection and opting out

Fishers know that if their organization tries to make and enforce regulations, many members will simply quit. Attitudes will not all be the same. But as their self-regulation becomes more onerous, each will eventually consider whether opting out would not be better. Some theorists, such as Sen (1977), have suggested that what is missing is the power to 'assure' all the members. Persons will not submit to a control unless they can be assured that others must also do so. If many defect, that assurance is lost. Because experienced fishers can anticipate defections, they have no wish to try cooperation.

Free riding

The problem of expected defection is similar to that of expected free riding. Fishers foresee that those who drop out will do so hoping that others will not drop out, but will persevere and succeed in cooperative management. Those who drop out can free ride on a managed stock. The expectation that those who persevered in self-regulation would see part of their pay-off captured by others will give everyone additional reason for not starting a self-regulation strategy.

The answer

The only answer to defection and free riding is compulsory universal membership of a cooperative scheme. This has been understood for centuries. The question it raises is, will fishers rate the expected benefits highly enough to ask government to make membership compulsory?

Cooperating with strangers: why heterogeneous fishers would resist cooperatives

No cooperation among strangers

In the absence of compulsory membership, a very few voluntary groups may survive being torn by defection and free riding. That these are small groups suggests that in such groups, fishers can trust each other and so agree on joint measures, where larger groups cannot (Ellickson, 1991, uses the term 'close-knit').

The pessimism about large groups implies that differences exist other than those engendered by personal preference. Examples are differences arising from social, economic or technical causes. Crews fishing the same waters, for example, differ in family size, where they live, their ports, languages, religions, gears, vessel types, days or hours of work and markets. They are strangers who exhibit too much diversity, heterogeneity or lack of a community of beliefs and norms. Diversity and heterogeneity are what prevents equilibrium according to those using game theory, those applying club theory and those placing emphasis on transaction costs. Ostrom *et al.* (1994) survey all this and Libecap (1980, Chapter 5) gives apt fishery illustrations (see also Hart, Chapter 17, this volume).

The game-theory reason for wishing that fishers were not heterogeneous is that when persons deal with strangers they have comparatively little reason to behave honourably toward them. Axelrod (1984) and others predict that when a self-interested person does expect to meet another frequently, he will not act as opportunistically as he would if he expected never to meet him again. Thus if persons live in the same community they will be less likely to try to gain by discriminating against each other. The more frequently they expect to meet, the more likely each is to behave in a trustworthy fashion. Even if there are many more than two persons, each can gain a reputation for behaving honourably toward each person with whom he deals in a small and closed community.

This abstract theory about contracting does not apply very neatly to our investigation of the acceptance of a system of organization for a fishery. But its emphasis on when a person will or will not expose himself to being harmed by others does suggest why many fishers would not want to get enmeshed in a cooperative. If the others will be strangers, all thrown together only by the accident of needing management of the same ocean fishery, each knows that mere membership will not resolve their old adversarial relationship. Each recognizes that those who use other gears, divide their time between this and other fisheries and live in different ports (provinces, countries) may harm him with few worries that he will ever deal with them again. Recognizing their incentives, the rational fisher will not place himself in a cooperative where they can affect him.

Fishers hoping to self-govern themselves can try some steps to prevent diversity from destroying their joint decision making. For example, a group could decentralize its administration to more homogeneous subgroups with more in common. A similar, and more promising, tactic for avoiding heterogeneity is to divide the larger group geographically, setting up fishing-zone, or area, subgroups. This approach is already behind the regionalization of fishery regulations. With the gaps lying between homogeneous groups rather than among individuals, the heterogeneity problem has been pushed one step up the ladder. When the management of a single stock used by all the different groups is in question, this may intensify disagreement. Furthermore the transaction costs of cooperation and compromise between groups could be higher than between unorganized individuals.

The main point is that many observers attribute any failure to cooperate to large size and diversity of tastes and preferences. They agree that just dividing a group into homogeneous subgroups is no answer.

The question and its answer

Can the benefits of cooperation be salvaged from these apprehensions? 'Share' is the key word. The typical fisher's share, under present conditions of government regulation, is more or less secure, thanks to his skill, work, capital, and vigilance about policy changes. Indeed, notwithstanding the instability of the fishery and the weather, data show that his percentage in the distribution of catches remains stable. Can the benefits of a fishery cooperative be obtained while keeping his percentage secure and stable? Can the fisher be protected against his fellow members' opportunism, block-voting and political manipulation, and their threats of defection and free riding?

When the question is seen this way, it follows that the fisher will support cooperation if his share is fixed in advance (just as a shareholder leaves his capital with his company as long as the companies act enforces his fixed share of dividends). Submitting to government that includes collective stock management, fishery improvement, rule enforcement, and harvesting will not drag him down from his previous income.

Is there evidence for this? There are few examples of heterogeneous offshore fishers successfully agreeing on self-regulation. Instead, to illustrate my contention we can look to other natural resources. There we find that wherever open-access or common-property resources are shared according to some fixed or external rule, the rule then serves as the percentage basis for later sharing of profits, and of joint activities. As evidence I mention the stinting formula on manorial or village common fields; the oil company's share of petroleum formations lying under more than one lot; and the farmer's right to irrigation water. Not all such examples include formal cooperatives. But they do show that users of a resource that is not fenced into individual spaces can work together without conflict because their percentages of the resource or its benefits have been fixed in advance.

15.4 ROLES FOR GOVERNMENT

Four provisions

What follows is very brief. Only government that manages the fishery now, can be the source of rules that prevent conflict and support a cooperative. It follows from what has been said that there are four provisions that fishers cannot provide for themselves without using force. Legislation would provide these:

1. compulsory membership, a new fishery requirement: this prevents free riding and threats of defection;
2. limited entry, an old requirement: this complements compulsory membership in preventing free riding;
3. voting rules: government makes these now for companies, societies, and farm cooperatives, and these rules would complement compulsory membership and safeguard minorities;
4. fixed shares: government already provides these in some fisheries (see below); these make support of and participation in fisheries risk-free.

Adapting fisher quotas (ITQs)

For a government to guarantee shares does not require strictly that percentage shares be fixed. Most fishers would be happy to be guaranteed an increasing absolute number of fish, even if their percentage share happened to be falling. But this is a technicality. In what follows, I will write as if fixed percentage shares were an essential.

The government could lay down members' exact percentage shares in advance. For example, it could mandate sharing equally; in proportion to initial subscription; in proportion to catch (though this would upset the whole scheme); or with default personal percentages as in patent and copyright laws. Customs in other natural-resource industries could also be the basis for mandatory catch sharing; the costs and benefits of joint activities are shared out according to the known division of other things to which the users also hold rights.

If government tries this in the fishery, looking around for an accepted traditional percentage sharing (such as the 'lay' aboard each vessel), it will fail. Instead of sharing, there has been a government-conducted race, in which the best catches usually go to the established champions, the fastest or best-skippered vessels. That this racing game has led to no customary peaceful sharing is widely recognized and is what in the first place induced government to introduce limited licensing and then individual transferable quotas (ITQs).

So, instead of traditional fishing percentages, the government can utilize those that are implicit in its ITQ scheme to become the basis of a set of accepted sharing percentages within the cooperative. No longer just a right to a share of the total allowable catch (TAC), the ITQs become a membership certificate evidencing the size of each holder's contribution to his organization's fishery wealth. The fisher adds his ITQ to the pool. For this the organization gives him a percentage of the fishery's managed yield. If the members only work one single-species fishery, this is of course just the same as their ITQ percentage of the TAC. If, however, the members pool their ITQs in more than one species, their various species' percentages must be weighted. And if they are in straddling stocks, they may hold memberships in two or more cooperatives. In spite of such complexities, each member's final percentage share of his own organization's pooled fishery benefits and costs is a single objective number.

The good news is that the toughest of the four requirements, to provide fixed shares, is something that governments everywhere have already shown they can do acceptably. They are handing out ITQs on historical principles. Hannesson (Chapter 19, this volume) shows how this has been accomplished, and what improvements it has brought about. My viewpoint goes beyond theirs, showing that ITQs are the invaluable, perhaps indispensable, foundation for self-government by fishers.

ACKNOWLEDGEMENTS

Thanks for discussion and suggestions from Rögnvaldur Hannesson, Philip Neher, Peter Pearse, Tony Pitcher and especially Ian Keay.

REFERENCES

Acheson, J. (1993) Capturing the Commons: Legal and Illegal Strategies, in *Political Economy of Commons and Culture* (eds T.L. Anderson and R.T. Simmons), Rowman & Littlefield, Lanham, MD, pp. 69–83.

Axelrod, R. (1984) *The Evolution of Cooperation*, Basic Books, New York.

Berkes, F. (1986) Marine inshore fishery management in Turkey, in *Common Property Resource Management*, National Academy Press, Washington, DC, pp. 63–83.

Brubaker, E. (1995) *Property Rights in the Defence of Nature*, Earthscan, Tornado.

Christy F.M. (1982) Territorial use rights in fisheries. *FAO Fish. Tech. Pap.* no. FIPP/T227.

Cruikshank, D. (1991) *A Commission of Inquiry: ... The Fisherman's Report*, United Fishermen, Victoria.

Ellickson, R.C. (1991) *Order without Law: How Neighbors Settle Disputes*, Harvard Univ. Press, Cambridge, MA.

Johannes, R.E. (1978) Traditional marine conservation methods in Oceania and their demise. *Ann. Rev. Ecol. Sys.* **9**, 349–364.

Johnson, R. and Libecap, G. (1982) Contracting problems and regulation: the case of the fishery. *Am. Econ. Rev.* **72**, 1005–1023.

Leal, D.M. (1995) Community-run fisheries, in *Taking Ownership* (ed. B.C. Crowley), Atlantic Institute for Market Studies, Halifax, NS, pp. 183–229.

Libecap, G. (1990) *Contracting for Property Rights*, Cambridge Univ. Press, New York.

MacClurg, T. (1996) *ITQs in New Zealand: Bureaucratic Management versus Property: The Score after 10 Years*, Treaty of Waitangi Fisheries Commission, Wellington, NZ.

Ostrom, E., Garner, R. and Walker, J. (1994) *Rules, Games and Common-Pool Resources*, University of Michigan Press, Ann Arbor, 370 pp.

Pinkerton, E. (ed.) (1991) *Co-operative Management of Local Fisheries*, UBC Press, Vancouver.

Schlager, E. (1990) Model specification and policy analysis: the governance of coastal fisheries, PhD dissertation, Indiana University, 247 pp.

Scott, A. (1995) The ITQ as a property right, in *Taking Ownership* (ed. B.C. Crowley), Atlantic Institute for Market Studies, Halifax, NS, pp. 31–98.

Scott, A. and Neher, P.A. (1981) *The Regulation of Commercial Fisheries in Canada*. Economic Council of Canada, Ottawa.

Sen, A. (1977) Rational fools: a critique of the behavioral foundations of economic theory. *Phil. Public Affairs* **6**, 317–344.

Chapter sixteen

Conflict, consent and cooperation: an evolutionary perspective on individual human behaviour in fisheries management

Paul J.B. Hart and Tony J. Pitcher

ABSTRACT

This chapter suggests that taking an evolutionary view of human behaviour may be helpful when designing or operating the management of humans involved in fisheries. Humans have evolved flexible, multipurpose mental organs through natural selection on genetic traits that maximized the survival of individuals and kin in small communities. We argue that management systems that foster behaviours that are congruent with evolutionary history of humans will be more likely to be successful and sustainable than those that rely either upon direct coercion, or upon purely economic incentives, or upon the emplacement of altruism as a social norm.

Reinventing Fisheries Management. Edited by Tony J. Pitcher, Paul J.B. Hart and Daniel Pauly. Published in 1998 by Kluwer Academic Publishers, London. ISBN 0 412 83410 3.

16.1 THE PROBLEM

Fishers are skilled at reducing the impact of fisheries regulations. For example, fishing effort has several interacting determinants (Wilen, 1979) and a control imposed on one, such as length of vessel, has stimulated fishers to change other factors, such as vessel width, to minimize the effect of the intended regulation on their catch and income. Ideal fishery regulations would manipulate a fisher's decision-making behaviour at a fundamental level, directly at the point where he/she evaluates the worth of carrying out some action. Properly done, this deeper level of control might lead to greater success for fisheries management measures, which are widely eluded with great skill and inventiveness in fisheries throughout the world. But a fresh approach of this kind requires a fundamental shift of perspective by fishery managers. It requires identifying the precise utility that an individual fisher maximizes when taking decisions. This chapter discusses the nature of utility for an individual fisher and attempts to highlight some misconceptions that have driven most fisheries management measures and their implementation.

The problem is illustrated by an example. Fowle (1993) describes the marine fish management regime in the United States. She points out that the fishery management regime in the US derives from the Magnuson Fishery Conservation and Management Act. This act was passed to 'Americanize' the fisheries off the US coast when 200 mile exclusive economic zones (EEZs) were established and to prevent overfishing, especially by foreign fleets, and to allow overfished stocks to recover. The Magnuson Act includes a number of national standards which set out the goals of the Act. Any regional fishery council must take note of the standards when it is drawing up a fishery management plan. The first standard states: 'Conservation and management measures shall prevent overfishing while achieving optimum yield', where optimum yield is defined as '...the catch from a particular fishery that will provide the greatest overall benefit to the United States in terms of food production and fishing opportunities. It is determined on the basis of Maximum Sustainable Yield from a fishery, as modified by any relevant ecological, economic, or social factors.' The key point about this example is that the utility maximizer is assumed to be the United States, or in more practical terms, the governing agency acting on behalf of the government or state. Similarly worded fishery legislation is in force in Canada, Europe, Australia and elsewhere. The problem is that, inevitably and not surprisingly, the collective interest expressed by government or state and the interests of individual fishers rarely match.

The history of fisheries biology and economics shows that practitioners of these disciplines usually took a global view of a resource, where fishing effort on a stock was accounted for as a single value, the sum of the effort exerted by all the fishers (Gordon, 1954; Beverton and Holt, 1957; Anderson, 1973). Despite pioneering work on individual fishers' strategies (McCay, 1978; Hilborn, 1985), from this viewpoint there was little room for individual specific fishing strategies and utility functions. Almost all fishery management goals have been explicitly based on a notion of collective societal good with which individuals are expected to conform, generally encouraged through some type of coercion. We suggest that such management is inherently doomed, is bound to cause extensive evasion and cheating with associated costly enforcement protocols, and is unlikely to lead to effective management. We consider that only when individual benefit is recognized as the driving force behind conflict among fishers, and between fishers and management, can conflict be turned into consent, fishers cooperate with management and fisheries become sustainably harvested.

The main points of this chapter will not be new to a section of the anthropological and sociological community (McCay, 1978; Durham, 1981; Ruddle, 1996), but the ideas we present have not been appreciated by fisheries biologists and managers. For that reason alone the ideas are worth presenting. We would claim also that even those anthropologists and sociologists who have perceived the importance of an individual approach to understanding fisheries have not emphasized how the viewpoint should permeate all fisheries management and not just that concerned with artisanal fisheries. They have also not appreciated how the approach can be used to create testable hypotheses that, if confirmed, provide sound evolutionary reasons for why fishers do what they do.

16.2 CANDIDATE DISCIPLINES

Several disciplines address this issue of individual versus collective benefit in building consent to be governed, but few have the theoretical rigour that is needed to address effectively the actions of individual fishers. Political science, sociology and anthropology are candidate social sciences. That they are largely preoccupied with the normative is not surprising, as past forays into social engineering have led to the horrors of Marxist, Nazi or fundamentalist regimes. These disciplines can describe in elegant detail issues in political, social or familial contexts, but deeper analysis is often avoided by their practitioners. These disciplines seem unable to use theoretical insight for prediction or to distinguish among alternative hypotheses. The politics of power, persuasion and exhortation offers a range of instruments of consent and cooperation that varies along a spectrum from raw coercion to cynical manipulation (e.g. Nazism, statist Marxism, state religions, commercial advertising, the hollow ring of synthetic bonhomie in politicians). All the points on this spectrum are so easily cheatable by individual humans, so expensive to maintain and so patently ineffective, that it is amazing that newspapers and other media dwell almost exclusively on such matters. Sociology offers to describe, but rarely dares to design, well-meaning, benevolent institutions that encourage consent and human welfare. When implemented, such arrangements invariably turn out to involve successively discovered onion-like layers of imperfections that need eternal adjustment, are costly to maintain in both money and human effort, must be administered by large bureaucracies and, in the last resort, are intrinsically ineffective. Only the discipline of anthropology offers a credible social incentive for consent and cooperation: the rosy glow engendered by working together with friends and kin, a human emotion that is fully congruent with the tribalism for which we humans are so well suited by our evolutionary history. The big problem here is that we do not cooperate with the tribe next door as they are invariably good-for-nothing scoundrels, and all humans know the tragedies to which that xenophobic pathway has led.

Alone among the social sciences (and often disavowed by them), only economics offers an internally consistent and rigorous mechanism for achieving cooperation. The power of the market-place has a strong claim to provide effective control measures for sustainable fisheries (see contributions in Munro and Pitcher, 1996): consent and cooperation is driven by individual economic monetary gain. This is a powerful argument (Hannesson, Chapter 19, this volume), but we will contend here that people do not necessarily act as though money were the sole measure of utility. In many instances humans choose a course of action because it provides intangible rewards such as happiness, delight at the beauty of nature, the contentment of a full stomach, relief in avoiding personal harm or social acceptance by other humans, none of which can be replaced by an amount of money.

The aim of this chapter is to suggest an alternative body of knowledge, founded in the Darwinian fitness-maximizing perspective of behavioural ecology, that may help to resolve the basic conflict of interest in fisheries. The advantage of this discipline is that it has developed a conceptual structure that makes it easier to devise hypotheses about behavioural strategies. An important advantage is that these hypotheses may be tested by experiment or observation.

16.3 EXPLAINING THE EVOLUTION OF SOCIAL BEHAVIOUR IN ANIMALS

In a now famous book, Wynne-Edwards (1962) argued that an individual animal in a dense population would limit its reproductive output to reduce intraspecific competition and to favour survival of the species. This thesis proposed that adaptations that favoured the group could arise through natural selection. Led by Williams (1966), subsequent work has shown that group selection is very unlikely and that most natural selection is through genes via the individual (R. Dawkins, 1976, 1982). Ultimately each individual animal acts to maximize its Darwinian fitness, defined as the number of offspring that survive to reproduce in the next generation (strictly, it is defined as female offspring that survive to reproduce in the grandchild generation). Inherent in this approach is that the variation in phenotypic properties of individual animals leads to individual-specific fitness-maximizing strategies (Parker and Sutherland, 1986). In the past 20 years, this paradigm has had stunning success in providing insight into the evolution of adaptations in behaviour, ecology, biochemistry and physiology throughout the animal, plant and fungal kingdoms (Trivers, 1985; M. Dawkins, 1986; Krebs and Davies, 1993; Alexander, 1996).

Using the gene's-eye view of selection has led to elegant theories explaining a range of animal behaviour which was hard to reconcile with the theory of natural selection, such as the cooperation of altruism (Hamilton, 1964a,b; Trivers, 1971), conflict and cooperation in fighting strategies (Maynard Smith, 1982), conflict and cooperation between the sexes and among generations (Parker *et al.*, 1972).

A key development in behavioural ecology was that many of the disputes seen between individual animals derive from a conflict of interest. Each individual, in trying to maximize its fitness, faces different demands. Where there are individual benefits to cooperative behaviour, individuals may act together (Axelrod, 1984) but the establishment of cooperation requires some special conditions (Hart, Chapter 17, this volume).

16.4 THE EVOLUTION AND MAINTENANCE OF SOCIAL BEHAVIOUR IN HUMANS

Some social scientists aver that animal behaviour can be no guide to human behaviour. Although some biologists also have taken this view (Rose *et al.*, 1984; Kitcher, 1985), we think it misguided. Because humans are one species of animal, human behaviour must have been moulded by the same evolutionary forces operating in the rest of the animal kingdom. Moreover, it is increasingly evident that taking the gene's-eye view of human behaviour can lead to valuable insights (Wright, 1994). Since the books of Morris (1967) popularized the concept, few (other than the creationist successors of Bishop Wilberforce) can doubt that humans are but one species of primate. Darwin himself assumed that there was no intrinsic difference between the emotions of animals and humans (Darwin, 1872). Provided that extrapolations from animal behaviour are made with care to distinguish general factors from factors specific to particular species (Huntingford and Turner, 1987), the gene's-eye view is proving valuable. Recent work on primate social behaviour (de Waal, 1982, 1996) has shown that social behaviours such as learning about tools, the use of language, a dynamic tribal politics based on individual expectations of gain tempered by kinship, and the exercising of moral scruples, all attributes long considered exclusively human characteristics, are emphatically present in our close animal relatives. These behavioural attributes in humans have arisen through natural selection, driven by ecological and social pressures (Byrne and Whiten, 1988). Hence, the evolution of the strategies used by humans to cope with the social environment must also have been driven by the need to maximize fitness. The vast weight of observation and evidence is now in support of this perspective: the obligation is now upon those who consider the human species in some way emancipated from such evolutionary forces, to prove their case. Many puzzles about human behaviour can be solved using evolutionary thinking. Moreover, we can expect evolutionary theory to shed light on the factors that lead to human conflict and cooperation.

The way in which ideas from behavioural ecology can explain human behaviour is discussed in detail in Wright's (1994) stimulating book. A key point that he makes is that many of the emotions and feelings of humans (Wright's so-called mental organs) were designed by natural selection to bring about fitness maximisation. Some analyses of the origin of the human species have argued that the evolution of mental organs for social discourse within and among tribes was the key event in human evolution, more so than upright posture, tool making or even language (Shreeve, 1995). Some of this selection occurred in an ancestral environment which was different from the way most humans live now. As a result, some of the behaviours with which humans are equipped may not bring about the effects for which they were designed. For example, our ancestral social environment was probably a small community where everyone knew everyone else, many near neighbours were kin and interactions between individuals were long-term. Many early humans made immense journeys on foot, and Shreeve (1995) argues that meetings with strangers were likely governed by behaviours that reduced the initial chance of conflict (even if warfare broke out later). Nowadays, at least in developed countries, we live in societies where we very often interact with strangers whom we will never see again and our contacts with kin are far fewer than of old. These differences can mean that people in modern societies tend to behave less cooperatively than they would do in small-scale societies (Axelrod, 1984). Modern management science recognizes many of these findings implicitly, advocates making the work-place a community, and tries to reduce the impact of conflict by mediation and facilitation skills.

Social scientists are also very aware of this problem and, for fisheries, are actively analysing management set in a community context (Jentoft, Chapter 12, and Hart, Chapter 17, both this volume) and the crucial role of kinship. But we think that these studies are not as instructive when lacking the concepts provided by the evolutionary perspective. An important putative mental organ that could have been a fitness maximizer in our ancestral environment is the feeling of satisfaction on the efficient completion of a task. People get endless satisfaction in finding shorter routes between two places. The spirit of invention is with many of us and there is nothing more satisfying than developing some simple widget to make a routine task easier. The technological history of humans is testament to this mental organ (Bronowski, 1973). For most of us today these achievements seem to have no fitness consequences. In the ancestral environment, being good at important tasks of the day, such as hunting or gathering food, could have had fitness consequences (Winterhalder and Smith, 1981; Borgerhoff Mulder, 1991). It is often ignored that an increasing proportion of the human population lives in poverty in huge urban centres with high pre-reproductive and adult death rates. An evolutionary biologist has to suspect that such human populations are unlikely to be emancipated from the genetic selection that has shaped all living creatures on our planet, but can only speculate at the outcome.

In conclusion to this section, there will always be someone to argue that humans and the rest of the animal world are too different to consider together and that animal behaviour can at best be only an analogy of the human condition. But this attitude ignores the great explanatory power of assuming that humans possess mental organs that have been designed by natural selection to maximize fitness. Given this assumption, we are able to understand better why men tend to be more promiscuous than women, why siblings alternate between fighting and friendship, why first-born children inherit, why status is so important to human societies, particularly to males, why people are more likely to cooperate with family and friends than with strangers, and why alliance and clever deception is a pervasive force in human social intercourse. Moreover, recent work has demonstrated the evolutionary origins of moral systems in primates and other mammals (de Waal, 1996).

16.5 FISHERIES MANAGEMENT AND AN EVOLUTIONARY PERSPECTIVE ON HUMAN SOCIAL BEHAVIOUR

Given the power of the evolutionary mode of thinking in explaining aspects of human behaviour, we argue that it is worth pursuing the same line when designing fisheries management regimes. Fishers, like the rest of us, will be motivated by those mental organs that natural selection designed to maximize fitness. As a result they will pursue mostly their own interests. They will not choose to sacrifice these interests for the good of the community, the government or the fish stock. Economic analysis takes a restricted view of human behaviour by assuming that a fishery can be treated just like a manufacturing company and overall efficiency assessed. If it is discovered that there are more fishers than required to harvest the quantity of fish available, then the argument is that the number should be reduced. This reasoning is fine from the point of view of the government scientist or the politician taking a global view. But the problem looks very different for each individual fisher. The fisher is possibly not only interested in economic efficiency but also values his activity as a way of living, as a vocation. Within a fishery there will also be different groups who have different ways of evaluating their activities and this may lead to several different strategies.

It is therefore likely that the global perspective taken by the government agency, as with the Magnuson Act, will lead to regulations that will not be compatible with the interests of individual fishers. This conflict of interest will mean that fisher and manager will be in a constant battle. A way to dissolve this conflict is to create management measures that are based on the assumption that fishers are driven by self-interest. The task then becomes one of identifying what it is that each fisher is trying to maximize, and then manipulate the cost–benefit function leading to that maximization so as to achieve the desired management goal. Some realized this point many years ago (Clark, 1981), and the trend towards ITQs (Grafton, 1996) is a recognition of the individual-specific factors motivating the decisions taken by fishers. Other economists are also beginning to take note of the individual nature of human action (Haltiwanger and Waldman, 1993).

The primary need of fishery management in relation to human social behaviour is to create conditions that promote cooperation between the fishers executing the fishery and the various management agencies. A knowledge of behavioural ecology encourages us to propose a list of methods that might encourage the establishment of cooperation. These are:

- establishing a role for fishers in gathering biological data for assessment;
- involving fishers directly in making decisions;
- encouraging self-policing of regulations to increase compliance;
- confining membership of the fleet fishing a stock to a fixed group of fishers;
- recognizing that fishers obtain more from their profession than money.

Moreover, in addition to prescribing systems that work better (anthropologists have an equal claim to be able to do that), theory allows evolutionary scientists to understand why such factors work and how each one may be improved in specific circumstances. A feeling of having a direct stake in the fish resource (e.g. through a property right), a feeling of being involved in making decisions about a resource with which the fisher has a long-term engagement, and a social structure for the fleet that encourages reciprocal altruism, should lead to greater cooperation with regulations. All these features derive from the evolutionary basis of human behaviour. The study of fisheries can profit from analytical tools developed in behavioural ecology that focus on conflict and cooperation among members of social groups. Examples of such concepts are the prisoner's dilemma (Clark, 1981; Hart, Chapter 17, this volume), game-theory models of animal contests (Maynard Smith, 1982) and optimization theory (Alexander, 1996).

In conclusion, this chapter suggests that using evolutionary theory about behaviour can help to design fishery management systems that have the consent of the fishers, encourage cooperation among them and between fishers and management, and reduce conflict. We contend that the insights and methods of behavioural ecology have as much to say as do the social sciences in a reinvented fisheries management.

REFERENCES

Alexander, R. McNeill (1996) *Optima for Animals*, 2nd edn, Princeton Univ. Press, Princeton, NJ.

Anderson, L.G. (1973) Optimum economic yield of a fishery given a variable price of output. *J. Fish. Res. Board Can.* **30**, 509–518.

Axelrod, R. (1984) *The Evolution of Cooperation*, Basic Books, New York, 241 pp.

Beverton, R.J.H. and Holt, S.J. (1957) *On the Dynamics of Exploited Fish Populations*. Min. Agr. Fish. Food, London, Fish. Invest. Ser. 2(19), 533 pp.

Borgerhoff Mulder, M. (1991) Human behavioural ecology, in *Behavioural ecology. An evolutionary approach* (eds J.R. Krebs and N.B. Davies), Blackwell Scientific Publications, Oxford, pp. 69–98.

Bronowski, J. (1973) *The Ascent of Man*, BBC Books, London, 448 pp.

Byrne, R. and Whiten, A. (1988) *Machiavellian Intelligence. Social Expertise and the Evolution of Intellect in Monkeys, Apes and Humans*. Oxford Science Publications, Oxford, 413 pp..

Clark, C.W. (1981) Bioeconomics of the ocean. *BioScience* **31**, 231–237.

Darwin, C.R. (1872) *The Expression of the Emotions in Man and Animals*, John Murray, London.

Dawkins, M.S. (1986) *Unravelling Animal Behaviour*. Longman, Harlow, 159 pp.

Dawkins, R. (1976) *The Selfish Gene*, Oxford Univ. Press, Oxford, 224 pp.

Dawkins, R. (1982) *The Extended Phenotype: the Gene as the Unit of Selection*. W.H. Freeman, Oxford, 307 pp.

Durham, W.H. (1981) Overview: optimal foraging analysis in human ecology, in *Hunter–gatherer Foraging Strategies. Ethnographic and Archaeological Analyses* (eds B. Winterhalder and E.A. Smith) Chicago Univ. Press, Chicago, pp. 218–231.

Fowle, S. (1993) *Fish for the Future. A Citizen's Guide to Federal Marine Fisheries Management*, Center for Marine Conservation, Washington, DC, 142 pp.
Gordon, H.S. (1954) The economic theory of a common-property resource; the fishery. *J. Polit. Econ.* **62**, 124–142.
Grafton, R.Q. (1996) Individual transferable quotas: theory and practice. *Rev. Fish Biol. Fish.* **6**, 5–20.
Haltiwanger, J. and Waldman, M. (1993) The role of altruism in economic interaction. *J. Econ. Behav. Organ.* **21**, 1–15.
Hamilton, W.D. (1964a) The genetical evolution of social behaviour, I. *J. Theoret. Biol.* **7**, 1–16.
Hamilton, W.D. (1964b) The genetical evolution of social behaviour, II. *J. Theoret. Biol.* **7**, 17–52.
Hilborn, R. (1985) Fleet dynamics and individual variation: why some fishers catch more than others. *Can. J. Fish. Aquat. Sci.* **42**, 2–13.
Huntingford, F.A. and Turner, A.K. (1987) *Animal Conflict*, Chapman & Hall, London, 448 pp.
Kitcher, P. (1985) *Vaulting Ambition: Sociobiology and the Quest for Human Nature*. MIT Press, Cambridge, MA, 456 pp.
Krebs, J.R. and Davies, N.B. (1993) *An Introduction to Behavioural Ecology*, 3rd edn. Blackwell Scientific Publications, Oxford, 420 pp.
Maynard Smith, J. (1982) *Evolution and the Theory of Games*. Cambridge Univ. Press, Cambridge, 224 pp.
McCay, B. (1978) Systems ecology, people ecology and the anthropology of fishing communities. *Human Ecol.* **6**, 397–422.
Morris, D. (1967) *The Naked Ape. A Zoologist's Study of the Human Animal*, Jonathan Cape, London, 252 pp.
Munro, G. and Pitcher, T.J. (eds) (1996) Individual Transferable Quotas in Fisheries. Special Issue. *Rev. Fish Biol. Fish.* **6**(1).
Parker, G.A. and Sutherland, W.J. (1986) Ideal free distributions when individuals differ in competitive ability: phenotype-limited ideal free models. *Anim. Behav.* **34**, 1222–1242.
Parker, G.A., Baker, R.R. and Smith, V.G.F. (1972) The origin and evolution of genetic dimorphism and the male–female phenomenon. *J. Theoret. Biol.* **36**, 529–553.
Rose, S., Kamin, L.J. and Lewontin, R.C. (1984) *Not in Our Genes*, Penguin, London.
Ruddle, K. (1996) Traditional management of reef fishing, in *Reef Fisheries* (eds N.V.C. Polunin and C.M. Roberts), Chapman & Hall, London, pp. 315–335.
Shreeve, J. (1995) *The Neandertal Enigma*. Morrow, NY, 309 pp.
Trivers, R.L. (1971) The evolution of reciprocal altruism. *Q. Rev. Biol.* **46**, 35–57.
Trivers, R.L. (1985) *Social Evolution*, Benjamin Cummings, Menlo Park, CA, 462 pp.
de Waal, F.B.M. (1982) *Chimpanzee Politics. Power and Sex among Apes*. Jonathan Cape, London, 223 pp.
de Waal, F.B.M. (1996) *Good Natured: The Origins of Right and Wrong in Humans and Other Animals*, Harvard Univ. Press, Cambridge, MA, 296 pp.
Wilen, J.E. (1979) Fisher behaviour and the design of efficient fisheries regulation programs. *J. Fish. Res. Bd Can.* **36**, 855–858.
Williams, G.C. (1966) *Adaptation and Natural Selection. A Critique of Some Current Evolutionary Thought*, Princeton Univ. Press, Princeton, 307 pp.
Winterhalder, B. and Smith, E.A. (eds) (1981) *Hunter–gatherer Foraging Strategies. Ethnographic and Archaeological Analyses*, Chicago Univ. Press, Chicago, 268 pp.

Wright, R. (1994). *The Moral Animal. Why We Are The Way We Are: The New Science of Evolutionary Psychology*. Vintage Books, New York, 466 pp.

Wynne-Edwards, V.C. (1962) *Animal Dispersion in Relation to Social Behaviour*, Oliver & Boyd, Edinburgh and London, 653 pp.

Chapter seventeen

Enlarging the shadow of the future: avoiding conflict and conserving fish

Paul J.B. Hart

In all of man's written record there has been a preoccupation with conflict of interest; possibly only the topics of god, love and inner struggle have received comparable attention.

Luce and Raiffa (1957)

ABSTRACT

Conflict of interest among fishers characterizes most fisheries and leads to competition for the resource. This puts a premium on present actions and discounts the future. As a result, most common-property fisheries inexorably move toward overexploitation. It is proposed here that the iterated prisoner's dilemma (PD) captures the essence of the conflict between present actions and the need for conservation. The iterated PD is outlined and the 'tit-for-tat' (TFT) strategy is explained. This proposes that players should cooperate on the first move and then do whatever the other player did on the previous move. It has been shown in computer tournaments that TFT is most often the highest score in the iterated PD, even though it does not win every time. To illustrate how the iterated PD may serve as a model for fisheries, it is used to interpret the interactions of crabbers,

Reinventing Fisheries Management. Edited by Tony J. Pitcher, Paul J.B. Hart and Daniel Pauly. Published in 1998 by Kluwer Academic Publishers, London. ISBN 0 412 83410 3.

trawlers and scallopers on the inshore trawling grounds of south Devon, UK. Mobile and fixed gears are separated in the inshore ground by voluntary partitioning arrangements called the Devon management system. Certain areas are reserved for crabbing only; the crabbers tend to live in small communities, and are related to one another. Small trawlers are also from local communities, but larger beam trawlers and scallopers, although based in south Devon ports, fish all around the UK. It is proposed that these differences in social and genetic structure predispose crabbers and small trawlers to respect the regulations of the partitioning agreement, whilst the larger vessels are more likely to break the agreement when it suits them.

17.1 INTRODUCTION

Conflict between fishers seeking different but overlapping resources is a recurrent problem for fishers and fisheries managers. Conflicts can be particularly acute when one fishery uses static gear that can be damaged by the gear used by fishers catching other species. An example is the static-gear crab fishery and the mobile-gear fishery for demersal fish and scallops in inshore waters off south Devon, UK. Crab fishers leave their gear on the bottom for long periods, making it vulnerable to destruction by trawls. Eventually the interference became a serious problem and a system was required to partition the fishing grounds to minimize contact between the two types of gear.

The closure of areas could lead to conflict between fishers using different gear types. Much of the conflict between fishers is driven by competition to obtain the resource before someone else does, which leads to the future being heavily discounted. In areas permanently closed to trawling, stock biomass may be expected to be more abundant than in frequently trawled areas (Pitcher and Pauly, Chapter 24, this volume). Consequently, fishers will be tempted to enter areas closed to trawling to obtain a higher catch rate, which will tend to have a destabilizing effect on management agreements involving closed or reserved areas.

Despite this, the partitioning system off Devon has remained stable for nearly 20 years, so there must be forces at work that make cooperation worthwhile among fishers. Understanding these forces is essential if new management regimes are to be devised (Jentoft, Chapter 12, this volume). I argue the case for using game theory as a basis for understanding and managing conflicts of interest in fisheries. In particular the prisoner's dilemma (PD) is described and discussed. Then, the Devon partitioning system is analysed in terms of the prisoner's dilemma.

17.2 GAME THEORY AND THE PRISONER'S DILEMMA

Conflicts of interest are best handled by game theory, developed first by von Neumann and Morgenstern (1944). Game theory deals with interactions between actors where an individual's pay-off depends on the action chosen by the other player. Any game theoretical approach assumes that the two or more players involved in an interaction will be able to choose from the same set of strategies and that the choice will be made to maximize individual pay-offs. The optimal solution to a game is the strategy that gives the highest and equal pay-off to all players and that cannot be bettered by any other alternative (Maynard Smith, 1982).

A fishery is a good candidate for the application of game theory in that the catch taken by one fisher is likely to be a function of the catches taken by others. Fisheries management measures attempt to regulate today's catches so that future catches can be assured. This poses a dilemma for the fisher. If individual fishers agree to limit their catch today in the expectation of continued future catches, they can never be sure that the catch they have just given up will not be immediately snapped up by competing fishers. In this sense, the present is given greater weight than the future. Managers need to find ways of making the needs of future fishers impinge on the day-to-day decisions of current fishers.

Within game theory, the prisoner's dilemma (PD) model of conflict and cooperation (Luce and Raiffa, 1957; Axelrod, 1984; Axelrod and Dion, 1988) captures most accurately this tension between the need to cooperate and the temptation to grab today's catch before someone else does. The suitability of the PD was recognized by Clark (1981), who wrote:

> The fundamental problem in common-property resource management is how to devise institutions that allow for decentralised decision making while at the same time circumventing the 'prisoner's dilemma', which if not resolved almost inevitably leads to resource depletion, economic impoverishment, or both.

The only surprising thing about this quote is that it did not lead to the application of the PD to fishery problems. Part of the reason might have been that when written, there was no general understanding of what constituted a 'best' strategy to play if a player was in a PD. It is only since the work of Axelrod (1984) that this knowledge has become available. The main points of the PD will now be described and then applied to understanding conflicts of interests in fisheries.

The prisoner's dilemma

A conflict of interest very often means that the two (or more) contestants in an interaction will not cooperate to achieve some common goal. The parties in a particular fishery each have different goals and there is little incentive for them to cooperate (Hart and Pitcher, Chapter 16, this volume). If fishers could be persuaded to cooperate with management agencies and with each other, then current catches could be cut back so as to ensure catches in the future. Under what conditions might cooperation develop? The PD is a theoretical construct that allows the costs and benefits of interactions to be evaluated and conditions to be determined for the development of cooperation. The name derives from the story that was used by Tucker (see Poundstone, 1992, p. 117) to set the model in a context that non-specialists could relate to. Here, only results from the theory relevant to this contribution are given; for more detail see Luce and Raiffa (1957), Maynard Smith (1982), Axelrod (1984), Lindley (1985), Rapoport (1985) and Axelrod and D'Ambrosio (1994).

Players of the game have two options: to cooperate with each other or to defect and look after their own interests. The cost or benefit to each player of the two options can be displayed in a pay-off matrix (Table 17.1). In the context of a fishery, the costs and benefits could be evaluated most simply in terms of income and expenditure. More generally, game theory models assume that players are trying to maximize their utility, a measure which summarizes their preferences for particular options. Pre-

Table 17.1 The prisoner's dilemma pay-off matrix. Scores are given for both players, subscript 1 for player 1 and subscript 2 for player 2. R_1, R_2 denote Reward for mutual cooperation; S_1, S_2 represent sucker's pay-off; T_1, T_2 are temptation to defect; and P_1, P_2 represent punishment for mutual defection. For this to be a dilemma, then $T > R > P \geqslant S$ and $R > (S + T)/2$ must be true (see text for further explanation)

	Player 2	
	Cooperate	Defect
Player 1		
Cooperate	$R_1 = 3$	$S_1 = 0$
	$R_2 = 3$	$T_2 = 5$
Defect	$T_1 = 5$	$P_1 = 1$
	$S_2 = 0$	$P_2 = 1$

ference is not always expressible only in terms of money and it is probably true of many fishers, particularly small-scale operators, that they fish for the lifestyle that fishing provides as well as for the income. This point of view is developed more fully by Hart and Pitcher (Chapter 16, this volume). When it is applied to animal interactions, biologists have assumed that contestants make choices so as to maximize their Darwinian fitness, usually expressed as number of offspring surviving to reproductive age. It could be argued that humans use rules that are still based on this index even though we think we are governed by costs and benefits of a more monetary nature.

In the PD, each player will be tempted to go for the option (defect) that maximizes the lowest score he or she could obtain (Table 17.1). This is a safe strategy. If both adopt this approach they will both play 'defect' and end up with a score of 1. Had they had greater faith in each other, they could have played 'cooperate' and achieved a score of 3. This is the dilemma.

There are two settings for the PD, each having its own outcome. If the contestants in a PD are to meet for only a fixed number of interactions, then the optimal strategy is always to defect (Axelrod, 1984). This outcome is not promising for fisheries management applications. However, in many local fisheries, the individuals involved always fish in a certain area and are likely to meet the same group of fishers repeatedly. The situation where such interactions occur over an indefinite period has been labelled an iterated PD and was studied in detail by Axelrod (1984). In an effort to determine which was the best strategy to play in an iterated PD, Axelrod (1984) used a computer tournament. Computer experts and game theorists were asked to write a computer program that would implement a strategy for playing the iterated PD. Each strategy was then played against every other one. The innovation introduced into his analysis by Axelrod was the variable w ($0 \leqslant w \leqslant 1$), where w is the probability with which contestants will meet again, providing a measure of how important the future is relative to the present. If the chance of meeting again the contestant with whom you are currently interacting is low then the best strategy will not necessarily be the same as it would be if you were very likely to meet again. As w approaches zero, the stable solution becomes defection for both. When w is close to 1, then a strategy called 'tit-for-tat' (TFT) turned out to be the one that most often won when played against many other strategies. TFT takes the form, 'cooperate on the first move and then do what the other player did on his/her previous move'. Axelrod (1984) shows that this strategy did well because it is:

1. a 'nice' strategy in that it is never the first to defect;
2. a strategy that is quick to punish uncooperative behaviour; and
3. a strategy that has a short memory, and thus forgives quickly.

17.3 APPLYING THE ITERATED PRISONER'S DILEMMA TO FISHERIES

Long-term interactions are possible only in populations where individuals are likely to meet again, to be able to recognize each other when they do and to remember what happened last time they met (Axelrod, 1984). These objectives can be achieved by acting in three ways:

1. make the future more important relative to the present; this is encapsulated by Axelrod (1984) in the evocative phrase 'enlarging the shadow of the future';
2. change the pay-offs to the players in the four possible outcomes of each move;
3. teach the players values, facts and skills that will promote cooperation.

I will analyse the Devon fishery to illustrate how the iterated PD could be applied to fisheries. However, before this analysis can be presented, it is necessary to give a brief outline of the fishery.

The structure of the south Devon inshore fishery

The fishery takes place off the south coast of west Devon, UK (Fig. 17.1). There are four types of vessel that use the area within the six mile territorial limit, between Brixham in the east and Plymouth in the west (Fig. 17.1); 'potters', small trawlers, 'beamers' and 'scallopers'. A key factor in understanding how these groups contribute to the fishery is that the smaller vessels are operated by people from closely knit communities with long histories in the area.

The crab, *Cancer pagurus*, fishery has been operated by local people for centuries (Firestone, 1972, 1976; Dickinson, 1987). Up until the Second World War, crews worked from small coastal villages such as Beesands, Hallsands and Torcross (Fig. 17.1) and many crew members were related (Firestone, 1972). Present-day boats are larger and operated from larger harbours such as Dartmouth and Salcombe, although crews still live in their original communities. There are approximately 36 crabbers, mostly vessels less than 15 m long but with some as small as 5 m. About 40% of the crab boats are owned and operated by people who have inherited the fishery from their family, most often their fathers (Firestone, 1972).

There are about 24–36 small trawlers. There is less information available to allow an estimate of what proportion are owned by fishers following in their father's footsteps. This information needs to be obtained. Most of the small trawlers are based in the same ports as the potters but some come from either Brixham or Plymouth.

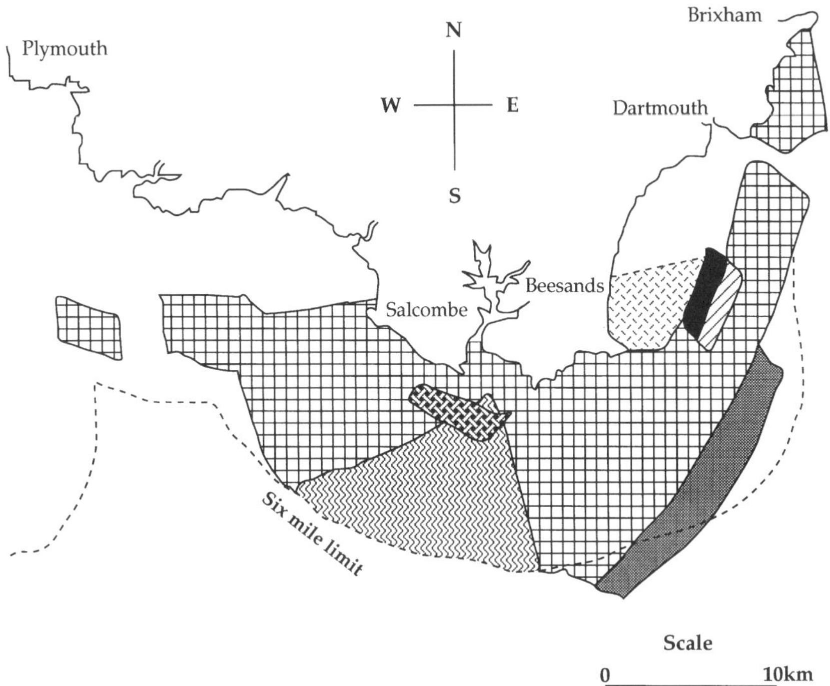

Fig. 17.1 A map of the English territorial sea between Plymouth and Brixham showing the agreed partitioning system of the fishing grounds between fishers using fixed and mobile gear. The six mile territorial limit is shown by the dashed line. ▦ Potting only areas; ▨ no potting or trawling; other shaded areas open to trawling at certain times of the year.

Beamers are larger vessels of about 15 m, and many are based in either Brixham or Plymouth. Vessels of this size and with beams less than 9 m can fish only outside three miles offshore. Beamers are not tied to the local area and can often be found anywhere around the British Isles. They are a seasonal component of the Devon fishery although many of the crew have their homes in the area.

The small group of scallop dredgers, or scallopers, visit the area sporadically. These vessels tow gear that can be destructive to the benthic community (Kaiser, 1996; Kaiser and Spencer, 1996). Like the beamers, scallopers are around 15 m in length and can operate over a wide area all around the UK. Some of these scallopers are large vessels operated by a hired crew for the owner. The crew's main task is to make money for the owner, or lose their job, and this leads to less than the usual regard for voluntary agreements.

In 1993 the south Devon crab fishery contributed 31% of the crab catch for England and Wales (MAFF, 1995). Fishing performance of the crab fleet off Devon is seasonal. The main period for crabs is between April and November, with greatest catches in October and November (Edwards, 1979), but 90% of the crab fleet keep their pots in the water for the whole year. Daily catches can be between 200 kg (winter) and 1250 kg (high season) per boat operating about 500–600 pots. Income for a day's fishing can be from £250 in winter to £1450 at the height of the season.

A day's fishing on a crab boat usually starts at about 0400–0430 h and finishes at 1530–1630 h. The physical labour exerted by the three crew is demanding. They work continuously, lifting about 22 tonnes of pots each day. One pot costs about £50 to replace and ship's running costs are approximately £200 a day at sea.

Outline of the management regime

To reduce conflict between fixed and mobile gear, the area between Plymouth and Brixham was partitioned in 1978 (Fig. 17.1). This is a voluntary agreement between the South Devon and Channel Shellfishermen Ltd, the Trawler Owners Association, Devon Sea Fisheries Committee and the South West Fish Producers Association. All trawling is prohibited in areas reserved for the setting of crab pots. Within these areas, a fisher's gear is set in a spot which becomes that fisher's 'territory'. 'Territories' are not necessarily contiguous. Other areas are open to trawlers at certain times of the year (Fig. 17.1). All local fishers' associations are party to the agreement, brokered by the Ministry of Agriculture, Fisheries and Food (MAFF) and the Devon Sea Fisheries Committee, and which I call the Devon management system (DMS). Because the partitioned area is entirely within British 10 km territorial waters, only British boats can use it. Reports from various sources indicate that the most frequent cheaters on the agreement are beam trawlers and scallop dredgers, while crabbers and small inshore trawlers generally comply with the DMS.

Interpreting the DMS in terms of an iterated prisoner's dilemma

Enlarging the shadow of the future can be achieved by increasing the frequency of interactions between the players. For crabbers and small trawlers off south Devon, this has already been achieved in that individuals from both groups meet regularly at sea, sometimes every day, and they may also meet socially when not at sea. In addition, the family ties between many crab fishers will promote regular meeting and cooperation. Scallopers and beam trawlers are different. These are more likely to enter the area occasionally, take a catch, leave and not return for months.

Crews of scallop boats, being employees, are particularly subject to discounting the future. Imminent threat of job loss must be a powerful incentive to emphasize present gains to the detriment of those that might be achieved in the future.

Because the DMS is a voluntary arrangement, the only method by which the utility of an option can be changed is through the threat of damage to boats or individuals. Large vessels entering the area can ignore such a threat for two reasons: (1) the intruder is bigger than resident fishing vessels, making any threat made on the fishing grounds harder to apply; and (2) the crew are harder to find when onshore, as they live in larger communities and are not so well known for their trade. An alternative way to manipulate the costs is to make the DMS legally binding so that fines and confiscations could be imposed by the authorities. The Devon Sea Fisheries Committee is currently elaborating a by-law to put the agreement on a legal basis, a move forced by the increase in damaging incursions by scallop dredgers and beam trawlers. A legal system might be hard to enforce. Defecting boats will know through contacts where the fisheries patrol vessel is, and this may not be able to maintain a sufficiently high rate of detection to make a sufficient threat to the defector (Hart, 1997).

Current research is working to identify the strategy employed by the various fishers in their interactions. The first stage has been to identify the factors that might lead groups to cooperate or defect and these are listed in Table 17.2. These show that crabbers and small trawlers show several characteristics identified by Axelrod (1984) as being essential if cooperation is to evolve. Apart from the features listed in Table 17.2 he suggests that players should behave in the following way for best results:

1. do not be envious;
2. do not be the first to defect;
3. reciprocate both cooperation and defection;
4. do not be too clever.

Possibly, the centuries of inshore fishing by the inhabitants of the Start Bay area have moved them to a position where they are already using the most effective strategy. Current research is modelling the PD to determine which features listed in Table 17.2 are essential for cooperation to become established and to be maintained.

The DMS as a model for fisheries management regimes

The DMS is a rare example of a conflict of interest in a fishery solved by a voluntary agreement that has worked for many years without a legal basis. For this reason it is valuable to try to understand the factors that

Table 17.2 A list of characteristics possessed by fishers using four different gear types. Each is grouped into those that might promote cooperation and those that might lead to defection. Trawlers have been divided into two groups to allow for the possibility that large beam trawlers have different interests. A question mark indicates doubt about the validity of the assessment

Gear type	Cooperation	Defection
Potting for crabs	Close kinship, live in close communities; few alternatives to areas worked	Different catch rates among 'territories'
Inshore trawlers	Close kinships? Close communities and contact with trawlers and crabbers	Catches higher in areas closed to trawlers
Larger beam trawlers	Damage to trawls from crab pots and loss of time?	Infrequently fish inshore; not much personal contact with small trawlers and crabbers; high catch areas closed to gear
Scallop dredger	Damage to dredge from crab pots, or at least, loss of time through entanglement?	Same as for larger beam trawlers; also, whole crew under threat of losing their job?

have led to success. It might then be possible to adapt the scheme for other fisheries that have properties conducive to cooperation. It might be possible also to devise management regimes that take account of the human behaviour and social structure necessary for them to work. My conjecture is that the DMS can be understood by using PD theory, which makes it possible to identify characteristics of the DMS that have led to success. The current task is to develop models that include the consequences for cooperation of the cost/benefit structure. Although the analysis is still at an early stage, it is possible to outline some general principles that will lie behind good management regimes.

Most modern commercial fisheries are composed of wide-ranging vessels with crews that have no bond with each other. National allegiances provide some grouping, but ultimately a British fisher is just as likely to cheat on a system devised in Britain as is a French fisher. The main deficiency in most management systems is that the players do not show the necessary characteristics outlined earlier: interacting individuals likely to meet again, recognizing each other when they do and remembering what happened last time they met (Axelrod, 1984).

I propose that the DMS works because it is built on the requirements necessary for successful cooperation. This proposition suggests that successful management regimes ought to restrict the participants in a fishery to a group of people who will interact frequently. The management regime could set the costs, through fines or other penalties, so that fishers did better by cooperating than by cheating. The 'shadow of the future' would be made sufficiently large by ensuring that the people engaged in the fishery all knew each other, met regularly on the fishing grounds and had good access to information on who had cheated. ITQs (individual transferable quotas) go some way towards generating such a system (Grafton, 1996).

As a corollary, those assessing the stocks and advising on quotas need also to be part of the community. The management methods proposed by Holling (1978) and Walters (1986) illustrate the way in which stock assessment, stock management and the fishery can be integrated to form a single community.

The analysis of the DMS shows that it is dangerous to imagine that a fishery can be regulated by assuming that each fisher is a free and rational agent who will eventually see the logic of quotas and closed areas. Fishers can only be made responsible to others if their activities are set into a social context and management regimes devised that recognize the often subconscious factors that determine fishers' decisions. New management measures should be devised with this in mind.

ACKNOWLEDGEMENTS

I acknowledge the assistance of the South Devon and Channel Shellfishermen Ltd, the Devon Sea Fisheries Committee and the fishers in the crab fleet who allowed me and Dr Chris Gilling to obtain first-hand experience of their working lives. I also thank Chris Venmore, Neil Downes, Michel Kaiser and anonymous referees for commenting on an earlier draft of this chapter. All views expressed are mine alone and do not represent the position of any fishery organization.

REFERENCES

Axelrod, R. (1984) *The Evolution of Co-operation*, Basic Books, New York (reprinted by Penguin Books, 1990), 241 pp.

Axelrod, R. and Dion, D. (1988) The further evolution of co-operation. *Science* **242**, 1385–1390.

Axelrod, R. and D'Ambrosio, L. (1994) Annotated bibliography on the evolution of cooperation. http://www.spp.umich.edu/spp/papers/axelrod/

Clark, C.W. (1981) Bioeconomics of the ocean. *BioScience* **31**, 231–237.
Dickinson, M.G. (ed.) (1987) *A Living from the Sea. Devon's Fishing Industry and its Fishers*, Devon Books, Exeter.
Edwards, E. (1979) *The Edible Crab and its Fishery in British Waters*, Fishing News Books Ltd, Farnham.
Firestone, M. (1972) The crew structure of the Start Bay crab fishery, in *North Atlantic Fishers: Anthropological Essays on Modern Fishing* (eds R. Anderson and C. Wadel), (Newfoundland Social and Economic Papers, No. 5), Inst. Social Econ. Res., Memorial Univ. Newfoundland, St John's, pp. 112–132.
Firestone, M. (1976) Crab fishers in south Devon. *The Devon Historian* **12**, 2–5.
Grafton, R.Q. (1996) Individual transferable quotas: theory and practice. *Rev. Fish Biol. Fish.* **6**, 5–20.
Hart, P.J.B. (1997) Controlling illegal fishing in closed areas; the case of mackerel off Norway, in *Proc. 2nd World Fisheries Congress* (ed. D. Hancock), Australian Society for Fish Biology, Collingwood, pp. 347–350.
Holling, C.S. (ed.) (1978) *Adaptive Environmental Assessment and Management*, Wiley, Chichester.
Kaiser, M.J. (1996) Starfish damage as an indicator of trawling intensity. *Mar. Ecol. Progr. Ser.* **134**, 303–307.
Kaiser, M.J. and Spencer, B.E. (1996) The effects of beam-trawl disturbance on infaunal communities in different habitats. *J. Anim. Ecol.* **65**, 348–358.
Lindley, D.V. (1985) *Making Decisions*, 2nd edn, Wiley, London.
Luce, R.D. and Raiffa, H. (1957) *Games and Decisions*, Wiley, New York.
MAFF (1995) *United Kingdom Sea Fisheries Statistics: 1993*, Her Majesty's Stationary Office, London.
Maynard Smith, J. (1982) *Evolution and the Theory of Games*, Cambridge Univ. Press, Cambridge.
von Neumann, J. and Morgenstern, O. (1944) *Theory of Games and Economic Behavior*, Princeton Univ. Press, Princeton.
Poundstone, W. (1992) *Prisoner's Dilemma*, Oxford Univ. Press, Oxford.
Rapoport, A. (1995) Prisoner's Dilemma: reflections and recollections. *Simulation and Gaming* **26**, 489–503.
Walters, C.J. (1986) *Adaptive Management of Renewable Resources*, Macmillan, New York.

Chapter eighteen

A political ecology of fisheries

Thomas R. McGuire

ABSTRACT

One of the central lessons harshly underscored by the collapse of northern cod stocks is that the relation between catch per unit of effort and stock abundance is problematic. It is confounded by local knowledge. But social scientists, with disciplinary training in the collection and interpretation of local knowledge, have made little contribution to this, and other problems in fisheries management. This commentary briefly reviews several contested issues in maritime social science: the skipper effect, fleet dynamics, folk management, scientific management, and adaptations to chaotic systems. It suggests that the debates, and much of the fine-grained empirical work underlying them, evolved in the context of largely academic contests over paradigms such as cultural ecology, political economy, and recently, political ecology.

An emergent approach seeks to combine prior concerns with individual and household adaptations to ecosystems and environments and with local knowledge (the domain of cultural ecology) with the problematic of political economy. These are the interrelations of production, class formation, the penetration of capital, the loss of local power and autonomy, all of which are essential antidotes to the limitations of cultural ecology. As such, the developing paradigm offers the potential for a close interface with the natural sciences, both theoretical and applied. However, political ecology, as it is emerging, is truer to its political economy roots than to its

Reinventing Fisheries Management. Edited by Tony J. Pitcher, Paul J.B. Hart and Daniel Pauly. Published in 1998 by Kluwer Academic Publishers, London. ISBN 0 412 83410 3.

cultural ecology ones: political ecologists know little about ecology. I suggest that the programme needs to be reinvented, to address the information requirements for effective fisheries.

18.1 INTRODUCTION

A cursory reading of the postmortems on the collapse of the northern cod *Gadus morhua*, in Canada in 1991 suggests obvious realms for collaboration between the natural and the social sciences. There was a strong reluctance to listen to 'local knowledge', particularly the protestations of inshore fishers that the stock was in trouble. And there was fundamental faith in assessment models which read – misread, most now realize – stock abundance from commercial catch data (Steele *et al.*, 1992; Finlayson, 1994; Hutchings and Myers, 1994; Dalhousie Law Journal, 1995; Maguire *et al.*, 1995; Walters and Maguire, 1996). Social scientists perhaps could have done more to articulate this local knowledge, as researchers are now doing (Neis, 1992). And social scientists could have observed and recorded those changes in technology and skill that apparently scuttled the stock (Acheson, 1988).

But it is perhaps asking too much of social scientists to assist in the shoring up of the stock assessment models and methods which many of us distrust on principle (McGuire, 1991). In addition, the local knowledge which we have been collecting for decades has seldom found its way into effective resource management (McEvoy, 1986; Davis *et al.*, 1991; Dyer, 1994; Nance *et al.*, 1994).

Nevertheless, there is an emergent field in the social sciences where natural and social interests overlap. In its simplest definition, this political ecology 'combines the concerns of ecology and a broadly defined political economy' (Blaikie and Brookfield, 1987, p. 17; see also Hjort, 1982; Atkinson, 1991; Bryant, 1992; Hershkovitz, 1993). In its most ambitious guise, it seeks to understand

> ...the social and economic processes that underlie environmental degradation and underdevelopment ... dealing explicitly with the fact that both processes occur in complex societies joined to one another by a global economy and ecology.
>
> Painter and Durham (1995)

In brief, political ecology hopes to connect ecological processes to the world economy, through the intervening links of individual, household, community, social class, region, and the state.

As an emergent field, however, it lacks definition, balance, and method. Straining to free itself from its two intellectual predecessors, cultural ecology (McCay, 1979, 1981; McCay and Acheson, 1987; Rappaport, 1979) and political economy (Wolf, 1972, 1982; McNally, 1981; Roseberry, 1988, 1989; Schneider and Rapp, 1995), political ecology, I fear, is simply restating the proposition of political economy: global economies drive local ecologies.

Political ecology's clearest expression to date is in the analysis of environmental destruction in tropical rain forests (Painter and Durham, 1995). The driving force for degradation is said to be market involvement. Only those rare communities based on purely subsistence economies are spared. As anthropologists Painter and Durham (1995) summarize,

> ...export production contributes to environmental degradation by aiding and abetting internal processes of capital accumulation whereby the wealthy appropriate natural resources without being accountable for the social and environmental consequences of their actions. Such actions displace large numbers of poor people who, denied access to the means of earning a living within those areas of the economy where accumulation is occurring, have little recourse but to behave in a predatory fashion toward an increasingly impoverished, unappropriated resource base

(Painter, 1995; see also Schmink and Wood, (1987). While there is substantial truth to this, it requires little ecological analysis. It is pure political economy. Contrast this to a relevant example, Hilborn and Walters' (1992, p. 189) enumeration of data requirements (after they have addressed such political economic factors as fleet investment and market demand) for reading stock abundance from catch per unit effort (CPUE), and thus for averting a collapse: (1) operational characteristics of fishing gear; (2) determinants of total fishing effort; (3) determinants of areas fished, such as quality of information and desirability of specific locations; (4) the 'concentration profiles' of localized stocks; and (5) the dynamics of movement and the diffusive behaviour of stocks (Wilen, 1979; Hilborn, 1985; Hilborn and Ledbetter, 1985).

Ecological factors, such as the behaviour of stocks under exploitation, that might have been addressed by the old cultural ecologists, appear to be of little import to the political economists. They are, however, critical factors in a new maritime political ecology. I want to briefly sketch my view about how and where this field should go.

18.2 PRINCIPLES OF POLITICAL ECOLOGY

To be true to its roots in cultural ecology and political economy, a new political ecology should adhere to the following six principles, the first two of which are controlling. It should pay simultaneous attention to social and ecological processes and it should determine empirically the interconnections between processes. Moreover, it should be actor-centred; it should be wary of the folk models of fishers and their managers; it should be sceptical of the panacea of 'folk management'; and it should nonetheless concern itself fundamentally with social justice in the distribution of resources.

1. Its defining characteristic is the simultaneous attention to scale and context in the social and the physical realms

Netting (1990, 1993), a prominent cultural ecologist, chose to study human–environment relations in autonomous, self-sufficient 'little communities'. He made few apologies for this:

> Ecological anthropologists with their commitment to gathering a wide range of non-cultural data and organizing these variables into systems models often focus on small units of interaction for both practical and theoretical reasons.

Theory drove practice in this case: the appropriation of a guiding ecosystem model from ecology and the insertion of human activity into that model. As a result, Netting (1990) noted that incorporating economic activities such as an exchange of goods into a systems model was hard to do adequately. After an initial period of excitement, this 'systems ecology' was attacked, both within and outside cultural ecology. The internal critique is well summarized by McCay (1978): (1) the presumption of system equilibrium, and the assumption, seldom effectively demonstrated, that human activities function to restore systems to equilibrium; (2) the teleological error that such functions therefore explain the existence of those human activities; and (3) the error in forgetting that ecosystems are simply analytic units, the boundaries of which seldom encompass the forces and constraints operating on human behaviour.

The political economists drove home this final critique. It is now more valid than ever. Their set of forces and constraints stemmed from the state and the world capitalist system. The heart of the issue, now, is how to link the analysis of ecosystem behaviour to the analysis of capital, class and power.

2. Causation across scale, context, the social and the physical, should be empirically determined

Vayda (1983, 1988), a cultural ecologist turned political ecologist, casts the linkage problem as one of 'progressive contextualization'. Two political economists turned political ecologists, Schmink and Wood (1987), review Vayda's approach and reject his empiricism. Vayda refuses to use a priori frameworks such as 'the concepts and relationships comprising the political economy perspective' (Schmink and Wood, 1987) to drive the contextualizing analysis.

My sympathies lie with Vayda, but my own strategic predilection is to start the process of setting an activity in context by beginning with key commodities such as cod, wheat, fur, sugar, shrimp and then determine which human activities affect their supply (Innis, 1940; Mintz, 1985).

For example, in the Gulf of California in north-west Mexico, the biggest game in town is shrimp *(Penaeus* spp.). To attempt to understand an apparent stock collapse in 1991 it is necessary to study the context of the fishery. This includes among other things, habitat conditions, growth rates, migratory patterns and interannual population variation of two commercial species (*P. stylirostris* and *P. californiensis*), individual and household production and consumption decisions, intra- and intersectional conflicts among the inshore and offshore fisheries, the behaviour of regulators, enforcers, avoiders and black marketeers, the science underlying management, and, of course, the 'global economy', or at least its manifestation in the international shrimp trade.

But the global economy, from which the political economists would drive the analysis, is only tenuously implicated in the collapse of shrimp stocks in the Gulf of California. Nor can the ecosystem consequences of the diminished freshwater flow from a river dammed decades ago shoulder the blame, although this was a popular target for local fishers and perhaps would have been highlighted by cultural ecologists. The primary culprit was overcapitalization, both inshore and offshore, fostered by the best available scientific advice that shrimp show no stock–recruitment relation (McGuire, 1991).

3. Political ecology should be actor-centred

In her critique of anthropological theory since the 1960s, Ortner (1984) observes of political economy (and its contentious cousin, 'structural Marxism') that they both

> ... assume, together with earlier anthropologies, that human action and historical process are almost entirely structurally or systemically determined. Whether it be the hidden hand of structure or the juggernaut of capitalism that is seen as the agent of things, it is certainly not in any central way real people doing real things.

In their critique of systems ecology, McCay (1978), Vayda (1983), Bennett (1969) and others trumpeted a people ecology in which individuals and groups make decisions (Hart and Pitcher, Chapter 16, this volume). These decisions result in adaptive strategies (Bennett, 1969). Such an orientation has now re-emerged in anthropology, but the model of the actor is rather dissimilar to that held by most microeconomists (for a 'bounded rationality' view, which I adhere to, see Simon, 1979). Choices and actions are shaped and constrained not so much by limited information but by asymmetrical and dominated social relations: the Marxian influence of the earlier political economy (Ortner, 1984).

Coming from a different tradition, Andersen (1972) offers an example of the insights an actor-orientated approach provides. A survey of patterns of deception and information-management among Newfoundland trawler skippers of company fleets in the late 1960s, showed that deception of one skipper by another exaggerated the disadvantages experienced by skippers with inferior gear. To skippers such behaviour is, in Bennett's (1969) terms, an adaptive strategy. It arises largely from the legal status of skippers and crews as 'co-adventurers' in vertically integrated fishing enterprises, 'wherein the major part of each individual's earnings comes from the catch of his vessel alone', not from overall company fleet performance (Andersen, 1972).

It is the cumulative results of such action, structured and constrained by social relations, that ultimately comprise fleet dynamics. More often than not, the aggregate behaviour of a fleet will be difficult to predict from formal models predicated on autonomous, all-knowing, rational skippers.

4. Political ecology should not assume that models mirror reality

Building upon the pioneering work of Barth (1966) and Andersen (Andersen and Wadel, 1972), anthropologists scrutinized the skipper effect in the 1980s. The statistical resolution to the question of how much variance in catch can be attributed to the skills and luck of a skipper or to equipment, vessel fishing power, abundance, and other factors went ultimately unresolved. The debate, though, was instructive on another matter: folk models, local belief systems regarding the nature of resources and their use, do not necessarily reflect reality.

In an exhaustive series of papers (Pálsson and Durrenberger, 1982, 1983, 1984; Durrenberger and Pálsson, 1983, 1985, 1986, 1987; Pálsson, 1988, 1989, 1991; Durrenberger, 1993), Pálsson and Durrenberger suggest that the 'folk model' of a strong skipper effect in Icelandic fisheries is a false ideology, imbedded in a particular mode of production: competitive (pre-quota) commercial fishing (for contrary views, see Gatewood, 1983, 1984a,b; McNabb, 1985; Thorlindsson, 1988). The peculiar Icelandic twist to the model, that a skipper's success lies in playing hunches and interpreting the dreams of others, serves in their argument to remove individual responsibility in cases of fishing failure.

Pálsson and Durrenberger (1990) draw obvious management implications from their analysis: effort reduction through the removal of 'good skippers' rather than successful boats may not achieve the intended results. But they make a larger point as well. Science itself may be imbedded in social dynamics which could influence the role the scientist plays in policy making (Rice, Chapter 14, this volume).

Finlayson (1994) makes a similar point about the assessors of the northern cod; they, like the policymakers, were caught up in the euphoria of Canada's extended jurisdiction declaration, and neither could nor would read signs that the stocks were not rebuilding. He overstates the case, but nonetheless he underscores the point that scientists and policymakers cannot be left out of the analytical contextualization. They, like fishers, are real people doing real things.

5. Political ecology should be sceptical of 'folk management'

The suggestion by Pálsson and Durrenberger that ideas have a social context should be applied to social scientists as well. A Rousseauian thread, always pervasive in anthropology, was bolstered by a systems ecology, the methods of which required small-scale research, and the findings of which typically pointed to equilibrium, homeostasis, and adaptation. This thread reappears in much of political economy, where premarket folks are sacred, those 'penetrated by capital' are profane. It also pervades the dominant and popular social scientific contributions to fisheries management: (1) the calls for 'parametric management' that have come out of the useful intrusion of chaos notions, reinforcing local patterns of when, how, and where to fish rather than imposing numerical quotas (Smith, 1990; Wilson and Kleban, 1992; Gilbertson, 1993; Wilson *et al.*, 1994); (2) the endeavour to study and engineer 'institutions for collective action' (Ostrom, 1990) to resolve common pool problems; and (3) the promotion of folk management (Felt, 1994).

But the literature typically illustrates the values of these suggestions by their absence. For example, a collection of case studies on local management (Dyer and McGoodwin, 1994) documents little but failure. The editors' introduction is an astonishing essay in counterfactual optimism, while Pinkerton's (1994) cautious and insightful summary suggests that folk management has a more limited contribution to make.

Speculating on the potential for co-management on the Great Northern Peninsula of Newfoundland, Sinclair (1990) suggests that co-management breaks down when resources become scarce.

There is substantial merit in recording local knowledge and promoting local management. But when things get tough in the fishery, local management may dissolve and local know-how may allow fishers to harvest the last fish.

6. Nonetheless, political ecology should be fundamentally value-driven toward social justice

Sinclair (1990) argues that we start the analysis of troubled fisheries with state policy. In his case, Canadian policy has systematically favoured the offshore trawler segment to the detriment of both the stocks and the traditional inshore fisheries. To Sinclair (1988), 'policy that deepens social inequalities or that necessarily deprives people of old rights is ... unjust' (see also Sinclair, 1985, 1988; Martin, 1990). The concern for social justice must not be lost in a new political ecology. Yet the issue is not quite so clear as Sinclair suggests. McCay (1993), in an argument against McGoodwin (1990), has written that people on large trawlers are just as much driven by cultural and social forces as are inshore fishers.

In short, social scientists should make value judgements about social justices and injustices. These should not be a priori judgements, based on presumptions about large-scale guilt and small-scale innocence. Such judgements should be applied to policies and practices that distort the distribution of income from natural resources and allow unequal access.

18.3 PRACTICAL IMPLICATIONS

The foregoing principles and polemics for a political ecology have been offered unapologetically in the spirit of empiricism. Cultural ecologists were led astray by presumptions of ecosystem equilibrium and homeostasis. Political economists eschewed such assumptions, but introduced their own about the inevitable and deleterious consequences of the expansion of capitalism. And when things go awry in the fisheries, a policy prescription flows, too easily, out of both of these traditions: folk management. I suggest, rather, that we return to the methodological programme proposed by Vayda (1983): '...to common sense, practical ways of seeing what is happening in the world'.

What is happening with alarming frequency is the decline of the world's fisheries. Political ecology, as an emergent field in the social sciences, will not arrest this process. Yet it may sharpen the analysis. It is concerned with real people doing real things in, and to, the environment. When these things are deleterious, they are of direct concern to the users of that environment and to the managers of those users. Political ecology, as sketched here, cannot and should not offer a prescriptive set of solutions. It can and should, however, seek to understand the problems. More often than not, the policy process is one of inappropriate answers applied to poorly understood questions. This, perhaps, is where a social science, focused on the causes and consequences of resource use, can assist.

ACKNOWLEDGEMENTS

Some of the thoughts expressed here were formulated during a sabbatical leave in the [Arizona] spring of 1995 in Atlantic Canada, Quebec, and Norway. I am deeply grateful to my hosts, colleagues, and friends at institutions thereabouts: Anthony Davis, Yvan Breton, Peter Holm, Dan MacInnes, John Phyne, Svein Jentoft, Knut Mikalsen, and Richard Apostle, among many others.

REFERENCES

Acheson, J.M. (1988) *The Lobster Gangs of Maine*. Univ. Press New England, Hanover, NH, 181 pp.

Andersen, R. (1972) Hunt and deceive: information management in Newfoundland deep-sea trawler fishing, in *North Atlantic Fishermen: Anthropological Essays on Modern Fishing* (Social and Economic Papers no. 5) (eds R. Andersen and C. Wadel), Inst. Social Econ. Res., Memorial Univ. Newfoundland, St John's, pp. 120–140.

Andersen, R. and Wadel, C. (eds) (1972) *North Atlantic Fishermen: Anthropological Essays on Modern Fishing* (Social and Economic Papers no. 5) Inst. Social Econ. Res., Memorial Univ. Newfoundland, St John's, 174 pp.

Atkinson, A. (1991) *Principles of Political Ecology*, Belhaven Press, London, 251 pp.

Barth, F. (1966) Models of social organization. *Royal Anthropol. Inst. Occ. Pap.* No. 23, London, 32 pp.

Bennett, J.W. (1969) *Northern Plainsmen: Adaptive Strategy and Agrarian Life*, Aldine, Chicago, 352 pp.

Blaikie, P., and Brookfield, H. (eds) (1987) *Land Degradation and Society*, Methuen, London, 296 pp.

Bryant, R.L. (1992) Political ecology: an emerging research agenda in Third-World studies. *Pol. Geog.* **11**, 12–36.

Dalhousie Law Journal (1995) After the Collapse (Special Issue). *Dalhousie Law J.* **18**(1), 167 pp.

Davis, A., MacInnes, D., and Jentoft, S. (1991) Social research and public policy formation in the fisheries: a report on a bilateral working seminar. *Marit. Anthropol. Stud.* **4**(2), 104–109.

Durrenberger, E.P. (1993) The skipper effect and folk models of the skipper effect among Mississippi shrimpers. *Human Organiz.* **52**, 194–202.

Durrenberger, E.P. and Pálsson, G. (1983) Riddles of herring and rhetorics of success. *J. Anthropol. Res.* **39**, 323–335.

Durrenberger, E.P. and Pálsson, G. (1985) Reply to Steven L. McNabb. *Am. Ethnol.* **12**, 544–545.

Durrenberger, E.P. and Pálsson, G. (1986) Finding fish: the tactics of Icelandic skippers. *Am. Ethnol.* **13**, 213–229.

Durrenberger, E.P. and Pálsson, G. (1987) Ownership at sea: fishing territories and access to sea resources. *Am. Ethnol.* **14**, 508–522.

Dyer, C.L. (1994) Proaction versus reaction: integrating applied anthropology into fishery management. *Human Organiz.* **53**, 83–88.

Dyer, C.L. and McGoodwin, J.R. (1994) Introduction, in *Folk Management in the World's Fisheries* (eds C.L. Dyer and J.R. McGoodwin), Univ. Press of Colorado, Niwot, CO, pp. 1–16.

Felt, L.F. (1994) Two tales of a fish: the social construction of indigenous knowledge among Atlantic Canadian salmon fishers, in *Folk Management in the World's Fisheries* (eds C.L. Dyer and J.R. McGoodwin), Univ. Press of Colorado, Niwot, CO, pp. 251–286.

Finlayson, A.C. (1994) *Fishing for Truth: A Sociological Analysis of Northern Cod Stock Assessments from 1977–1990* (Social and Economic Papers no. 52), Inst. Social Econ. Res., Memorial Univ. of Newfoundland, St John's, 176 pp.

Gatewood, J.B. (1983) Cooperation, competition, and synergy: information-sharing groups among southeast Alaskan salmon seiners. *Am. Ethnol.* **10**, 350–370.

Gatewood, J.B. (1984a) Deciding where to fish: the skipper's dilemma in southeast Alaskan salmon seining. *Coastal Zone Manage. J.* **10**, 347–368.

Gatewood, J.B. (1984b) Is the 'skipper effect' really a false ideology? *Am. Ethnol.* **11**, 378–379.

Gilbertson, N. (1993) Chaos on the commons: salmon and such. *Marit. Anthropol. Stud.* **6**, 74–91.

Hershkovitz, L. (1993) Political ecology and environmental management in the Loess Plateau, China. *Human Ecol.* **21**, 327–353.

Hilborn, R. (1985) Fleet dynamics and individual variation: why some people catch more fish than others. *Can. J. Fish. Aquat. Sci.* **42**, 2–13.

Hilborn, R. and Ledbetter, M. (1985) Determinants of catching power in the British Columbia salmon purse seine fleet. *Can. J. Fish. Aquat. Sci.* **42**, 51–56.

Hilborn R. and Walters, C.J. (1992) *Quantitative Fisheries Stock Assessment: Choice, Dynamics and Uncertainty*, Chapman & Hall, New York, 570 pp.

Hjort, A. (1982) A critique of 'ecological' models of pastoral land use. *Nomadic Peoples* **10**, 11–27.

Hutchings, J.A., and Myers, R.A. (1994) What can be learned from the collapse of a renewable resource? Atlantic cod, *Gadus morhua*, of Newfoundland and Labrador. *Can. J. Fish. Aquat. Sci.* **51**, 2126–2146.

Innis, H.A. (1940) *The Cod Fisheries: The History of an International Economy*, Univ. Toronto Press, Toronto, [1978 edition] 522 pp.

McCay, B.J. (1978) Systems Ecology, People Ecology, and the Anthropology of Fishing Communities. *Human Ecology* **6**(4), 397–423.

McCay, B.J. (1979) 'Fish is scarce': fisheries modernization on Fogo Island, Newfoundland, in *North Atlantic Maritime Cultures: Anthropological Essays on Changing Adaptations* (ed. R. Andersen), Mouton Publishers, The Hague, pp. 155–188.

McCay, B.J. (1981) Optimal foragers or political actors? Ecological analysis of a New Jersey fishery. *Am. Ethnol.* **8**, 356–382.

McCay, B.J. (1993) Fisherwomen, fisheries policy, and maritime anthropology. *Rev. Anthropol.* **22**, 77–89.

McCay, B.J. and Acheson, J.M. (eds) (1987) *The Question of the Commons: The Culture and Ecology of Communal Resources*, Univ. Arizona Press, Tucson, 439 pp.

McEvoy, A.F. (1986) *The Fisherman's Problem: Ecology and Law in the California Fisheries, 1850–1980*, Cambridge Univ. Press, Cambridge, 368 pp.

McGoodwin, J.R. (1990) *Crisis in the World's Fisheries: People, Problems, and Policies*, Stanford Univ. Press, Stanford, CA, 235 pp.

McGuire, T.R. (1991) Science and the destruction of a shrimp fleet. *Marit. Anthropol. Stud.* **4**, 32–55.

McNabb, S.L. (1985) A final comment on measurement of the 'skipper effect'. *Am. Ethnol.* **12**, 543–544.

McNally, D. (1981) Staple theory as commodity fetishism: Marx, Innis and Canadian political economy. *Stud. Polit. Econ.* **6**, 35–63.

Maguire, J.-J., Neis, B. and Sinclair, P.R. (1995) What are we managing anyway? The need for an interdisciplinary approach to managing fisheries ecosystems. *Dalhousie Law J.* **18**, 141–153.

Martin, C. (1990) Comment on Sinclair's 'Fisheries management and problems of social justice: reflections on northwest Newfoundland'. *Marit. Anthropol. Stud.* **3**(2), 103–110.

Mintz, S.W. (1985) *Sugar and Power: The Place of Sugar in Modern History*, Viking, New York, 274 pp.

Nance, J.M., Garfield, N.H. and Paredes, J.A. (1994) Studying the social impact of the Texas shrimp closure. *Human Organiz.* **53**, 88–92

Neis, B. (1992) Fishers' ecological knowledge and stock assessment in Newfoundland. *Newfoundland Stud.* **8**, 155–178.

Netting, R. McC. (1990) Links and boundaries: reconsidering the Alpine village as ecosystem, in *The Ecosystem Approach in Anthropology: From Concept to Practice* (ed. E.F. Moran), Univ. Michigan Press, Ann Arbor, pp. 229–245.

Netting, R. McC. (1993) *Smallholders, Householders: Farm Families and the Ecology of Intensive, Sustainable Agriculture*, Univ. California Press, Berkeley, 389 pp.

Ortner, S.B. (1984) Theory in anthropology since the sixties. *Comp. Stud. Society Hist.* **26**, 126–166.

Ostrom, E. (1990) *Governing the Commons: The Evolution of Institutions for Collective Action*, Cambridge Univ. Press, Cambridge, 280 pp.

Painter, M. (1995) Introduction: anthropological perspectives on environmental destruction, in *The Social Causes of Environmental Destruction in Latin America* (eds M. Painter and W.H. Durham), Univ. Michigan Press, Ann Arbor, pp. 1–21.

Painter, M. and Durham, W.H. (eds) (1995) *The Social Causes of Environmental Destruction in Latin America*, Univ. Michigan Press, Ann Arbor, 274 pp.

Pálsson, G. (1988) Models for fishing and models of success. *Marit. Anthropol. Stud.* **1**, 15–28.

Pálsson, G. (1989) The art of fishing. *Marit. Anthropol. Stud.* **2**, 1–20.

Pálsson, G. (1991) *Coastal Economies, Cultural Accounts: Human Ecology and Icelandic Discourse*, Manchester Univ. Press, Manchester, 191 pp.

Pálsson, G. and Durrenberger, E.P. (1982) To dream of fish: the causes of Icelandic skippers' fishing success. *J. Anthropol. Res.* **38**, 227–242.

Pálsson, G. and Durrenberger, E.P. (1983) Icelandic foremen and skippers: the structure and evolution of a folk model. *Am. Ethnol.* **10**, 511–528.

Pálsson, G. and Durrenberger, E.P. (1984) Reply to Gatewood. *Am. Ethnol.* **11**, 379–380.

Pálsson, G. and Durrenberger, E.P. (1990) Systems of production and social discourse: the skipper effect revisited. *Am. Anthropol.* **92**, 130–141.

Pinkerton, E.W. (1994) Summary and conclusions, in *Folk Management in the World's Fisheries* (eds C.L. Dyer and J.R. McGoodwin), Univ. Press of Colorado, Niwot, CO, pp. 317–337.
Rappaport, R.A. (1979) *Ecology, Meaning, and Religion*, North Atlantic Books, Berkeley, CA, 259 pp.
Roseberry, W. (1988) Political economy. *Ann. Rev. Anthropol.* **17**, 161–185.
Roseberry, W. (1989) *Anthropologies and Histories: Essays in Culture, History, and Political Economy*, Rutgers Univ. Press, New Brunswick, NJ, 278 pp.
Schmink, M. and Wood, C. H. (1987) The 'political ecology' of Amazonia, in *Lands at Risk in the Third World: Local-level Perspectives* (eds P.D. Little and M.M. Horowitz with A.E. Nyerges), Westview Press, Boulder, CO, pp. 38–57.
Schneider, J. and Rapp, R. (eds) (1995) *Articulating Hidden Histories: Exploring the Influence of Eric R. Wolf*, Univ. California Press, Berkeley, 400 pp.
Simon, H.A. (1979) *Models of Thought*, Yale Univ. Press, New Haven, CT, 524 pp.
Sinclair, P.R. (1985) *From Traps to Draggers: Domestic Commodity Production in Northwest Newfoundland, 1850–1982* (Social and Economic Papers no. 31), Inst. Social Econ. Res., Memorial Univ. Newfoundland, St John's, 271 pp.
Sinclair, P.R., (ed.) (1988) *A Question of Survival: The Fisheries and Newfoundland Society* (Social and Economic Papers no. 17), Inst. Social Econ. Res., Memorial Univ. Newfoundland, St John's, 254 pp.
Sinclair, P.R. (1990) Fisheries management and problems of social justice: reflections on northwest Newfoundland. *Marit. Anthropol. Stud.* **3**, 30–47.
Smith, M.E. (1990) Chaos in fisheries management. *Marit. Anthropol. Stud.* **3**(2), 1–13.
Steele, D.H., Andersen, R. and Green, J.M. (1992) The managed commercial annihilation of northern cod. *Newfoundland Stud.* **8**(1), 34–68.
Thorlindsson, T. (1988) The skipper effect in the Icelandic herring fishery. *Human Organiz.* **47**, 199–212.
Vayda, A.P. (1983) Progressive contextualization: methods for research in human ecology. *Human Ecol.* **11**, 265–281.
Vayda, A.P. (1988) Actions and consequences as objects of explanation in human ecology. *Env., Technol., Soc.* **51**, 2–7.
Walters, C.J. and Maguire, J.-J. (1996) Lessons for stock assessment from the northern cod collapse. *Rev. Fish Biol. Fish.* **6**, 125–137.
Wilen, J.E. (1979) Fisherman behavior and the design of efficient fisheries regulation programs. *Can. J. Fish. Aquat. Sci.* **36**, 855–858.
Wilson, J.A. and Kleban, P. (1992) Practical implications of chaos in fisheries. *Marit. Anthropol. Stud.* **5**(1), 67–75.
Wilson, J.A. Acheson, J.M., Metcalfe, M. and Kleban, P. (1994) Chaos, complexity and community management of fisheries. *Mar. Policy* **18**, 291–305.
Wolf, E.R. (1972) Ownership and political ecology. *Anthropol. Q.* **45**, 201–205.
Wolf, E.R. (1982) *Europe and the People Without History*, Univ. California Press, Berkeley, 503 pp.

Chapter nineteen

The role of economic tools in redefining fisheries management

Rögnvaldur Hannesson

ABSTRACT

This chapter discusses the cause of the present crisis in world fisheries, i.e. overexploitation of fish stocks and excessive fleets and labour. The cause is identified as the absence of ownership of fish resources, and the main remedy proposed is the introduction of use rights such as transferable fish quotas or fishing licences. The reason why traditional regulations have not been adequate is traced back to the political control of fisheries management; rather than being profit-driven and trying to correct for the basic market failure that stems from free access to the fish, management has aimed at serving political goals by catering to special interests in the fishing industry at the expense of economic efficiency.

19.1 INTRODUCTION

The fisheries of the world are often described as being in crisis. This is not just hype. One fish stock after another has been depleted, although seldom if ever beyond ultimate recovery but certainly for long periods in which fishing has been severely reduced or halted altogether. Pelagic stocks appear to be particularly vulnerable, as illustrated by the California

Reinventing Fisheries Management. Edited by Tony J. Pitcher, Paul J.B. Hart and Daniel Pauly. Published in 1998 by Kluwer Academic Publishers, London. ISBN 0 412 83410 3.

sardine, the Peruvian anchovy, and the Atlanto-Scandian herring. But demersal stocks are not immune. The once abundant northern cod off Newfoundland and Labrador has also been depleted to the point that fishing has been banned (see also Buckworth, Chapter 1, this volume).

Another aspect of the crisis in world fisheries is that the cost of catching fish is much higher than it need be. A report by the FAO has alleged that the cost of fishing is in fact higher than the value of the catch (FAO, 1993). This is a no less serious problem than stock depletion. Sacrificing future catches for short-term gains means less nutrition for a hungry world and lower incomes for those who make their living by fishing, but using more labour and boats than necessary is no less deplorable. The excess boats and labour are resources that could be put to better use than making zero or perhaps a negative contribution to the total production value in the fishing industry.

In this chapter I will briefly discuss the background of the fishery crisis, why traditional attempts at dealing with it have failed, and what can be done to get out of it.

19.2 WHY FISH STOCKS ARE OVEREXPLOITED

The reasons for the fishery crisis are economic and organizational. Fishing (except for sport fishing) is an activity people engage in to make a living. They do so to the extent they see their interests being better served by fishing than by doing something else. Clearly, if there is too much fishing going on, people do not face appropriate incentives, they are encouraged to engage in fishing on a greater scale than is appropriate.

Why this occurs has been known for a long time (Warming, 1911; Gordon, 1954; Scott, 1955; Andersen, 1983). To put it in perspective, let us ponder a question that I owe to my colleague Alf Erling Risa at the University of Bergen, 'why is the pig not an endangered species?' Pig farming is a sustainable undertaking because farmers own their pigs, they raise pigs because they are entitled to whatever price their pigs fetch in the market-place, and they grow their pigs to an optimal size. Fishers (as opposed to fish farmers) are not farmers but hunters, they do not raise fish but hunt them in the wild, and it makes no sense for a fisher to save a fish for breeding or for growing to a more appropriate size, as long as other fishers have the right to catch it in the meantime. Unfortunately it can be quite profitable for fishers to search for fish until stocks have been depleted to unproductive and perhaps irreversibly low levels. It is this destructive mechanism that is responsible for the crisis in world fisheries, both the depletion of fish stocks and the use of capital and human resources beyond what is needed to take the fish that it is advisable to take.

Would it be possible to solve this problem by making fishers the owners of the fish that swim around in the sea? For all but the most sedentary fish stocks, this would not appear to be a feasible solution. To be effective, the ownership would have to extend to an entire stock. This is not a particularly appealing solution; letting one individual or company own, say, the northern cod stock (assuming it recovers) or the Barents Sea cod would give that entity an enormous influence in the labour market in Newfoundland or northern Norway and possibly also over the world fish market. Possibly the ownership could be shared by a limited number of agents who would agree, implicitly or explicitly, to limit their hunting for fish, in their mutual self-interest. Such solutions become less and less likely to work, however, as the number of agents sharing a resource increases. The losses from playing foul are shared by all and thus become less of a deterrent as the number of players increases.

On the other hand, exclusive fishing rights are a distinct possibility and probably the only method that will ensure an economically efficient use of fish resources, as will be further discussed below. The 200 mile exclusive economic zone made such rights-based management possible, although this is increasingly being undermined by the accessibility of fish stocks outside the zone. A further extension of the zone is probably the only effective method to deal with this problem. Fisheries management by corrective taxes, popular among some economic theorists, would not work, because there is a need to respond to short-term, environmental fluctuations (Hannesson, 1993). The information requirements and administrative flexibility required for using taxes to deal with such problems are daunting. This is not to say that special taxes on fisheries could not be used to skim off some of the excess profits arising in well-managed fisheries. Such taxes would even be helpful, if insufficient, to achieve an optimum level of investment in fishing fleets.

19.3 WHY FISHERIES MANAGEMENT HAS FAILED

What, then, can be done? Some authority has to step in and regulate the industry, to limit fishing and investment in fishing boats. The problem is not, however, the absence of regulations. Regulations are very much a fact of life in most if not all fisheries. Rather, the problem is inappropriate regulations. Despite regulations, fish stocks have been depleted, and the industry is typically characterized by overcapitalization and excess labour. There are basically two faults with most fishery regulations in effect: (1) inappropriate objectives, and (2) inappropriate incentives faced by the industry.

Objectives

If we regard the fish resources as sources of food only, the appropriate objective of fisheries management is to utilize these resources as efficiently as possible. Each additional unit of labour allocated to the fishing industry should produce a value equal to what it otherwise could produce in its best alternative use, and the same goes for each additional unit of money spent on fuel or fishing gear, or invested in fishing boats. Fish are not the sole source of food, and food is not the sole source of material well-being. To maximize material well-being, a marginal unit of labour or unit of money invested should produce the same value irrespective of how it is used or where it is invested. In that situation, there is nothing to be gained by reallocating human resources and other factors of production between industries.

This would occur if the rent of the fishery (revenue less all necessary costs) were maximized at the appropriate discount rate and provided that all prices reflect the opportunity cost of goods, services and factors of production. The general validity of this objective is debatable, however. It would not be very appropriate for recreational fishing, but the major world fisheries can hardly be characterized as recreational, being conducted in some of the roughest waters of the world and at a relatively high risk for loss of life and limb. Then there are environmental objectives such as preservation of biodiversity and avoiding too radical a disturbance of the ecological balance (but some disturbance cannot be avoided if there is to be any fishing at all). To what extent these objectives interfere with the maximization of rent will vary from case to case, but maximization of rent would appear to leave fish stocks and their environment in a more healthy condition than other approaches, except for species with a low rate of growth and low costs of fishing, and it would certainly outscore free access.

Maximization of rent, or maximum economic efficiency, is seldom, however, a goal of fisheries regulation, neither explicitly nor implicitly. Instead one finds objectives such as provision of job opportunities or preservation of rural communities. Superficially these objectives may appear as unassailable as God, motherhood and apple pie but on closer inspection they are not. It is not immaterial what sort of jobs are being created; our appetite for a high and rising material standard of living makes it important, to say the least, to ensure that jobs contribute as much as possible to the creation of wealth. Similarly it would appear important to ensure that rural and other communities are economically vibrant on their own and not a drain on public finances, which in all developed welfare states are coming under increasing strain.

Then where do these objectives come from? Typically they are formulated by elected politicians and civil servants acting on their behalf. This origin would seem to ensure the legitimacy of these objectives. The problem is, however, that collective decision making often fails to attain results that serve the interests of the public at large, even if such decisions are taken by elected representatives and in accordance with democratic principles. The problem is the overriding importance that people typically attach to the status quo and to group interests rather than the general interest. Most of us are more occupied with our interests as wage and salary earners, producers or owners of real estate in a certain location than with such things as economic efficiency. It would be great to be able to buy milk and bread a little more cheaply, but more important still to have a higher salary. The farmer is likely, therefore, to win the ear of the politician when lobbying for restricted trade in milk or grain, and so would the fisher when arguing for subsidies or for restrictions on the imports of fish.

Not infrequently, the goal of preserving or increasing job opportunities is pursued at the expense of the fish resources; fish quotas are stretched a little further than recommended, so as to avoid throwing people out of work. Elsewhere (Hannesson, 1996) I have discussed the experiences of four countries and provinces around the North Atlantic rim with respect to management of their cod fisheries. That is not an uplifting story. Despite the facts that they have a high degree of control over their stocks and, no less importantly, that these resources are quite or even extremely important to them, their experiences range from outright disaster to no more than a moderate success. For decades the Newfoundland fishing industry has been regarded as an employer of last resort. This policy, aided by faulty methodology of stock assessment (Walters and Maguire, 1996), reached an impasse with the virtual extinction of the northern cod off Newfoundland, which threw tens of thousands of people out of work for an indefinite period. The Faeroe Islanders brought a major economic crisis upon themselves by overinvesting in fishing vessels and processing plants and by overharvesting their resources. Norway has, together with Russia, managed to rebuild the Barents Sea cod stock, despite placing a low priority on the profitability of its fisheries. Even Iceland, which has the most productive fishing industry and the most sensible management system of all four, has overexploited its resources in the past.

Incentives

More often than not, fisheries regulations have been designed without paying much attention to economic incentives, or for the purpose of making the industry less efficient than it would otherwise have been. This approach is, needless to say, related to the neglect of economic efficiency as an objective; if economic efficiency is of little or no importance, it does not matter whether the management measures themselves promote economic inefficiency. A typical example is the absence of controls of access to fisheries and investment in fishing equipment even if the total catch is being limited. Crutchfield and Pontecorvo (1969) used the phrase 'irrational conservation' when dealing with the intricacies of the Pacific salmon fisheries.

The recognition that fish stocks need to be saved from depletion appears, fortunately, to be widely acknowledged while the need to limit the use of labour and capital in the fishing industry is less generally appreciated. This has led to situations bordering on the absurd, such as the two six-hour openings in a year that once occurred in the Alaskan halibut fishery, an activity that could have been pursued at a more even rate for months if there had been fewer boats participating (in 1995, individual fish quotas were introduced in this fishery). Other examples are restrictions on the design of fishing boats, where length regulation, for example, can end up making them about as long as they are wide, sometimes at a considerable cost in the form of less seaworthiness and comfort for the crew. A humorous example of such wasteful regulation was reported in the Icelandic newspaper *Dagblaðið* on 2 February 1996. Recently the rules for fishing for shrimp in a certain area were changed and boats more than 39 m long were not allowed to fish there. This excluded the boat *Dagfari*, which measured 40 m. The owners had the tip of the prow removed, which satisfied the regulators. The excess metre is now welded on and removed again, depending on whether the boat is fishing in the coveted area or not.

Some of these apparently irrational rules can be traced back to a perceived need to divide the resources between competing claimants. By requiring that only boats below a certain size can participate in a given fishery, that fishery is effectively being reserved for those who own small boats. From the point of view of resource conservation, it hardly matters whether fish are caught by small or large boats. The economic efficiency problem would be taken care of by selling or leasing fishing rights to whoever can pay most for them, with appropriate measures being taken to limit those rights to whatever the resources can support.

This illustrates another weakness of the political process in economic decison making; it tends to become preoccupied with rules, procedures and distribution rather than wealth creation. In fact, fisheries managers do not typically have much regard for the cost increases that their regulations impose upon the industry. This disregard is, perhaps, not surprising, as the economic interests of the managers themselves are in no way at stake.

The neglect of the economic aspects of management can be highly undesirable even if only the biological health of the fish stocks is a matter of concern. The larger the overcapacity in the industry, the greater is the pressure for keeping the wheels running and making ends meet by catching as much as possible and perhaps a little more than one should, as appears to have happened with the northern cod. Furthermore, it typically becomes more difficult to limit the activities of the fleet the bigger it is.

19.4 AN ALTERNATIVE MANAGEMENT APPROACH

What can we do about this? We need a management system that emulates a perfect market and avoids replacing market failure with government failure, or perhaps even adding the two. Competitive markets fail to utilize common resources efficiently, because a common resource has no price even if it is scarce; there is no one who can ration the access to the resource and charge an appropriate price for its use. Competition, it may be added, is neither good nor bad in itself: it produces advantageous results when applied within the appropriate framework but can be disastrous otherwise. Competition to catch fish from a common stock can be and has been disastrous. Government regulations that at the best of times preserve the fish but waste productive resources through excess labour and overcapitalization of the industry are hardly any better. Without them the industry would break even and the resources would be depleted, but we would be spared the costs of the regulations.

A system of this kind would need built-in incentives to maximize the rent from the fishery. Examples of such systems are individual transferable fish quotas (ITQs) or fishing licences. Of these, the ITQs appear much superior. The landings taken from any particular stock are only indirectly controlled by controlling fishing vessels. Restricting the number of fishing vessels is insufficient for controlling total investment, and regulating their design has proven to be difficult and counterproductive. An upper limit to the total catch will, if appropriately set, protect a fish stock from depletion, while dividing it up among the vessels will eliminate the incentive to race for the fish and the overinvestment in fishing capacity undertaken to win the race. Transferability of quotas (or fishing licences) provides incentives to minimize costs. In the short term some will lease their quotas to more efficient operators and lay up their boats when there is too little fish for all to operate profitably, so saving on operating costs, while in the long term people will have incentives to invest in fishing equipment that is appropriate for their quota holdings, which would minimize capital costs. Some countries have indeed adopted management systems of this kind, the most comprehensive ones being found in Iceland and New Zealand (these systems, as well as ITQs in general, are discussed in Munro and Pitcher, 1996).

As with any other system, ITQs are not without drawbacks, and the experience with ITQ systems in place has led to heated discussions about their pros and cons (for different points of view on the Icelandic system, see Árnason, 1995, and Pálsson and Helgason, 1995). Some alleged drawbacks are doubtful. In Iceland the ITQ system has caused some concentration of quotas among firms and made it difficult for prospective boat owners to enter the industry. This is an unavoidable and indeed a desirable effect. Getting rid of excessive fishing boats and labour must necessarily mean some concentration of quota holdings. How much concentration is desirable can be debated, and it is in any case possible to put a limit on quota holdings, as has been done in New Zealand. It should be noted, however, that firms in the fishing industry often compete with giants in the food industry and are, by comparison, dwarfs. To the extent there are economies of scale and scope in the food industry, the firms in the fishing industry would improve their competitive edge by getting bigger. Doing away with open access, the main cause of depletion of fish stocks and overinvestment in fishing fleets, must by necessity mean that it becomes more difficult to get into the industry. ITQs can, however, be associated with a lopsided distribution of benefits. If quotas are initially allocated for free, the economic rent emerging as a result of a more efficient industry will become capitalized as a value of the quota and accrue as a capital gain for those who initially got the quota for free.

Among real problems caused by ITQs are stronger incentives to discard fish of low value and incidental catches in multispecies fisheries where one particular species can be targeted only imperfectly (Árnason, 1994). There is also an incentive to underreport landings, which biases landings data and stock assessment. In general the costs of monitoring and enforcement are likely to be much higher in quota-based systems than in licence-based systems; fishing vessels can be seen and counted while landings can be in the dark, even literally speaking. The control of quotas would appear to be most difficult in small-scale fisheries where the catches are landed without any elaborate equipment and used for human consumption in local communities. Group discipline among small-scale fishers could, however, mean some self-policing, to the extent that everyone is aware of excessive fishing endangering the future possibilities for the group as a whole. There is little or no experience, however, of ITQs in such fisheries.

An interesting question is to what extent ITQ systems would enable the industry to manage the resources on which it is based for its common benefit, or to what extent government regulations would still be needed. Could the industry, for example, set the total allowable catch in an ITQ system? An individual fisher or a fishing firm would dearly like a higher catch quota, particularly in years with a small total allowable catch, but the only way to accomplish this under an ITQ system based on individual shares in the total allowable catch is to buy quotas from others or to push for a higher total allowable catch. Would individual fishing companies push for a higher total allowable catch for the sake of increasing their individual allocation when they know that this would increase the allocations of others and deplete the resource stock too far, perhaps putting its viability in danger? This would not appear to be rational behaviour. It is true, however, that different firms may have different time preferences, some having paid off their debts while others will put a high premium on avoiding bankruptcy. The effect of restricting catches in the near term on future catches is uncertain, so opinions on the desirable degree of resource conservation may differ among individuals. Having the industry set the total catch limit by some form of collective decision making may not, therefore, be satisfactory. Furthermore, the industry is not likely to take into account any value that might be attached to fish stocks as such (biodiversity, ecological relationships), over and above what affects the profits and losses of the industry itself. Some public agency may be necessary, therefore, to set the total allowable catch.

19.5 CONCLUSIONS

The role of economic tools in redefining fisheries management is, in my view, essential. Up to now, economic goals and incentive mechanisms have largely been ignored, both in design and execution of fisheries management. Not only is this unfortunate, it is also self-defeating. The ultimate purpose of the fishing industry is to provide its customers with food and its members with incomes and a congenial way of life. It is for this purpose that we are concerned with fish stocks, and we need to be concerned because we know from bitter experience that the industry is unable to take care of its resource base, in any case as long as it is fragmented and governed by the rule 'first come, first served.' But fisheries management has largely been limited to protecting fish stocks, disregarding the economic perspective. It has tried to limit fishing by putting all sorts of restraints on the industry without addressing the problem of fishing capacity that far outweighs the carrying capacity of fish stocks. Even if the knowledge of fish stocks and their environment is far from perfect, I would contend that inappropriate management is a much greater obstacle and a more important cause of the resource collapses that we have seen in recent times.

The role of economic tools in redefining fisheries management is, as I see it, the following. First, economic objectives, summarized as maximizing the net present value of fish catches over an indefinite time horizon, need to be made the primary goal of management. Second, rights-based incentive mechanisms have to be put in place as management tools, instead of the restrictions on fishing activity such as number of days at sea, limits on vessel size, etc. These measures have proved counter-productive in the end, as they do not address the fundamental problem of too many fishers chasing too few fish. At the best of times they may save the fish, but at the expense of wasting human and man-made productive resources.

REFERENCES

Andersen, P. (1983) On rent of fishing grounds [translation of Jens Warming's 1911 article, b.v. with an introduction]. *Hist. Polit. Econ.* **15**, 391–396.

Árnason, R. (1994) On catch discarding in fisheries. *Mar. Resource Econ.* **9**, 189–207.

Árnason, R. (1995) *The Icelandic Fisheries*. Fishing News Books, Oxford.

Crutchfield, J.A. and Pontecorvo, G. (1969) *The Pacific Salmon Fisheries: A Study in Irrational Conservation*, Johns Hopkins Press, Baltimore.

FAO (1993) Marine fisheries and the law of the sea: a decade of change. Special chapter (revised) of *The State of Food and Agriculture 1992. FAO Fish. Circ.* no. 853.

Gordon, H.S. (1954) The economic theory of a common property resource: the fishery. *J. Polit. Econ.* **62**, 124–142.

Hannesson, R. (1993) *Bioeconomic Analysis of Fisheries*. Fishing News Books, Oxford.

Hannesson, R. (1996) *Fisheries Mismanagement. The Case of the North-Atlantic Cod*, Fishing News Books, Oxford.

Munro, G. and Pitcher, T.J. (eds) (1996) Individual transferable quotas (special issue). *Rev. Fish Biol. Fish.* **6**(1), 1–116.

Pálsson, G. and Helgason, A. (1995) Figuring fish and measuring men: the quota system in the Icelandic cod fishery. *Ocean Coastal Manage.* **21**, 117–146.

Scott, A.D. (1955) The fishery: the objectives of sole ownership. *J. Polit. Econ.* **63**, 116–124.

Walters, C.J. and Maguire, J.-J. (1996) Lessons for stock assessment from the northern cod collapse. *Rev. Fish Biol. Fish.* **6**, 125–137.

Warming, J. (1911) Om grundrente af fiskegrunde. *Nationaløkonomisk Tidsskrift* 499–505 (in Danish: see Andersen, 1983).

Chapter twenty

Social regime formation and community participation in fisheries management

Craig K. Harris

ABSTRACT

Recent developments in the fisheries of Lake Victoria in East Africa provide a paradigmatic case study of an emerging pattern in the management of fisheries in developing countries. Not only does the history of the Lake Victoria fishery exemplify changes that have occurred in fisheries management more generally; it is those changes that have brought actors in the Lake Victoria fishery system to the point where they feel that fisheries management needs to be reinvented. Current developments in the management of the Lake Victoria fisheries illustrate more general themes in the reinvention of fisheries management – ecosystem-based management, adaptive management, multilevel multiparty management, stakeholder participation, individual quotas, environmental and economic and social sustainability, regime formation, chaotic systems, and the precautionary principle. The immediate tasks of fishery management are to define the resource, to identify which groups speak for the resource, and to determine what is in the interest of the resource. The social sciences can contribute significantly to each of these tasks.

Reinventing Fisheries Management. Edited by Tony J. Pitcher, Paul J.B. Hart and Daniel Pauly. Published in 1998 by Kluwer Academic Publishers, London. ISBN 0 412 83410 3.

20.1 INTRODUCTION

The fisheries of Lake Victoria in East Africa provide a paradigmatic case study of an emerging pattern in the management of fisheries in developing countries. In this chapter I will first sketch the recent developments in the Lake Victoria fishery. In the second part, I will suggest that the history of this fishery exemplifies changes that have occurred in fisheries management more generally; I will propose that it is those changes that have brought us to the point where we feel that fisheries management needs to be reinvented. In the third part of the chapter, I will indicate some of the ways in which those developments illustrate more general themes in the reinvention of fisheries management. I will conclude with some thoughts on the future of fisheries management.

20.2 THE LAKE VICTORIA FISHERY

During the 1970s and 1980s, a fairly rich and complex trophic community shifted to a simpler and more sparse trophic structure dominated by an introduced top predator (Ogutu-Ohwayo, 1995), and a small-scale artisanal fishery was largely replaced by an intermediate-scale industrial fishery (Yongo and Wiley, 1997). The causes of these changes include the introduction of the exotic species (Pitcher and Bundy, 1995), increases in siltation-borne and precipitation-borne nutrients (Hecky *et al.*, 1994), land-use changes in the riparian ecotone (Bugenyi, 1994), and the incorporation of the Lake Victoria region into the global fish trade system (Harris *et al.*, 1997).

These changes in the fisheries of Lake Victoria have generated high levels of concern in lakeside communities, in the riparian nations, and in international bodies. The topic has been addressed in at least seven international forums. The World Bank, in conjunction with the Global Environmental Facility of UNDP and UNEP, is currently considering an 80 million dollar programme for Lake Victoria (World Bank, 1995). Funding to support research concerning the Lake Victoria aquatic ecosystem has been allocated by at least ten donor nations, two international scientific bodies, and the European Union (Harris *et al.*, 1995).

These various programmatic and scientific efforts relate to the management of the Lake Victoria fisheries through the establishment of a social regime (Haas, 1989, 1990). A regime is established by a coalition of actors which gains control of the authority for fisheries management. Because the management of a fishery involves, at least in part, the resolution of claims on the fishery that are not fully mutually compatible, the establishment of the regime almost always entails a contest among the various stakeholders for control of the relevant authority.

The past ten years has been a period of significant conflict over the management of the Lake Victoria fishery. At least three distinct viewpoints have been expressed during that conflict: species conservation, commodity production, and riparian well-being (Harris *et al.*, 1997). The contest for the establishment of the Lake Victoria fisheries management regime has involved many different actors advocating one or more of these viewpoints – international scientific organizations, global organizations, transnational corporations. That contest is now coming to a resolution with the establishment of the trilateral Lake Victoria Fisheries Organization.

In establishing a fishery management regime, the dominant actors attempt to structure relevant institutions so as to ensure the actors' continued influence. At the same time, for these institutions to be perceived as legitimate (Dryzek, 1988), they must include a significant element of democratic openness, often referred to as community participation (Pinkerton, 1989; Brown, Chapter 13, this volume). This interplay between regime formation and community participation establishes a dynamic for the development of the new Lake Victoria Fisheries Organization.

20.3 THREE VIEWS OF FISHERIES MANAGEMENT

The sequence of changes in the Lake Victoria fishery exemplifies changes that have occurred in fisheries management more generally and we can see three distinctly different views of fisheries management. I refer to these as the traditional, modern and postmodern views.

Traditional view

In the traditional view of fisheries management, authority was originally invested in local communities. This authority was exercised by the community as a collective, often including regard for future generations and for non-human species (Martin, 1978). Regulation was accomplished by communal processes of influence and suasion (McCay and Acheson, 1987), and was based on the accumulation of indigenous knowledge (Brokensha *et al.*, 1980; Warren *et al.*, 1989). While it is generally assumed that these management systems existed across the millennia of human existence (McEvoy, 1986), contemporary examples of such traditional management systems have been recorded for Cat Harbour in Newfoundland (Faris, 1972), the lobster fishery in Maine (Acheson, 1970), the inland fisheries of Mexico (Pomeroy, 1994), and minor lake fisheries in Uganda (Ogutu-Ohwayo, 1995).

Modern view

Gradually, over the past four centuries, the traditional view of fisheries management has been supplanted by what one could call the modern view of fisheries management. In this view, authority over the fisheries is vested in the nation-state, which, depending on the type of government, may exercise this authority for what it perceives as some common good. When, in the judgment of the national government, the common good is best served by transferring authority over the fisheries to international treaty organizations, that is done (Peterson, 1993). In the modern view of fisheries management, authority is structured hierarchically and is exercised in a top-down manner; the mode of regulation is generally command and control. Science is called upon to determine best management practices for any particular fishery, and to advise fishery managers accordingly. While it is probably a common occurrence, in some situations it can be shown empirically that the nation-state destroys the effective authority of the local community and displaces the local community's management of the fishery. Faris (1972) and Durrenberger (1987) showed how civil society displaced the traditional fishery management in Newfoundland and Iceland, respectively. Fallers (1965) and Vercruijsse (1988) showed how colonial administrations displaced the traditional fishery management in Uganda and Ghana, respectively.

Postmodern view 1

The tension between regime formation and community participation noted above for the Lake Victoria fishery illustrates the conflict between two postmodern views of fishery management. In the first, 'regime formation', at least three major goals of management compete for primacy. One widely heralded goal is the conservation of the 'natural' fishery; management should seek to restore the fishery as it is believed to have existed at some earlier time (Pitcher and Pauly, Chapter 24, this volume). Equally prominent is the goal of economic rationalization of the fishery; management should determine the sustainable yields of the target species and should allocate those allowable catches among various stakeholders so as to accomplish objectives of economic efficiency and societal well-being (Hannesson, Chapter 19, this volume). Less widely publicized is the goal of local well-being; advocates of this view argue that the priority for the management of the fishery should be employment, income, nutrition and well-being of people in the riparian countries and especially those in the vicinity of the Lake.

In Lake Victoria, each of these three positions is advocated by powerful transnational and national actors. The conservation view is advanced most forcefully by scientific organizations such as the International Limnological Society (SILS), by international agencies such as UNESCO and UNEP, and by transnational organizations such as the World Wide Fund for Nature (WWF) and the International Union for the Conservation of Nature (IUCN). The economic rationalization position is advocated by transnational corporations which trade in fish, by international agencies such as UNDP and FAO, and by regional development banks. The local well-being position is advanced by various non-governmental organizations (NGOs) such as Oxfam, and international agencies such as the World Health Organization (WHO). As noted above, each of these views attempts to enroll local actors in its position (Callon, 1986), through such activities as graduate training and conferences. Each of the three positions sees some potential advantage in the establishment of formal organizations of control, such as the Lake Victoria Fisheries Organization (Ssentongo, 1991), and attempts to attain control over the emergent organization.

Postmodern view 2

In the other postmodern view of fishery management, 'community participation', the conflict between disparate goals is submerged in a conflict over management process. While many divergent interests can see community participation in fishery management as potentially beneficial, working out the details of participatory management can be very complex (Pinkerton, 1989). In particular, an effort to establish community participation is likely to founder on two major questions.

The first troublesome question is: who is the community that is to participate in the management process? In contrast to the traditional view of fishery management, the community is no longer just local actors. This expanded view of the community has led analysts to talk of stakeholders (Decker *et al.*, 1996). In the case of Lake Victoria, stakeholders are found at ten levels or realms.

1. global level: biodiversity, common heritage of humanity;
2. transnational level: traders;
3. society level: national resource;
4. citizen level: food security, nutritional needs for protein;
5. national government level: foreign exchange;
6. recreational realm: tour operators;
7. industrial realm: processors and marketers;
8. fishers: boat owners, operators, crew;
9. family and neighbours: employment;
10. lakeside people: subsistence consumption.

Assuming that some agreement can be reached on who is the community, the second problem of community participation is how to facilitate participation in fisheries management (Pinkerton, 1996).

First, community participation must be seen as a process rather than as an event. It is not a one-time consultation; it is a developmental process. This process is still emergent in the case of Lake Victoria. While some consultations were held in the development of the Lake Victoria Environmental Management Plan (World Bank, 1995), it remains to be seen whether the participation process will continue during the implementation of the Plan.

Second, it is important that the diverse groups of stakeholders have some parity of authority; the groups must be able to participate as equals in the process. The views of one group must not be privileged with respect to those of other groups. In some cases this has been accomplished by providing each group with access to its own expertise in fishery science; to date no efforts of this sort have occurred on Lake Victoria.

Third, the groups of stakeholders must be treated as equals in their claims to legitimacy. Legitimacy can have many different bases – historical, monetary, basic human needs, human capital, social capital. Each of these bases can be claimed by one or more of the stakeholder groups on Lake Victoria; for example, while the lakeside communities can base their claims on social capital, traders can base their claims on monetary value.

20.4 REINVENTING FISHERIES MANAGEMENT

In the preceding section I have suggested that the history of the Lake Victoria fishery exemplifies changes that have occurred in fisheries management more generally (Crean and Symes, 1996). We see the effects of these changes in the widespread debates about the existence, causes and remedies of overfishing (Ecologist, 1995; Schurman, 1996). I would suggest that it is those changes that have brought us to the point where we feel that fisheries management needs to be reinvented. In this section of the chapter I want to indicate some of the ways in which the developments on Lake Victoria illustrate more general themes in the reinvention of fisheries management. My underlying notion in this section is that the challenge in reinventing fisheries management is to reconcile the two postmodern views of fishery management: regime formation, with its inherent conflicts over alternative goals, and community management, with its inherent conflicts over power and equity.

The first general theme in the reinvention of fisheries management is ecological management or ecosystem-based management (Larkin, 1996).

For a start, this certainly means that it is a whole fishery that is managed, not just one species, because most species have significant ecosystem interactions. But then including all the significant species means including humans as a variable component of the system, because it is their motivations and responses which significantly affect the other components of the system (Ludwig *et al.*, 1993). Eventually ecological-based management comes to mean entire ecosystem approaches (Griffis and Kimball, 1996). Managing a fishery requires managing and monitoring with respect to ecosystem units, not political boundaries. It acknowledges the need to address anthropogenic inputs and to influence policy and governance processes, and the need for stakeholder input from the beginning to define desired ecosystem outcome. It was this realization that led initial efforts to manage the Lake Victoria fisheries to become a Lake Victoria Environmental Management Plan, including components covering soil erosion and watershed pollution.

Ecosystem management implies that management activities will occur at multiple levels and will involve multiple parties. While this adds to the difficulty of the management project, the difficulties are not insuperable. The three countries which border the Wadden Sea off western Europe have established a multilevel, multiparty management scheme (Miller, 1996). The three countries have different arrangements at the national level for the management of the sea area, and each of the three countries has various arrangements at local levels for involvement in the management of local sea territory. Miller suggests that intergovernmental mechanisms can be minimal if local and national institutions and practices are well established, and that states can retain sovereignty and culture if national and local mechanisms can accomplish multicountry goals. For this to happen, the agenda discussed by the multiple levels and parties must include programme goals, means, evaluation and compliance. Miller notes that it has been found necessary to establish a secretariat for implementation, information collection, and sensitization. Likewise, the collection and publication of information was the original task of the Great Lakes Fishery Commission in North America, which involves the United States and Canada and the riparian states and provinces. In a similar fashion, the harmonization of the collection of information is seen as one of the main purposes of the Lake Victoria Fisheries Organization.

Another example of multilevel, multiparty management is the Great Barrier Reef in Australia (Miller, 1996). Although the Reef is under state jurisdiction, the national government has coordinated the management effort. The management functions by allocating areas to different (multiple) uses. The largest areas permit all legal uses, including trawling and commercial netting; the second largest areas include commercial netting but not trawling. According to Miller, the biggest problem is monitoring and analysing the impacts of the uses on the different areas.

The problem of collecting sufficient data to make management decisions was considered by a conference convened by the US National Academy of Sciences in 1996 in Monterey, California. One of the conclusions of that conference was that the monitoring of fishery parameters and the collection of fishery data by multiple actors, including fishing outfits, is both feasible and efficient (forthcoming in a special issue of *Ecological Applications*). While the involvement of multiple actors in the collection and analysis of monitoring data is certainly an important component of reinventing fisheries management, efforts to move toward ecosystem management imply that data need to be gathered also on parameters of the aquatic ecosystem other than the fisheries (Gutting, 1996), and on the human components of the ecosystem.

Considering the possibilities of multilevel, multiparty management raises the question of the optimal levels for different kinds of management activities. Ostrom *et al.* (1993) suggest that, rather than trying to locate all fishery activities at one level or another (e.g. the international organization, the nation-state, or the local community), the different activities should each be allowed to devolve to optimal levels. This would suggest that, for example, one level might authorize the enforcement of fishery regulations, another level would establish the structures for enforcement, and a third level might provide the actual enforcement services. While this would seem to have many advantages for a situation like Lake Victoria (Wilson, forthcoming), little effort has been made to experiment with such arrangements.

One of the problems which needs to be solved in reinventing fisheries management is covering the costs of management. Currently the management activities which are carried out by public agencies are generally supported by general governmental funds, most of which come from general taxes. An alternative is to develop ways in which the fishery can support its own management directly. Internally supported management could be accomplished either by user fees, or by collecting some sort of rent for the exploitation of the fishery. While this seems straightforward on the surface, deciding which income stream(s) will be tapped for the fee or rent almost certainly has the effect of redistributing the costs and benefits of the fishery from some stakeholders to other stakeholders. Although an assessment on the production of each industrial fish processing plant was originally proposed in the Lake Victoria Environmental Management Plan as a way of paying for the costs of fishery management, this proposal was dropped from the final plan. This direct support of management costs becomes especially problematic if the activities of fishery management include management of other non-aquatic, distant parts of the ecosystem (e.g. upland factories) without those components contributing to the management cost.

The modern view of management, as described above, believed that fisheries were sufficiently stable that, once science identified the best management practices for that fishery and the correct management regime was in place, it could be relatively permanent (Scheiber, 1995; Wooster, 1988; Wooster and Miller, 1988). The postmodern view of management, in contrast, sees fisheries as sufficiently complex and emergent that it is not possible to specify an optimal management scheme for even a few years into the future. This suggests that reinventing fisheries management will have to include adaptive management regimes. While active adaptive management arrangements have benefits of flexibility and responsiveness, Abramovitz (1996, 42–43) notes that they may do more harm than good if they allow rapid intensification of effort or switching between target species. Supplementing native stocks with plantings or exotics, and increasing effort in aquaculture are some of the responses which adaptive management can use, but Abramovitz argues that none of these should be used without long and careful study of the environmental and social impacts. The need for adaptive management suggests that both decision tree analysis (neural nets) and simulation analysis will have a role in reinventing fisheries management (Bundy, Chapter 26, this volume).

As fishery management seeks to be adaptive, it will find it advantageous to work with combinations of both tried and true, and new and unfamiliar, management devices. Perhaps the most unfamiliar of these is the assumption of closure; this reverses the customary understanding in many areas that access to a fishery is open until restricted in some way. The assumption of closure, on the other hand, means that no human exploitation is allowed until management permits it. While this approach has some superficial attractiveness, it ignores the fact that almost all fisheries in the world are exploited by some human group (Weber, 1995). In that situation, assuming closure will deprive those human users of the goods and services they are currently obtaining from the fishery – in some cases, their subsistence. To the extent that the users thus deprived are politically effective, the government depriving them of their fishery may pay a political cost. The Lake Victoria Environmental Management Plan implies the possibility of an assumption of closure when it talks about allowing fishing only from gazetted landings. Because the political systems of the three riparian countries are characterized by rather loose internal integration, it might be possible for the state to impose this assumption without high political cost. At the same time, that loose internal integration would also make it difficult for the closure of significant areas to be enforced.

While area-based management is not as novel as the assumption of closure, it also is likely to be a feature of reinventing fisheries management (Roberts, 1997). Other chapters in this volume discuss closed areas or refugia in detail (Walters, Chapter 21, and Sumaila, Chapter 23), so I will simply note that the imposition of closed areas and the creation of refugia entail the redistribution of the benefits of the fishery from some stakeholders to other stakeholders. To the extent that this redistribution imposes significant economic and social costs on some stakeholders, the management scheme might want to include some compensation for their burdens. One way that has been suggested for dealing with the unequal burden produced by area closures is to rotate the closures among the various areas; while this may appear to distribute the burden equally, it benefits the stakeholders with the greatest mobility, and in the short to middle term it benefits the stakeholders whose open area happens to coincide with an abundance of fish. Closed areas are being considered by the Lake Victoria Fisheries Organization, and are advocated in the Lake Victoria Environmental Management Plan. At the moment not enough is known about where boats go on the Lake to harvest fish, to be able to evaluate the social and economic impacts of different closure schemes.

The third somewhat new device available to the reinvention of fisheries management is the individual quota (IQ), and especially the individual transferable quota (ITQ) (Scott, Chapter 15, and Hannesson, Chapter 19 both this volume). While quota may be held by individual persons or families, by processing firms, or by communities, it is the transferability which makes it financially useful and leads to more efficient fisheries. While quota-based management schemes have been criticized in the past for being ineffective, Wilder (1995) argues that it is important to distinguish the quota system *per se* from the management regime implementing a quota. He argues that quotas can accomplish environmental protection if the management regime has accurate information and analysis, effective enforcement, and international cooperation. While IQs and ITQs solve the problems of high grading and derby fishing (Dewees, 1990), issues remain concerning the allocation of quotas among the various stakeholders (Pauly, 1996). While fisheries biologists have argued for the imposition of total quotas on the harvest from Lake Victoria, little thought has been given to which stakeholders would get quota or how the quotas would be allocated.

The various components of reinventing fisheries management mentioned above all entail some sort of stakeholder participation in fishery management. As noted, such participatory schemes are generally referred to as cooperative management (Pinkerton, 1989, 1996). Indeed, Griffis and Kimball (1996) suggest that the ecosystem approach facilitates the identification of stakeholders and their incorporation in management. Pido (1995) has argued that both politicians and government administrators must be included among the stakeholders, rather than being accorded a privileged position outside the participatory process.

The diversity of stakeholders implies that fisheries will be managed for multiple objectives. Onal (1996) shows that a management regime relying on closure to accomplish its aims will have to vary its closure policies according to the objective(s) it wants to achieve – efficiency, aggregate net income, net income distribution, consumer welfare, enforcement cost. As noted above, the consideration of enforcement cost in the calculus of optimization will depend on which actor(s) is currently paying the bill for enforcement activities. Although enforcement is not currently effective on Lake Victoria, the expenses of the enforcement effort that does exist are being borne by the three riparian nations.

All of these considerations point to the idea that, to be effective, fisheries management must be environmentally, economically and socially sustainable. Griffis and Kimball (1996) argue that sustainability needs to be measured at the level of the ecosystem, and Worster (1993) argues that environmental sustainability must go beyond the anthropocentrism of focusing on one or a few resources, and focus instead on inherent ecosystem properties like biodiversity. While much of the literature sees the environmental dimension as the highest priority, Worster (1993) argues that social equity will lead to sustainability on the other two dimensions. The frequency of appeals to sustainability leads some authors (Ludwig *et al.*, 1993) to be distrustful of claims for sustainable management; they challenge the claimants to show how a new plan will avert or overcome the difficulties encountered by past plans. Worster (1993) argues that the ambiguity of 'normal yield' or MSY or 'optimum yield', and the inability unambiguously to determine carrying capacity, imply that environmental sustainability cannot be determined a priori; for this reason, it becomes important to create a management structure that will be socially and economically sustainable, because that structure will be most likely to produce a management process that ensures environmental sustainability.

One aspect of social sustainability concerns the process of social regime formation. If a regime is to be sustainable, its formation must be consensual and participatory (Pinkerton, 1996). As noted above, the development of the Lake Victoria Environmental Management Plan showed some of these characteristics. On the one hand, most of the international meetings leading to the formation of the Plan were held outside Africa, and most of the influential actors in the planning process were international agencies and transnational associations. On the other hand, the final document resulted from a consultative process which engaged over 100 scientists, officials and business leaders in each country, and the formal proposals were assembled by task forces of officials from the three countries. Although this process included the industrial fish processors and marketers, it excluded any participation by the boat owners and operators and crew, or by the leaders of fishing communities.

Another general theme in the reinvention of fisheries management is the precautionary principle (O'Riordan, 1994). This suggests that management decisions and actions should err in the direction of maintaining the greatest future flexibility and potential. Although it is often necessary to act before scientific consensus is achieved (Ludwig *et al.*, 1993), the precautionary principle offers some guidance for such actions. Apart from the formation of scientific consensus, it is often necessary to make decisions under uncertainty (Ludwig *et al.*, 1993; Gordon and Munro, 1996; Lauck, 1996); again, the precautionary principle suggests criteria for such decisions. Worster (1993) points to the difficulty of taking any action if the natural system is chaotic; the precautionary principle encourages management to base its decisions and actions on assumptions and models that allow for chaotic behaviour.

The final theme in the reinvention of fisheries management, then, concerns our efforts to adapt our activities to chaotic systems. Worster (1990) suggests that this development in fisheries management parallels a shift in biological ecology from a concern with order to a concern with chaos. What makes this change promising for fisheries management is that a concern with chaos may promote a less hierarchical view of life, and may foster new relations between humans and nature (cf. Prigogine and Stenger). In this view, effective management requires that humans be respectful of the complexity, unpredictability and turbulence of nature. Whether this shift in fisheries management will occur is yet to be seen, and indeed, the shift in ecological thinking is somewhat ambiguous. Emphasizing the chaos of nature may lead to a *laissez-faire* approach to management; if there is no inherent order in nature, then humans can do whatever they want to the environment. This line of thinking has been expressed by Botkin (1990). Not only does he argue that we should reject nature as a norm or standard for human civilization; in his view, humans have both the need and the right to give order and shape to nature.

20.5 CONCLUSIONS

To conclude this discussion of the reinvention of fisheries management as it applies to Lake Victoria, I suggest that the management of fisheries (and indeed of all natural resources) involves at least three tasks. The first is defining the resource. As noted above, the resource may be conceptualized as the stock of a species, the aquatic ecosystem in which the species exists, or the larger ecosystem of which the stock and its habitat and human communities are parts. The second task is deciding who speaks for the resource; this requires being able to hear the different voices which represent themselves as speaking for the interests of the resource. The third task involves deciding what is in the interest of the resource; this decision may be made processually, or substantively.

It is important to realize that various groups claim to speak for the resource, and in doing so represent their own particular conceptualization of the resource. Among the groups that claim to articulate the interests of the resource are resource managers. They may be located in various agencies (e.g. Marine Fisheries Service, Fisheries and Wildlife Service) in different realms of government (state, federal, executive, judicial), and they may come from different intellectual disciplines (e.g. biology, economics), but when they speak, they emphasize the authority which has been invested in them to act as the caretakers of the resource. Another set of groups that claim to speak for the resource are scientists of various disciplines (ichthyology, ecology, economics, sociology). When they speak, they emphasize the truth of their statements, which comes from the objectivity of their methods. A third set of spokespersons represent the various uses and users of the fisheries; when they speak, they emphasize the consequences for psychological, economic and social well-being of different management decisions. The fourth set of groups that claim to speak for the resource are environmentalists, but even among environmentalists there are variations in worldview (e.g. egocentric, anthropocentric, biocentric). But whatever their worldview, when they speak, they emphasize the consequences of different management actions for elements of the biophysical environment which are most relevant to their worldview.

I suggest that the social sciences (anthropology, economics, geography, political science, psychology, sociology) have several contributions to make to the decision about what is in the interest of the resource, and thus to the management of fisheries and other natural resources (see also Jentoft, Chapter 12, this volume). The first contribution they can make is in developing models of the fishery with specific parameters. Here their contribution may be both to the identification of biological–social interactions, and to the specification of interactions between social components. The second contribution they can make is to measuring the current values of the parameters of those models. The third contribution is the identification of the critical indicators of the social and economic subsystems in those models; these critical indicators are useful for managers and policy makers. Fourth, social scientists can measure the current state of those indicators, and design monitoring systems for the routine measurement of their states. Fifth, social scientists can analyse the processes of policy formation and management decision making, so that government and management systems can be reflective about their effectiveness and efficiency. Finally, the application of social science can contribute to the revision of the management and policy processes in ways that increase the capacity for hearing and respecting the multiple contending stakeholder voices, and thus of increasing the sustainability of fisheries management.

In conclusion, I would illustrate this perspective on the reinvention of fisheries management with a question: what if an authentic stakeholder participation process results in a decrease in the target fish population? I would argue that this occurrence would not represent a failure of the process, for the following three reasons. First, this negative impact gives the stakeholders an opportunity to take responsibility for the consequences of their own decisions. It is to be hoped that they will be capable of saying 'we diminished the stock'; at least it will be difficult for them to say that some other decision maker diminished the stock. Second, they will have to manage the stock at a lower level (i.e. they will have to make allocation decisions among their highest priorities, including rebuilding). And third, they will have an opportunity to cooperate in rebuilding the population. In the ability of all stakeholder groups to participate in the reinvention of fisheries management lies the prospect for the sustainability of the world's fisheries.

REFERENCES

Acheson, J.A. (1975) The lobster fiefs. *Human Ecol.* **3**, 183–207.

Botkin, D.B. (1990) *Discordant Harmonies: A New Ecology for the Twenty-first Century*, Oxford Univ. Press, New York.

Brokensha, D., Warren, D.M. and Werner, O. (1980) *Indigenous Knowledge Systems and Development*, Univ. Press of America, Washington, DC.

Bungenyi, F.W.B. (1994) Ecological Influences On the Fisheries of Lake Victoria. Paper presented at the Working Conference on Managing the Ecology of Lakes Malawi and Victoria, Michigan State University, 25–27 February 1994.

Callon, M. (1986) Some elements of a sociology of translation: domestication of the scallops and the fishermen of St. Brieux Bay, in *Power, Action and Belief: A New Sociology of Knowledge* (Sociological Review Monograph) (ed. J. Law), Routledge and Kegan Paul, London, pp. 196–229.

Crean, K. and Symes, D. (eds) (1996) *Fisheries Management In Crisis*, Fishing News Books, Oxford.

Decker, D.J., Krueger, C.C., Baer, R.A. Jr, Knuth, B.A. and Richmond, M.E. (1996) From clients to stakeholders: a philsophical shift for fish and wildlife management. *Human Dimensions of Wildlife* **1**(1), 70–82.

Dewees, C.M. (1990) *Multispecies Fishery Challenges With Individual Transferable Quotas (ITQs). Effects of Different Management Schemes On Bycatch, Joint Catch and Discards*, California Sea Grant College, La Jolla.

Dryzek, J. (1988) *Discursive Democracy: Politics, Policy, and Political Science*, Cambridge Univ. Press, New York.

Durrenberger, P. (1987) The economy of a Lisu village. *Am. Ethnol.* **3**, 633–644.

Ecological Applications. forthcoming. 'Special Issue on Gathering Fisheries Data'. Harold Mooney (ed.)

Ecologist (The) (1995) Overfishing: causes and consequences. *The Ecologist* **25**(2/3).
Fallers, L.A. (1965) *Bantu Bureaucracy: a Century of Political Evolution among the Basoga of Uganda.* Univ. Chicago Press, Chicago.
Faris, J.C. (1972) *Cat Harbour: a Newfoundland Fishing Settlement* Inst. Social Econ. Res., Memorial Univ. Newfoundland, St John's.
Gordon, D.V. and Munro, G.R. (eds) (1996) *Fisheries under Uncertainty: A Precautionary Approach to Resource Management*, Univ. Calgary Press, Calgary.
Griffis, R.B. and Kimball, K.W. (1996) Ecosystem approaches to coastal and ocean stewardship. *Ecol. Appl.* **6**, 708–712.
Gutting, R. (1996) Management challenges facing U.S. fisheries. *Forum Appl. Res. Public Pol.* **11**(2), 35–38.
Haas, E.B. (1989) Do regimes matter? *Int. Org.* **43**, 377–403.
Haas, E.B. (1990) Reason and change in international life: justifying a hypothesis. *J. Int. Aff.* **44**, 209–240.
Harris, C.K., Wiley, D.S. and Wilson, D.C. (1995) Social and economic impacts of the introduction of the Nile Perch in Lake Victoria, in *Impact of Species Changes In African Lakes* (eds T.J. Pitcher and P.J.B. Hart), Chapman & Hall, London, pp. 496–537.
Harris, C.K., Wiley, D.S. and Wilson, D.C. (1997) 'Tensions in the Management of the Lake Victoria Fishery'. African Rural and Urban Studies.
Hecky, R.E., Bugenyi, F.W.B., Ochumba, P., Talling, J.F., Mugidde, R., Gophen, M. and Kaufman, L. (1994) Deoxygenation of the deep-water of Lake Victoria, East Africa. *Limnol. Oceanogr.* **39**, 1476–1481.
Larkin, P.A. (1996) Concepts and issues in ecosystem management. *Rev. Fish Biol. Fish.* **6**, 139–164.
Lauck, T. (1996) Uncertainty in fisheries management, in *Fisheries under Uncertainty: A Precautionary Approach to Resource Management* (eds D.V. Gordon and G.R. Munro), Univ. Calgary Press, Calgary, pp. 91–105.
Ludwig, D., Hilborn, R. and Walters, C. (1993) Uncertainty, resource exploitation, and conservation: lessons from history. *Science* **260** 17, 36.
McCay, B.J. and Acheson, J.A. (eds) (1987) *The Question of the Commons: the Culture and Ecology of Communal Resources*, Univ. Arizona Press, Tucson.
McEvoy, A. (1986) *The Fisherman's Problem: Ecology and Law In California Fisheries, 1850–1980*, Cambridge Univ. Press, New York.
Martin, C. (1978) *Keepers of the Game: Indian–animal Relationships and the Fur Trade*, Univ. California Press, Berkeley.
Miller, K.R. (1996) *Balancing the Scales: Guidelines for Increasing Biodiversity's Chances Through Bioregional Management*. World Resources Institute, Washington, DC.
Ogutu-Ohwayo, R. (1995) Refugia for endangered fishes from an introduced predator in Lake Nabugabo, Uganda. *Cons. Biol.* **10**, 554–561.
Onal, H. (1996) Optimum management of a hierarchically exploited open access resource: a multilevel optimization approach. *Am. J. Agric. Econ.* **78**, 448–459.
O'Riordan, T. (1994) *Interpreting the Precautionary Principle*. Earthscan Publications Ltd, London.
Ostrom, E., Schroeder, L.D. and Wynne, S.G. (1993) *Institutional Incentives and Sustainable Development: Infrastructure Policies in Perspective*, Westview Press, Boulder.
Pauly, D. (1996) ITQ: the assumptions behind a meme. *Rev. Fish Biol. Fish.* **6**, 109–112.
Peterson, M.J. (1993) International fisheries management, in *Institutions for the Earth: Sources of Effective International Environmental Protection* (eds P.M. Haas, R.O. Keohane and M.A. Levy), MIT Press, Cambridge, MA, pp. 249–305.

Pido, M.D. (1995) Towards improved linkage between research and management in marine fisheries. *Naga, ICLARM Q.* **18**(3), 16–17.

Pinkerton, E. (ed.) (1989) *Co-operative Management of Local Fisheries: New Directions For Improved Management and Community Development*. Univ. British Columbia Press, Vancouver.

Pinkerton, E.S. (1996) *Participatory Management*. David Suzuki Foundation, Vancouver.

Pitcher, T. and Bundy, A. (1995) Assessment of the Nile perch fishery in Lake Victoria, in *The Impact of Species Changes in African Lakes* (eds T.J. Pitcher and P.J.B. Hart), Chapman & Hall, London, pp. 163–180.

Pomeroy, C.S. (1994) Self-Regulation In a Traditional Fishery. Am. Fisheries Soc., Portland.

Roberts, C.M. (1997) Ecological advice for the global fisheries crisis. *Trends Ecol. Evol.* **12**, 35–38.

Scheiber, H.N. (1995) Success and failure in science–policy interactions: cases from the history of California coastal and ocean studies, 1945–1973, in *Improving Interactions Between Coastal Science and Policy* (ed. National Res. Council), National Acad. Sci., Washington, DC, pp. 121–147.

Schurman, R. (1996) Overfishing. *CNS March* **7**(1), 131–137.

Ssentongo, G. (1991) *Report on the harmonisation of fishery regulations on Lake Victoria*. Committee On Inland Fisheries of Africa, Bujumbura.

Vercruijsse, E.V.W. (1988) *The Penetration of Capitalism : a West African Case Study*, Zed Books, London.

Warren, D., Michael, L., Slikkerveer, J. and Titilola, S.O. (1989) *Indigenous Knowledge Systems: Implications for Agriculture and International Development*, Technology and Social Change Program, Iowa State Univ., Ames, IA.

Weber, M. (1995) *The Oceans Are Emptying*. WorldWatch Press, New York.

Wilder, M. (1995) Quota systems in international wildlife and fisheries regimes. *J. Env. Devel.* **4**(2), 55–104.

Wilson, D.C. forthcoming. Participation and economic equity in fisheries management, *Soc. Nat. Resour.*

Wooster, W.S. (ed.) (1988) *Fishery Science and Management: Objectives and Limitations*. Springer-Verlag, Berlin.

Wooster, W.S. and Miller, M.L. (1988) On fishery science and management, in *Fishery Science and Management: Objectives and Limitations* (ed. W.S. Wooster), Springer-Verlag, Berlin, pp. 289–294.

World Bank (1995) *Lake Victoria Environmental Management Plan*, The World Bank, Washington, DC.

Worster, D. (1990) The ecology of order and chaos. *Env. Hist. Rev.* **14**, 1–18.

Worster, D. (1993) The shaky ground of sustainability, in *Global Ecology: A New Arena of Political Conflict* (ed. W. Sachs), Zed Books, London, pp. 132–145.

Yongo, E.O. and Wiley, D.S. (1997) A short history of the exploitation of the Lake Victoria fishery. *African Rural Urban Stud.*

Part Four

Coping with Ecology in a Reinvented Fisheries Management

Chapter twenty-one

Designing fisheries management systems that do not depend upon accurate stock assessment

Carl Walters

ABSTRACT

Fisheries management systems are coming to rely more on quota and allowable catch regulation, which in turn depends for success on accurate stock size assessments. But we cannot provide accurate assessments for most fish stocks, and probably never will. Instead of treating the seas as open to fishing with small exceptions (marine refugia), we will only safely limit harvest rates if we reverse this view and treat the seas as closed to fishing with small exceptions (limited fishing areas and times). The will to make this basic change in attitude will come only if we recognize deep uncertainty in management capability, and a public commitment to manage our resources safely no matter how we must reorganize fishing industries to make that happen.

Reinventing Fisheries Management. Edited by Tony J. Pitcher, Paul J.B. Hart and Daniel Pauly. Published in 1998 by Kluwer Academic Publishers, London. ISBN 0 412 83410 3.

21.1 INTRODUCTION

It is a sad comment on the state of world fisheries that we can now develop fairly elaborate taxonomies for the causes of collapse. Some fisheries have apparently failed simply because environmental or habitat change has been too severe for the stocks to handle. Others have collapsed through overfishing or a combination of overfishing and habitat change, combined with failure of fisheries governance systems to react in a timely fashion. Still others have collapsed because, in spite of adequate governance systems and the will to act wisely, scientists have provided incorrect advice concerning stock sizes and potential production. So nature abuses us, governments react too slowly, and our scientific authorities often aim governments in the wrong direction.

It is clear that the reinvention of sustainable management systems for fisheries must involve far more robust ways of dealing with natural variation and scientific uncertainty than have been employed in the past. It is no solution to simply assert that we should press for a 'precautionary principle' in fisheries management; cautious management can still be utterly destructive if it is based on assumptions and analyses that are not even in the right general ball park in the first place. In this chapter, I propose that the key starting point for designing sustainable fisheries systems is to look at our few success stories, and to learn from these that the basic structure of a successful system will involve severe limitation of how much damage we can do in the first place. In practice, this severe limitation will generally involve making sure that a high percentage of every fish stock is not even exposed to the risk of capture during any fishing season. From that basis in ecological safety, we can then begin to build more elaborate and risky management systems as our knowledge and regulatory ability improves.

In the following sections I first argue that successful fisheries are likely to be those that are managed with conservative fixed exploitation rate policies. I then describe why fisheries assessment scientists cannot count on implementing such policies by the currently popular approach of assessing abundance and then setting allowable catches or quotas on the basis of such assessments. This leads to a discussion of why a fundamental component of successful fisheries management in the future will be to have very large marine protected areas, where we view areas open to fishing as the exceptional places, rather than marine protected areas as the exceptional places. I close with some comments about why the public should seriously consider such drastic measures to protect *our* fisheries, even in circumstances where these measures lead to considerable hardship for today's fisheries stakeholders.

21.2 A COMMON FEATURE OF SUCCESSFUL FISHERIES: LARGE SPACE–TIME REFUGES

There is a startling contrast between the east and west coasts of Canada, which have the same basic legal and institutional structures for conservation, but have had very different fisheries outcomes: many major salmon fisheries are healthy and even increasing, while there has been collapse after collapse on the east coast. A naïve view might be that salmon management is just easier than groundfish management, because it is relatively easy to count spawning salmon and salmon are very productive anyway. This view is dead wrong – at the time salmon managers have to make harvest decisions, they have no better stock size assessments than do groundfish managers (the salmon manager maybe gets a bit better hindsight, but that is not really helpful), and many salmon stocks are not very productive.

Success at managing salmon in British Columbia (BC) and Alaska does not result from better knowledge, but rather a reversal in the way managers think about regulation compared with trawl fisheries (Figs 21.1, 21.2). With salmon, it has long been known that concentration of fish during spawning migrations makes them vulnerable to overharvesting, and managers have acted accordingly by closing most of the ocean to fishing for most of the time (salmon fishing areas near rivers are characteristically very small, and open for only a few hours or days at a time). BC's very successful herring roe fishery (on spawning concentrations of Pacific herring) is operated in the same way as that for salmon (Fig. 21.2). In contrast, Canadian trawl (and salmon recreational) fisheries are operated with the opposite perspective, namely that the world should be open to fishing except perhaps for limited time/area protection of spawning aggregations. With salmon, extremely efficient commercial fishing fleets cut 'holes' in the migrating runs (Fig. 21.3), with openings resulting in virtual destruction of whatever part of a stock happens to be migrating through a fishing area at the time when the area opens. But these areas/times are short enough that a substantial percentage of the fish never see a net: exploitation rate of salmon and herring is limited not by knowing how many fish there are, but rather by allowing substantial percentages of the fish to pass through the coastal zone unexposed to the risk of harvest. Stocks have been sustained by following a very simple rule: 'if you do not want to catch more than *X* per cent of the stock each year, do not expose more than *X* percent of the fish to harvest'. Salmon managers have learned what *X* should be through hard adaptive management experience, during which stocks have been driven down enough for their productive limits (*X*) to be directly evident.

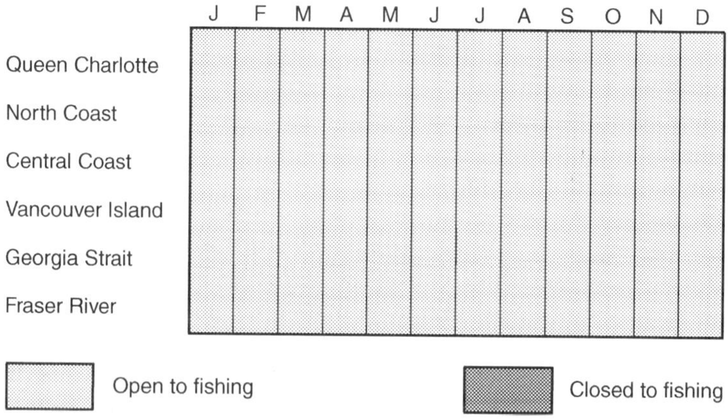

Fig. 21.1 One philosophy of management is that the British Columbia coast (vertical axis of graph) should be open to fishing at all times of year (horizontal axis of graph), with management acting to restrict harvesting by setting total allowable catches over the space–time grid but with fishers free to move over the grid to minimize harvesting costs and maximize product quality. Compare Fig. 21.2.

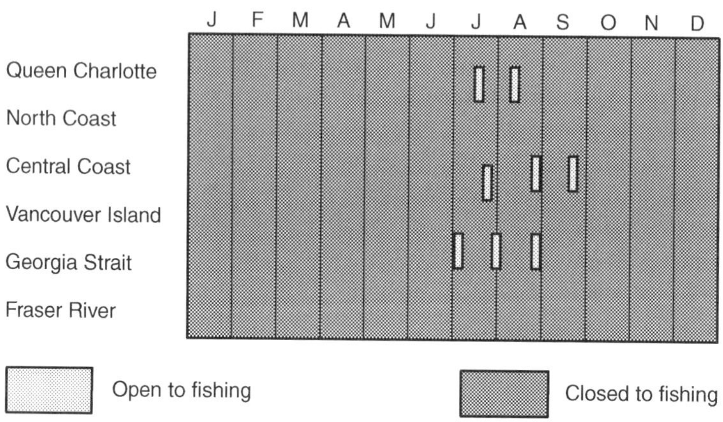

Fig. 21.2 The philosophy of Fig. 21.1 has been reversed in the case of Pacific salmon and herring fisheries. The coast is viewed as closed to fishing at all times and places, and managers then establish a few short openings to take migrating fish. Total exploitation rate is limited by the size and duration of these openings, not by setting an allowable catch based on accurate stock size estimates.

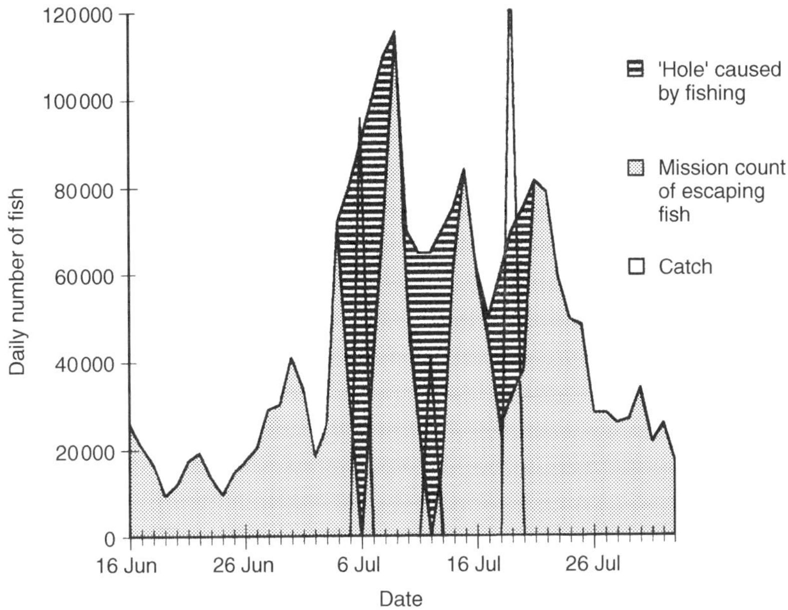

Fig. 21.3 Safe harvesting of Pacific salmon cuts 'holes' in the migration timing, as in this case showing number of sockeye salmon, *Oncorhynchus nerka*, making it past (just upstream from) the last major commercial fishery they face along their spawning migration, to Mission in the Fraser River, in 1992. Note how virtually all fish are removed from the portions of the run available during the three short fishery openings. This removal is largely independent of the number or efficiency of fishing vessels, and thus can be done safely providing the fishing areas/times are small enough.

It is not difficult to show that such 'fixed exploitation rate strategies' (constant percentage *X* harvested each year independent of stock size) are a very good way to manage in the face of long-term, unpredictable changes in natural productivity and carrying capacity owing to environmental changes (Walters and Parma, 1996). In essence, allowing a fixed percentage of the stock to escape each year allows the stock to 'track' changes in environmental conditions such that it will increase almost optimally when the environment is favourable, and decrease when the environment is unfavourable. Various twists on the basic idea can be introduced to avoid extreme low and high population sizes (so-called 'cut-off' rules for stopping fishing if stock size goes below a critical minimum, etc.), and these can be implemented with only crude information on relative stock size.

Looking beyond salmon and herring, to other fisheries that appear to have been sustained over long periods without resorting deliberately to large space/time closures, we find in virtually all such cases some natural or bioeconomic refuge that has protected large parts of the stock from fishing. Until recently, tuna fisheries were concentrated in coastal areas, and there were large offshore stock components that buffered against overfishing. Cod off Newfoundland had large offshore spawning components that were not accessible to the inshore fishery over most of its history; trouble started immediately when foreign trawlers began to trawl along the continental shelf edge. Lobster fisheries have enjoyed offshore, deep-water refuges for larger reproductive animals (Walters *et al.*, 1993), owing to offshore migration of these animals and economic impracticality of setting traps in deep water over scattered bottom habitats (until GPS systems became available).

21.3 WHY STOCK ASSESSMENTS FAIL, AND WILL CONTINUE TO DO SO

Many of the world's major fisheries are now managed by setting total allowable catch (TAC) or annual quotas as a fixed percentage of the estimated stock size. This approach would work fine as a way of implementing a fixed exploitation rate strategy indicated as good for sustainability in the previous section, if assessment scientists were indeed good at estimating stock size each year. But the sad fact is that most stock assessment systems are just not up to the task of providing accurate enough stock size estimates to be safe in the face of fishing technologies that can get the quota even if stock size is much lower than estimated; expected management performance degrades sharply as the average error in stock size estimation increases (Walters and Parma, 1996). Proponents of quota management should heed well the results presented in Fig. 3 of Walters and Parma (1996), as these results imply that typical stock size assessment errors (CV of estimates 0.5 or larger) are large enough to more than wipe out the economic gains to be expected from quota management (Walters and Pearse, 1996).

Why are fisheries scientists so bad at estimating stock size on a timely basis for management? We are good at reconstructing what happened long ago (using virtual population analysis of fish that were subsequently caught or observed to spawn), but we have failed badly at estimating current abundance, even in some very data-rich situations such as the cod off Newfoundland (Hutchings and Myers, 1994; Walters and MacGuire, 1996). Indeed, the basic problem that we have encountered in stock assessment is to obtain good data on trends in relative abundance. We must have such trends to establish the effects of fishing, and hence total current stock size, in all fish stock assessment situations where the fish cannot be directly counted (i.e. almost all fisheries). Usually, the relative abundance measure we have available is catch per unit fishing effort (CPUE), and fishers are seldom so stupid as to let their catches fall in proportion to fish abundance (i.e. CPUE indices usually stay high during fishery collapses, making us think that stocks have not fallen as rapidly as they really have). We have also got into trouble in some cases when we have assumed constant relative vulnerability of young fish, when in fact fishers have started targeting such fish during fishery collapse (and their high catches of these young fish then make us think there are lots of young fish out there if we do not account for the targeting change).

Difficulties with relative abundance data in stock assessment could be solved in principle by investing more in fishery-independent surveys. But such surveys are extremely costly, even when done by, or in close co-operation with, commercial fishers. There is little prospect that scientists will develop magical new technologies for making fish counting and surveys much cheaper: the ocean is an inherently costly place to work. So particularly for smaller fisheries, scientists often must admit that the costs of better stock assessment would be greater than the value of the fishery that they are trying to assess! This is perhaps acceptable when the costs of management are 'hidden' or borne by the public through fisheries management agency budgets, but it is coming more to the fore as the public begins to demand that fishers shoulder much or all of the costs of regulation and management. It would not be at all surprising to find many fishers willing to accept radical restriction in fishing areas/times, should the only alternative for insuring sustainability be a costly survey system that they are required to fund.

21.4 DENOMINATING FISHING RIGHTS IN SAFE TERMS

Suppose there is a movement toward the 'salmon' management approach in Fig. 21.2, by closing very large areas to fishing so as to insure that exploitation rates remain safely low. Will this not create severe competition among fishers, and unpredictability in fishing opportunities, for more typical fisheries where the fish are widely distributed and fishers must search hard for economical fishing sites? This is certainly a possible outcome, and few would disagree that some of the BC salmon and herring fisheries are indeed pathological in terms of competition on the fishing grounds. Small fishing areas and short openings create scramble competition for the best fishing sites, and this competition can drive wasteful investment in technologies that do not fish better, but just get fishers to the best fishing places quicker.

Economic theory about how to deal with the wasteful competition problem has focused almost entirely on individual, transferable quota management systems (Hannesson, Chapter 19, this volume). It is obvious that the predictability associated with knowing how much a fisher will be allowed to catch before each season can allow that fisher both to optimize fishing activities and to achieve the highest price for the fishing 'right'. But if managers cannot in fact assure such predictability without placing the resource at risk, or if the hidden cost of achieving predictability is in fact greater than this predictability is really worth, then it is completely misleading to denominate the fishing right in such precise terms.

There is in fact nothing to stop the public from denominating fishing rights in ways that assure *opportunity* to fish without guarantee of a particular catch each year. For example, the fishing right can be denominated as a guarantee of exclusive access to a particular place at a particular time each year. This approach has been discussed for years with BC salmon: let each licence consist of a set of 'access cards', each guaranteeing the holder a fishing site or open area on a particular date; then let fishers freely trade these cards to define fishing plans that best fit personal preferences and beliefs about the best places to be. Then manage the fishery for improved ecological and overall economic performance by limiting the number of cards issued each year.

Twenty years ago such an elaborate fishing rights system, with many small space–time fishing opportunities, might have been seen as too complex and cumbersome for both fishers and public managers to administer effectively. But modern information management and communication technologies (e.g. the Internet) can make this administration (and trading system) almost trivial. And other information-gathering technologies, like satellite surveillance and required vessel location transponders, can make it possible and economical to enforce the overall system of space–time fishing restrictions. If we can assure fishers that such technologies will indeed protect their fishing opportunities, we may obtain even more co-operation from them in monitoring and enforcement than has been achieved through quota management systems.

A further objection to using a salmon-like approach for other marine fisheries is that spatial distributions and migration patterns are seldom as predictable as they are for BC salmon and herring, so that expensive distribution surveys might be needed each year to define appropriate locations for fishery openings. Even a small trawl fishing area could have devastated the spawning stock of cod off Newfoundland near the end of the stock decline, should this area happen to have been exactly where the last spawning aggregations formed. There is no simple answer to this objection for fisheries in general, because we have little experience with design of surveys intended only to rapidly map distribution patterns without a more ambitious aim of estimating stock size. Existing survey methods are often too expensive for wide and routine use in mapping out fishing areas for each fishing season, but it is entirely possible that scientists could develop much cheaper, rapid survey methods for situations where the only aim is to define locations for openings that will insure limitation in total exploitation rate; indeed, the pre-season herring surveys conducted by DFO and fishers in BC now represent just such an approach. Locations of herring openings are adjusted on small space–time scales every year in relation to shifts in locations of spawning aggregations. Surely such cooperative arrangements between managers and fishers can be exploited in many other fisheries.

21.5 GETTING THE PRIORITIES STRAIGHT: *OUR* FISH, *OUR* LEGACY, *OUR* RESPONSIBILITY

Much discussion about reinventing fisheries has centred on the development of more effective institutional structures for management (Preikshot, Chapter 7, this volume). Much has been said about rights-based fishing, quota management systems, community-based management, and a lot of other ideas that are based on the fundamental presumption that cooperation among fisheries stakeholders will lead to more responsible and risk-averse harvesting practices and to initiatives for resource enhancement. Hiding behind all these discussions is the fundamental assumption that fishers do in fact have an established right of access to the fish, that in some sense they 'own' the productive capability or natural surplus that the stocks produce.

The public needs to seriously question this hidden assumption (Charles, Chapter 6, this volume). In fact, most states still claim public ownership of fisheries resources. The public has not knowingly and willingly relinquished that ownership. Yet our public management agencies, whose salaries we pay, still insist on treating fishers as their 'clients' and on negotiating with fishers in co-management committees, public meetings, and other forums as though the fishers did have some special and permanent entitlement. In many cases, our public managers are thrust into meetings with industry that are every bit as threatening as it would be for prison guards to have to sit with their charges at every meal. It is little wonder that these managers are notorious for bowing to the pressures of industry for a little more time, a few more fish, even in situations where there is a clear conservation problem and the fishery should simply be closed.

So until we, the citizens of fishing states, explicitly decide to give away *our* resources to the fishing interests, my most important recommendation for reinventing fisheries is to reassert the public right to see safe and sustainable management practices, no matter how such practices may affect those fishers to whom we extend the privilege of access to our fish. From this reference point it is not radical or unacceptable at all to assert that we should have large closed areas and short seasons; indeed, these become the most prudent initial steps toward a sustainable fishery for the future.

REFERENCES

Hutchings, J. and Myers, R.A. (1994) What can be learned from the collapse of a renewable resource? Atlantic cod, *Gadus morhua*, of Newfoundland and Labrador. *Can. J. Fish. Aquat. Sci.* **51**, 2126–2146.

Walters, C.J. and MacGuire, J.-J. (1996) Lessons for stock assessment from the northern cod collapse. *Rev. Fish Biol. Fish.* **6**, 125–137.

Walters, C.J. and Parma, A. (1996) Fixed exploitation rate strategies for coping with the effects of climate change. *Can. J. Fish. Aquat. Sci.* **53**.

Walters, C.J. and Pearse, P.H. (1996) Stock information requirements for quota management systems in commercial fisheries. *Rev. Fish Biol. Fish.* **6**, 21–42.

Walters, C.J., Hall, N., Brown, R. and Chubb, C. (1993) Spatial model for the population dynamics and exploitation of the Western Australian rock lobster, *Panulirus cygnus*. *Can. J. Fish. Aquat. Sci.* **50**, 1650–1662.

Chapter twenty-two

Fish production, food webs and simple trophic models

William E. Neill

ABSTRACT

Fisheries science and management have traditionally focused on direct physiological effects of environmental variation on fish production. But as consumers near the top of aquatic bioenergetic pyramids, fish depend on the productivity and transfer efficiencies of lower trophic levels to fuel their production. Lindeman's trophodynamic model of ecosystem productivity and energy transfer is the predominant conceptual framework underlying most ecosystem-based management of fish production. The biomass, spatial and temporal distribution, availability, and productivity of prey at each trophic level establish the fluxes of energy/carbon into each consumer level. Management consists either of monitoring those fluxes (and advising the fishery of expected upturns or downturns in production) or of deliberately manipulating those fluxes to enhance production of desirable species. Nutrient additions or deletions, habitat improvements, and introductions of exotic prey to facilitate trophic transfers are common bottom-up manipulations. However, as mean fish production increases, so too does the variance in production.

Much of this added variance can be accounted for by a top-down trophic cascade. This model argues that while energy/carbon flows up the trophic pyramid, the control of that flow into various biomass pools at each trophic level comes down from higher trophic levels. Thus, piscivores control the types, activities, and production of planktivores. Planktivores, if

Reinventing Fisheries Management. Edited by Tony J. Pitcher, Paul J.B. Hart and Daniel Pauly. Published in 1998 by Kluwer Academic Publishers, London. ISBN 0 412 83410 3.

not regulated by piscivores, similarly control zooplankton. In short, predation at the top of trophic pyramids can have substantial indirect influence on the rates of energy/carbon transfer by lower levels, and effects appear to be strongest where ecosystems are relatively productive.

While still controversial, this model implies that fisheries managers must be cognizant of potentially complex interactions between bottom-up and top-down forces that continually change in aquatic food webs. Much theory and empirical work is required before management of food webs to enhance fish production can be done reliably.

22.1 INTRODUCTION

Fisheries scientists have long recognized that the productivities of exploited stocks are sensitive to environmental variability. Environmental factors may directly affect physiological properties of individual growth, survival and reproduction, and assessing these direct effects is a well-established part of modern fisheries science. But environmental variation may also *indirectly* affect productivity by altering the properties of biotic linkages among organisms in a species' food web, sometimes several steps removed from fisheries near the top of the trophic pyramid. While forecasting the occurrence, timing and magnitude of environmental change is at best stochastic, forecasting subsequent indirect effects on stock productivity is much more difficult than forecasting direct physiological effects. The range of possibilities for indirect effects is potentially enormous. What, for example, will be the effects on yellowfin tuna production and distribution of a three-degree temperature rise and 50% increase in phytoplankton production stimulated by elevated atmospheric CO_2 from combusted fossil fuels? Will the direct effects of warmer temperatures and altered growth efficiency override the indirect effects of changes in competitors, prey abundance and composition, or predators of juveniles? Much of the uncertainty stems from contingent factors in the physical, chemical and biological environment that alter the magnitude, timing and persistence of effects. Such contextual concerns have prompted calls for intensive research on 'ecosystems approaches to fisheries management' in the 21st century.

While it is at least hypothetically possible to undertake ecosystems-scale research and management on a case-by-case basis, greater efficiency, insight and generality are likely if driven by one or more robust conceptual models of system function. The strength (and weakness) of a 'conceptual framework' approach is that only certain kinds of explanations are included a priori as probable agents of change. For the past 50 years, the conceptual basis of nearly all work on (indirect) food web mechanics in marine and much of freshwater fisheries has been a donor-controlled energy/carbon flow model of ecosystem function and fish productivity. This is the Lindeman trophodynamic model, in which external environmental variation affects rates of carbon uptake, incorporation and transmission from prey and to consumers in series. Upper trophic levels receive the largesse of whatever fixed energy from primary production is passed to them by intermediate trophic levels. Consumption by organisms in any trophic level is seen to act only to acquire the biomass of prey and has no regulatory impact on the abundance, composition or productivity of any lower level. It is thus a model of predation on 'excess' biomass produced by the level below. While there is no debate about the general validity of energy flow from primary producers to fish, the contentious issue is whether energy flow is strictly donor controlled.

Management under a 'bottom-up' model of limited energy flow usually takes one of two forms: (1) prescriptions for *reactive* responses by fisheries from forecasts of the occurrence, timing and magnitude of inherently unmanageable environmental variation or (2) *proactive* manipulations of the environment to enhance the rate of energy/carbon transfer at each level to maximize productivity of top trophic levels. The spatial extent of the system to be managed generally determines whether the approach must be reactive or can be proactive. In marine fisheries, donor-controlled food web theory is deeply entrenched and is reactive. Fisheries oceanographers expend enormous effort pursuing ever-finer-grained measurements of primary production, temperature, and water movement across large spatial scales to try to anticipate probable variations in food web production. At the smaller scales of many freshwater fisheries, bottom-up trophodynamics constitute the theoretical basis for enhancing the production of harvested species by manipulating the flow of nutrients and energy through the underlying trophic web (Hyatt and Stockner, 1985). Recreational fisheries at the scale of individual streams or lakes are centres of proactive, manipulative management at the base of the food web.

A second conceptual model for focusing fisheries research and management in the next century was codified in theory only about a decade ago, although it had existed less formally prior to this time (Hairston *et al.*, 1960). In contrast to the 'bottom-up' trophodynamic model, a 'top-down' perspective of food webs such as the trophic cascade model (Carpenter *et al.*, 1985) holds that the biomass, composition and productivity of lower trophic levels not only underwrite higher trophic levels energetically (as above), but also are limited by them. Fish production is still considered to be limited on average by the thermodynamics of energy fixation and inefficient energy transfer, but the structure and function of the trophodynamic web are modified by the predation process, thereby altering fish production itself. In theory, such predation-driven variations in prey, if large, generally reduce trophic transfer efficiency, so management must optimize the influence of top-down against bottom-up processes. Changes may be only transient, varying with fluctuating predator densities or behaviours. As such they may require continual hands-on tuning. Management is proactive. It involves manipulating the types and intensities of predation to direct the pattern of predatory mortality, maximize useful production at each level and adjust the efficiency of energy transfer through the food web. It is potentially much more complex than trophodynamic bottom-up management because of the multiplicity of feedbacks involved. Consequently it requires a much more complete map of trophic connectedness and its regulation.

As is frequently the case at early stages of theoretical development, these two models of ecosystem organization have been hotly debated during the past decade (Carpenter and Kitchell, 1992; DeMelo *et al.*, 1992), particularly after previous large-scale bottom-up management schemes had resulted in either mixed success (e.g. Hyatt and Stockner, 1985) or spectacular failures (e.g. Spencer *et al.*, 1991). In this short chapter, I discuss some implications for fisheries of adopting one or the other of these two models of aquatic food web organization.

22.2 BOTTOM-UP DYNAMICS

Figure 22.1 is a simplified depiction of the bottom-up perspective of production across lake or oceanic ecosystems with four trophic levels. Note that predators do not suppress prey in this model so biomasses of all consumer levels increase with algal biomass. At low algal densities, total production in any trophic level is low because of insufficient biomass to generate the productivity. Secondary production follows increasing primary production and is correlated with factors that enhance algal production (e.g. nutrients, light), although not necessarily at the same rate in all trophic levels. At very high levels of algal biomass, production falls as a result of habitat degradation, disease or density-dependent self-suppression of production (competition for light, habitat space, etc.). In short, maximum production of each consumer trophic level occurs at intermediate (to high) producer and consumer biomasses.

Production of the whole food web – and from a fisheries perspective, of the piscivorous and planktivorous fish in particular – is maximized when the peaks in production are aligned as in Fig. 22.1. The theoretical expectation from a strictly bottom-up perspective is that *on average*, trophic webs will tend toward this aligned arrangement as a result of organisms at each trophic level maximizing their own production on the available production of their prey. However, environmental vicissitudes that affect survival,

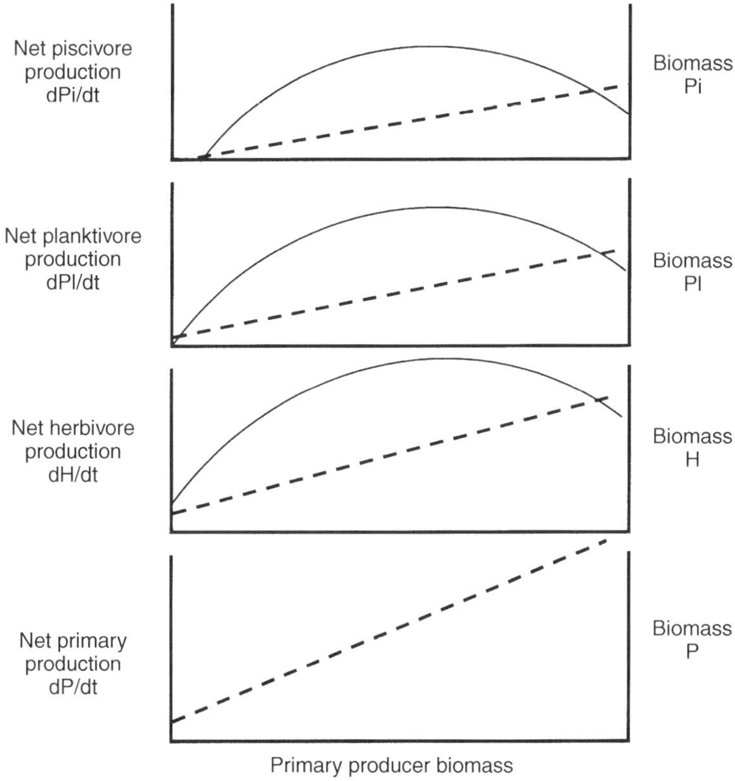

Fig. 22.1 Simple 'bottom-up' trophodynamic model of a four-trophic-level aquatic food web. Biomasses (dashed lines) of each trophic level increase as a function of prey availability which is determined only by abiotic factors and the food availability from the previous level. Total production (solid line) assumes density-dependent resource limitation at high biomass concentrations at each trophic level. Energy flow through the system is maximized when peaks in the production curves are aligned.

growth or reproduction might reduce the height of a trophic level's production or displace the location of its maximum on the biomass axis, particularly if community taxonomic composition changes with environmental conditions. Distortions of these sorts may be more likely at high biomasses where the scope for changes is greater. In all cases, these kinds of changes reduce the quantity and perhaps quality of organic matter transmissible to higher levels, and pelagic fish production falls.

In the absence of long-term, within-lake production data for all trophic levels, comparisons among lakes can be used with caution to evaluate

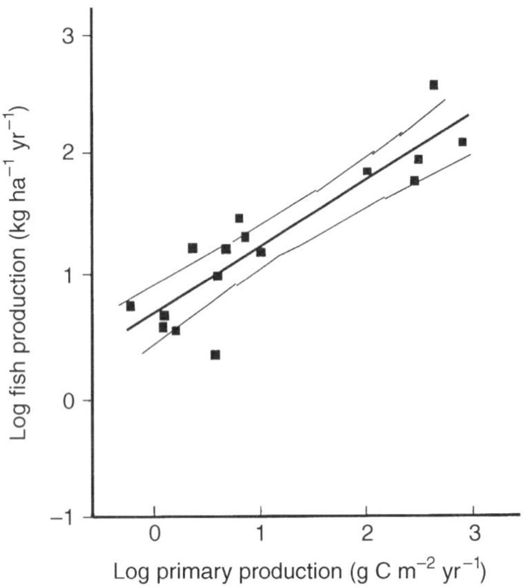

Fig. 22.2 Relationship between annual total primary production and annual fish production among lakes; 95% confidence interval around the regression line is shown. Modified from Downing *et al.* (1993).

model predictions of energy flow and relative variability in primary or secondary production. Comparisons among lakes (Fig. 22.2) indicate that the logarithm of annual fish-community production is strongly positively correlated ($r^2 > 0.75$) with the logarithm of limnological properties related to lake fertility. These factors include lake morphometry (Rawson, 1952), annual mean concentration of (algal) chlorophyll *a* (Oglesby, 1977), annual mean nutrient supply (Hansen and Leggett, 1982) and annual primary production (Oglesby, 1977, Downing *et al.*, 1990, Fig. 2). Similarly, in a literature survey of 100 fish populations, Downing and Plante (1993) found the logarithm of annual population production to be well correlated with the logarithms of nutrient supplies and primary production, after the effects of population biomass and individual body size had been accounted for. These relationships, coupled with similarly strong correlations between (log) primary production and (log) annual invertebrate production (Plante and Downing, 1989), reinforce the bottom-up, trophodynamic paradigm.

Logarithmic transformation of both axes in plots like Fig. 22.2 reduces the density dependent curvature that is anticipated from Fig. 22.1. It also shrinks the ballooning variance in production that occurs commonly at higher productivities. By not logarithmically transforming the production data of Downing *et al.* (1990), or of similar data of Hanson and Leggett (1982) or of Plante and Downing (1989) and Downing and Plante (1993), one sees modest, density-dependent curvature in annual fish production at high values of the independent variables, though not so strongly as in Fig. 22.1. Limited density dependence in fish production probably occurs because interannual comparisons *among lakes* are not appropriately scaled to detect the density-dependent mortality, growth and reproduction of fish that occur at seasonal and monthly time scales *within lakes*. Variability, on the other hand, is probably exaggerated in such among-lakes relationships, because of differences in the physical-chemical environments of lakes. Variance in (untransformed) fish production increases with mean concentration of algal biomass, nutrients or primary production, suggesting that the bottom-up signal becomes less predictable across richer, more productive lakes.

Watson *et al.* (1992) show that much of this variation seems to originate at the primary producer level. For example, they found that chlorophyll *a* concentration (primary production) at moderate to high nutrient concentrations (10–200 $\mu g\,l^{-1}$ total phosphorus) is 10–25 times more variable than at low nutrient concentrations. They attributed almost all of this added uncertainty to variable growth during summer by certain algal and cyanobacterial species that can be resistant to grazers. Scarcity of grazer-resistant forms promotes large-bodied suspension feeders among the zooplankton that are in turn readily consumed by planktivorous fish. A preponderance of ungrazable primary producers shifts the competitive balance among zooplankton toward small taxa that are able to glean useful particles from the ungrazable matrix. However, small grazers are less productive and much less attractive to planktivores than are large grazers and result in relatively reduced growth, survival, and production of fish. Consequently, year-to-year and lake-to-lake variability of fish production is substantially higher when limnological conditions favour potential blooms of mainly grazer-resistant primary producers. In the extreme, lake or stream managers who manipulate nutrients or other limnological properties so that algal and invertebrate production are inappropriately packaged to supply energy for pelagic fish production may shift the location of production into non-target benthic food chains. Desirable fish species may or may not be able to take advantage of the resulting benthic production of decomposers that exploit decaying ungrazable primary production. Less desirable taxa may be favoured, especially if water quality deteriorates. Further, losses of organic matter to sediments and decomposer metabolism may be high, so overall fish production may be substantially reduced.

In summary, the bottom-up model of ecosystem production describes the maximal rate of energy flow possible from primary producers to fish. It explains departures from that rate to be the result of environmental variations which depress survival, growth and reproduction of some portion of the community, mostly at lower trophic levels. In short, variability in fish production is seen as a consequence of variable bottom-up dynamics. Under this model, the fisheries manager must assess particulars of bottom-up limitation for given circumstances and plan ameliorative actions to nutrients, habitat or food supplies.

In contrast, many fisheries biologists and systems ecologists have argued that this trophodynamic perspective has missed a fundamental property of food web processes, namely that many consumers at each trophic level in aquatic ecosystems strongly affect their prey's abundances, composition, distributions, and productivity. They argue that, while energy and nutrients flow up a trophic pyramid, feedbacks of predators on prey are common and powerful modifiers of transfer rates and production efficiencies. Further, the magnitude of these feedbacks may be equivalent to the magnitudes of bottom-up environmental variations. Failure to recognize such processes when they are significant weakens managers' abilities to achieve desired fisheries outcomes.

22.3 TOP-DOWN DYNAMICS

Cognizant of the direct effects of predators on prey populations, Carpenter *et al.* (1985 and ff.) argue that much of the variability in lower trophic levels stems not (only) from direct environmental disturbances, but from variations in functionally linked predator–prey sequences starting near the top of the pyramid. This variability is generated by changes in reproductive output, fishing mortality, over-winter survival, disease, predation and other variations in the abundance and behaviour of fish. Proponents of top-down models suggest that changes in web structure and function may be largest in systems that are relatively productive because replacement taxa that are rare or exotic prior to food web disturbance are least likely to be resource limited. That is, the scope for change is potentially larger in more productive systems, though the realized magnitude of response may be more variable as well. Accordingly, the variable colonization and proliferation of some productive lakes by ungrazable algal taxa (see previous section; Watson *et al.*, 1989) is almost certainly a direct adaptive response to changes in the intensity of grazing pressure by zooplankton, the composition and abundance of which varies strongly with planktivory. Likewise, it is probably not coincidence that the exotic invertebrate predator *Bythotrephes* successfully invaded the North American Great Lakes and radically altered the plankton community just after the collapse of planktivorous alewife (*Alosa*) populations under sustained heavy predation by introduced salmon (Lehman *et al.*, 1993).

Several discussions of trophic cascades have commented on the magnitudes of changes at the top of the trophic web that are necessary to detect effects two or more trophic levels away (Carpenter *et al.*, 1988, DeMelo *et al.*, 1992, Power, 1992). Small, short-lived changes in piscivores are probably dissipated quickly in stochastic variations within existing prey populations of a lake. Larger changes are believed to be capable of penetrating the web more completely, altering biomasses and productivity, shifting relative abundances, and even causing local exterminations of taxa several steps removed from the initial changes (Carpenter and Kitchell, 1992). In general, the consensus seems to be that large changes are required to propagate a trophic cascade that substantially alters the food web structure. Failure of the cascade to appear in the Great Lakes until very large numbers of Pacific salmon were introduced is consistent with this assessment. But it is also apparent that even large changes in piscivory do not necessarily cascade through every food web. If prey are dynamically decoupled from their predators (e.g. by refuges), top-down effects are dissipated. High variance in fish production in more productive lakes (Fig. 22.2) is at least partially driven by variable transmission of top-down signals as a result of highly variable species composition, physical-chemical constraints, and vulnerabilities to predation at various life history stages.

In terms of overall web productivity, and especially productivity of fish near the top of the trophic pyramid, functioning trophic cascades may seriously reduce the capacity of the ecosystem to produce fish biomass, compared with bottom-up trophodynamic models. In a trophic cascade, predatory suppression of a prey trophic level causes the next lower level to be at least partially released from limitation by predation. This means that adjacent levels will fail to 'provide' and 'harvest' newly produced biomass at comparable rates, resulting in accumulation of biomass at some trophic levels and depletion at others. Inefficiencies of trophic transfer are the result. In general, the important consequences in the food web are non-alignment of the maxima of production among trophic levels, reduced transfer efficiencies in ecosystem trophodynamics and ultimately lower production of harvestable fish.

These effects can be seen in Fig. 22.3, in which the productivity distributions are distorted from the well-aligned configuration of Fig. 22.1 as a result of cascading trophic effects. At low piscivore biomass, biomass tends to accumulate in algae and planktivores because their production is not harvested readily by their scarce consumers. At the other extreme, high piscivore biomass directly suppresses planktivore biomass and production, and indirectly, primary producers and their production. The latter

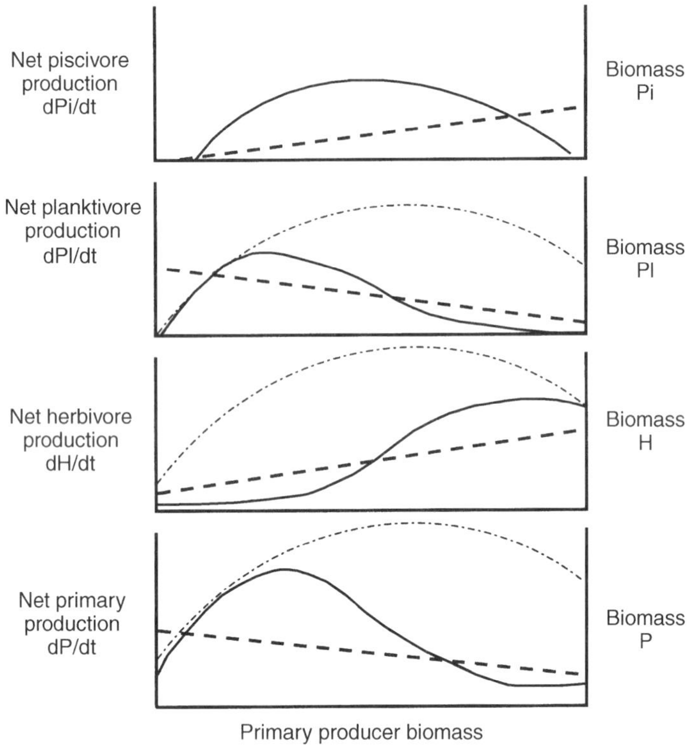

Fig. 22.3 Modified food web relationships under 'top-down' influence of a trophic cascade from top predators. Biomasses (dashed lines) and total production curves (solid lines) for each trophic level below the piscivore level have been altered by predation compared with the bottom-up model in Fig. 22.1. For reference, production curves (dot-dashed lines) for the donor-controlled trophodynamic model are shown but do not relate to the biomass functions.

occurs because herbivores are released from limitation by their consumers and then suppress algal abundance. At both extremes, at least one trophic level acts as a bottleneck to energy transmission, so overall efficiency of fish production declines. Piscivory maintained at intermediate intensities should facilitate energy flow, though overall efficiency will probably be lower than if bottom-up forces alone pertained. Management decisions regarding piscivore abundance can have major indirect consequences on the composition of the food web, water quality and piscivore growth, survival and reproduction. To the degree that managing piscivory is an inexact science, top-down uncertainty has the potential for generating considerable variance in energy flow.

The important distinction between a traditional trophodynamic perspective and a top-down trophic cascade is that predators in the latter can potentially limit (and probably regulate, *sensu strictu*) the abundance and production of their prey. None can in the donor-controlled trophodynamic model. The empirical validity of this potential predatory control is by no means obvious for all trophic interactions, but especially at zooplankton and phytoplankton levels (DeMelo *et al.*, 1992). Nevertheless, evidence is strong from experimental enclosures and many small to medium-sized lakes (and a few large ones such as Lake Michigan) that enhancement of consumers commonly results in large fluctuations in the abundance of sensitive prey and then their prey. Any particular outcome is not always observed everywhere, because effects clearly depend on local characteristics of the habitat, predators and prey, as they do in all predator–prey relationships (Holling, 1959). Outcomes may also be confounded by other indirect effects of changes in predator abundance (e.g. interference behaviours, avoidance behaviours by prey, altered nutrient availability via recycling) that mask effects on prey biomass. But many results recur sufficiently frequently that pure bottom-up control cannot be the sole or even predominant organizing principle for these links.

From a resource manager's perspective, recognizing that top-down signals are potentially very significant structuring agents in lakes does not necessarily lead to simple prescriptions for action. Complex non-linear dynamics are possible, and even likely, depending on the forms and magnitudes of predatory functional and numerical relationships. Consequences for production of desirable fish taxa may be difficult to predict quantitatively, given our incomplete knowledge of the functional interconnectedness of many food web components. The difficulties in forecasting fisheries effects under such complexity can be readily illustrated with an example drawn from small, modestly productive sport-fishing lakes in British Columbia.

A common management problem in trout (*Oncorhynchus*) lakes is anglers' release of unused bait fish (usually small cyprinids) that may proliferate, consume large-bodied zooplankton and competitively suppress growth and survival of trout fry, juveniles and even planktivorous adults. Trout, most bait fish and many other temperate zone planktivores prefer to eat *Daphnia* and other common, large-bodied zooplankton because of their size and ease of capture. If bait fish numbers are reduced under management (e.g. by trapping or enhancing piscivory), large-bodied zooplankton typically rebound rapidly and remove several times the standing crop of grazable algae in the manipulated lake each day. There can be three effects of this release of *Daphnia*, depending upon various contingent factors.

1. Algal biomass and production may be rapidly depleted, strikingly improving water clarity, producing a 'clearwater' period. Vulnerable zooplankton, benthic invertebrates and small fish rapidly adjust their behaviours to lessen elevated risk of predation in the transparent water. *Daphnia* populations collapse under insufficient primary production, and fish growth and survival fall.
2. Less competitive (small) zooplankton taxa may be excluded from much of the lake by highly efficient *Daphnia*, drastically reducing the diversity of foods for fry and juveniles of certain fish taxa specialized on non-daphnid prey. Survival and growth of these fish decline.
3. Grazer-resistant algae and cyanobacteria proliferate under altered grazing conditions and changes in nutrient fluxes. These ungrazable forms may come to dominate the primary producers within a few weeks, minimizing grazing harvest, potentially releasing toxins to fish and invertebrates, stimulating bacterial growth, and depleting O_2 at depth. Invertebrate production declines, fish alter their foraging patterns, and survival and growth are compromised.

Daphnia is undoubtedly the best-studied zooplanktonic taxon in fresh water. Yet I doubt anyone would be willing to predict the consequences to fish production and water quality of indirectly manipulating their biomass in such systems. In fact, the most risk-averse management procedure to improve game fisheries in small lakes with excessively high bait fish populations in British Columbia has been to reset the food web to a more desirable species configuration. Intensive trapping or even use of toxins can remove nearly all fish. Desirable fish can then be restocked after the altered food web has been permitted time for recovery. Intensive manipulations of the food web on this scale approach aquaculture.

The point here is that, especially in more nutrient-rich lakes, a simple reduction of planktivory by bait fish can generate enormous consequences for food web function and fish production. Our ability to accurately forecast these contingent effects is very limited. Undoubtedly, organisms other than *Daphnia* play similar multidimensional roles in lake communities, and probably in marine systems as well. Research priority must be placed on identifying and evaluating consequences of alterations in such organisms prior to employing half-understood top-down theory to lake fisheries management.

22.4 CONCLUSIONS

Manipulation of aquatic ecosystems to modify the energetics of fish production is still in its infancy. Even classical bottom-up approaches such as nutrient enrichment of lakes or introductions of exotic species as conduits of energy from one trophic level to the next are fraught with problems. Understanding the types and strengths of regulatory processes governing food web dynamics, including bottom-up and top-down mechanisms, is a goal to which ecosystem-level fisheries management must strive. Neither the theory nor data are yet sufficient to achieve desired fish production reliably, except perhaps under oligotrophy where the scope for unwittingly stimulating alternative bottom-up or top-down responses is relatively small (but see Hyatt and Stockner, 1985). Initially, reliable theory and data are likely to be developed for systems in which there is strong dominance of food web dynamics by relatively few persistent taxa (or many taxa that are functionally interchangeable). In functionally diverse food webs, where regulatory processes governing each of many food chains are weak or highly interactive, management manipulations have high potential for inadvertently stimulating some pathways, depressing others and generating unpredicted fisheries outcomes.

The most instructive courses for research in the medium term are likely critical analyses of existing case histories, modelling, and adaptive experimental management. Unfortunately, some basic properties of ecological processes in aquatic food webs are still quite vague (e.g. energy flow in spatiotemporally patchy environments), so much basic research is needed. If fisheries are to be reinvented from an ecosystem-management perspective, a complex task lies ahead of us. Recognizing that predatory impacts may cascade down a trophic pyramid and accordingly can have substantial organizational consequences for energy flow up the pyramid is a good start. But it is just a start.

REFERENCES

Carpenter, S.R. and Kitchell, J.F. (1992) Trophic cascade and biomanipulation: interface of research and management – a reply to the comment of DeMelo *et al.* *Limnol. Oceanogr.* **37**, 208–212.

DeMelo, R., France, R. and McQueen, D.J. (1992) Biomanipulation: hit or myth. *Limnol. Oceanogr.* **27**, 192–207.

Downing, J.A. and Plante, C. (1993) Production of fish populations in lakes. *Can. J. Fish. Aquat. Sci.* **50**, 110–120.

Downing, J.A., Plante, C. and Lalonde, S. (1990) Fish production correlated with primary productivity, not the morphoedaphic index. *Can. J. Fish. Aquat. Sci.* **47**, 1929–1936.

Hairston, N.G., Smith, F.E. and Slobodkin, L.B. (1960) Community structure, population control, and competition. *Am. Nat.* **94**, 421–425.

Hanson, J.M. and Leggett, W.C. (1982) Empirical prediction of fish biomass and yield. *Can. J. Fish. Aquat. Sci.* **39**, 257–263.

Holling, C.S. (1959) The components of predation as revealed by a study of small mammal predation of the European pine sawfly. *Can. Entomol.* **91**, 293–320.

Hyatt, K.D. and Stockner, J.G. (1985) Responses of sockeye salmon (*Oncorhynchus nerka*) to fertilization of British Columbia coastal lakes. *Can. J. Fish. Aquat. Sci.* **42**, 320–331.

Lehman, J.T. and Caceres, E.E. (1993) Food-web responses to species invasion by a predatory invertebrate: *Bythotrephes cederstroemii* in Lake Michigan. *Limnol. Oceanogr.* **38**, 879–891.

Oglesby, R.T. (1977) Relationship of fish yield to lake phytoplankton standing crop, production, and morphoedaphic factors. *J. Fish. Res. Bd Can.* **34**, 2271–2279.

Plante, C. and Downing, J.A. (1989) Production of freshwater invertebrate populations in lakes. *Can. J. Fish. Aquat. Sci.* **46**, 1489–1498.

Power, M.E. (1992) Top-down and bottom-up forces in food webs: do plants have primacy? *Ecology* **73**, 733–746.

Rawson, D.S. (1952) Mean depth and the fish production of large lakes. *Ecology* **33**, 513–521.

Spencer, C.N., McClelland, B.R. and Stanford, J.A. (1991) Shrimp stocking, salmon collapse and eagle displacement: cascading interactions. *Bioscience* **41**, 14–21.

Watson, S., McCauley, E. and Downing, J.A. (1992) Sigmoid relationships between phosphorus, algal biomass and algal community structure. *Can. J. Fish. Aquat. Sci.* **49**, 2605–2610.

Chapter twenty-three

Protected marine reserves as hedges against uncertainty: an economist's perspective

Ussif Rashid Sumaila

ABSTRACT

Most of the problems we see today in world fisheries emanate from two broad sources – the lack of sufficient and correct information about how fisheries work, and the inadequacy of current institutional arrangements to deal with the problems at hand. I trace recent proposals in the fisheries literature on the use of protected marine reserves to hedge against uncertainty. Finally, I give an economist's perspective on marine reserves in fisheries management, and suggest how economics may help to bring about their timely implementation in some of the world's fisheries.

23.1 INTRODUCTION

A number of policy recommendations have been advanced in the fisheries management literature to address issues related to the achievement of viable and sustainable fisheries. Most of these can be classified into four groups: (1) control of allowable catch or effort levels; (2) use of socioeconomic

Reinventing Fisheries Management. Edited by Tony J. Pitcher, Paul J.B. Hart and Daniel Pauly. Published in 1998 by Kluwer Academic Publishers, London. ISBN 0 412 83410 3.

instruments such as individual transferable quotas (ITQs) and decentralization; (3) more intense scientific research including 'adaptive' management; and (4) use of ecosystem-based management models.

The majority of these recommendations hinge on the recognition of the common pool (= common property) nature of capture fisheries and the difficulties it normally imposes on any attempts to optimally manage these resources (Gordon, 1954; Munro and Scott, 1985).

Three questions are posed and addressed here.

- Why is it that even for fisheries where some or all of the above suggestions have been applied, efforts at optimal management still fail? As a concrete case, why was there a dramatic collapse of Atlantic Canada's cod fishery, given all the optimism that followed Canada's claim of sole property rights to the resource?
- What efforts are being made by scientists to address the new issues that arise as a result of such failures?
- What role can economics play in helping to move the current debate from the theoretical level to the practical application of the new ideas arising from the ongoing debate in the management of real fisheries?

It may be argued that the main sources of problems for fisheries management, and indeed the driving force for the evolution of fisheries management concepts and practices, are twofold: first, the lack of adequate and correct information about how fisheries work; and second, the inadequacy of current institutional arrangements to deal with the problems at hand. Included in the first category are issues such as the inadequacy of current biological models of the fishery to capture what is actually going on under the sea, to the problems of getting proper hold of the socio-economic dynamics of fishing communities, right down to the problems of uncertainty, both biological and otherwise. Simply put, the complexity, and perhaps chaotic nature, of the biological (and the socio-economic) environment creates a massive, intractable information problem for fisheries management (Wilson *et al.*, 1994). On the other hand, the second category covers such issues as the ability or inability of society to put in place incentive and policy structures to implement what are thought to be optimal strategies for the management of the fishery. This chapter concentrates on the former problem; in particular, it discusses the burning issue of uncertainty.

23.2 UNCERTAINTY IN FISHERIES MANAGEMENT

Uncertainty is clearly one of the biggest obstacles to the achievement of the desired amount of information about a fishery that would, at least, provide a basis for efficient management. Sources of uncertainty may include the dynamic nature of fish populations in the wild and the variability and complexity of the marine ecosystems of which they are a part, the impact of fishing activity upon the resources, and the fact that perfect monitoring and control of harvesting in marine capture fisheries will forever be problematic.

Uncertainty has been classified in various ways in the literature. Here, I will classify uncertainty into two broad categories, that is, *first degree uncertainty*, and *second degree uncertainty*. First degree uncertainty consists of 'random effects whose future frequency of occurrence can be determined from past experience' (Walters and Hilborn, 1978). Hence, it is possible to construct objective probability distributions to capture this class of uncertainty. Further research has a great deal to offer here. Indeed, most stochastic economic models of fisheries incorporate only this type of uncertainty (Andersen and Sutinen, 1984; Clark, 1990).

Second degree uncertainty, usually termed true uncertainty, covers uncertain events that cannot be predicted, and for which objective probability cannot be estimated (C.W. Clark *et al.*, pers. comm.). It is possible to reduce this class of uncertainty through further research, but to eliminate it completely is but a dream: there will always exist an irreducible level of uncertainty. Ludwig *et al.* (1993) argue that the complexity of fisheries systems precludes predictability of the sort required to exercise the numerical control envisioned by current theory. So what is to be done, given this reality?

23.3 PROTECTED MARINE RESERVES AND THE MITIGATION OF UNCERTAINTY

The establishment of protected marine reserves (PMRs) has been promoted by many as a viable alternative where other forms of fisheries management are impracticable or unsuccessful (Wallis, 1971; Davis, 1981; Bohnsack, 1990; Lauck *et al.*, 1997). Bohnsack (1990) gives a comprehensive list of the potential benefits that may be expected from the establishment of PMRs, including (1) protection of spawning biomass, (2) providing a recruitment source for surrounding areas, (3) supplemental restocking of fished areas through emigration, (4) maintenance of natural population age structure, (5) maintenance of areas of undisturbed habitat, and (6) insurance against management failures in fished areas.

Recent work by Lauck *et al.* (1997) has explicitly linked PMRs to the mitigation of uncertainty in fisheries management. Lauck *et al.* contend that diversification and risk mitigation may be achieved by establishing protected reserves, a contention which is easy to subscribe to, given the potential benefits of PMRs enumerated above.

Most biological models produce results that tend to be positive about the biological usefulness of PMRs. For example, Bohnsack (1990) came to the conclusion that PMRs would be superior to other management strategies, in terms of maintaining productive populations of fish and ensuring sustainable harvests. Using a simple two-patch logistic model, Quinn *et al.* (1993) suggest that the establishment of a harvest refuge is necessary to prevent collapse and to maintain sustainable catches at high levels of harvesting effort. Roberts and Polunin (1991) list a number of studies that compare protected and unprotected areas for different fisheries with a view to investigating the benefit of PMRs with respect to various biological variables. The overwhelming majority of the studies tend to show that PMRs generally lead to positive biological effects. Indeed, a number of official and non-official initiatives such as the Marine Stewardship Council initiative (Sutton, Chapter 8, this volume) are positive to the use of PMRs to improve fisheries management.

To realize any or all of the potential benefits listed above, biologists correctly argue that PMRs must be designed appropriately: the location, size and shape of the reserve must be carefully chosen to reflect the realities of the habitat and the fishes to be protected (Rowley, 1994). A number of scientists have looked specifically at the question of the optimal size of protected reserves from different perspectives. Polacheck (1990) asserts that for the effects of protection to be realized, a reserve must be big enough to retain a large proportion of the protected individuals. Quinn *et al.* (1993) claim that the long-term persistence of the population and the long-term sustainable catch can benefit with increasing numbers and decreasing spacing of reserves, even up to half of the total area. Man *et al.* (1995) conclude that sustainable yield is maximized when 50% of all available patches are occupied. Clark *et al.* (pers. comm.) show, among other things, that a single non-age-structured population is quickly terminated in the face of true uncertainty when less than half of the habitat is protected. On the other hand, when 80% of the population is protected, the mean time for extinction exceeds 10^6 generations. Similarly, models developed by Mangel (pers. comm.) predict that the use of substantial protected reserves, that is, over 50% of the habitat, can greatly increase the long-term sustainability of fisheries subject to uncertainty and uncontrollability.

23.4 THE ROLE OF ECONOMICS

There seems to be agreement that to meet both the concerns for biology and uncertainty, a significant size of a given fishery must be protected. By significant, it is meant that the size of the reserve should be between 50% and 80% of the fishery. A practical and relevant question to ask here is, is it politically, socially, and economically possible to protect such large proportions of the fishery completely from exploitation (Walters, Chapter 21, this volume)? Because I am an economist, I will focus the preceding question more sharply: does it make economic sense to protect such large proportions of the fishery completely from exploitation for a long time? The answer to this question is probably no, particularly because it is likely that such closures would result in a 'decades-long transient period in which catches are depressed before the long-term benefits of the refugia on sustainable catch are realized' (Quinn *et al.*, 1993).

The role of economics in all of this is to help operationalize the principle of reserves by using bioeconomic criteria to determine the optimal size. What is advocated here is similar to the maximum sustainable yield (MSY) and the maximum economic yield (MEY) criteria, where the former is biological while the latter is bioeconomic. Hence, one can envisage a situation where, based purely on biological concerns a certain size of PMR is considered optimal for a given fishery, while based on bioeconomic concerns a different optimal size of PMR for the same fishery may be derived. The reader is referred to Farrow (1996) for a discussion of some of the economic considerations in the creation of marine reserves.

A bioeconomic analysis of PMRs will entail the maximization of discounted economic rent from the resource, subject to the stock dynamics of both the protected and the non-protected parts of the habitat (Holland and Brazee, 1996). The basic requirement for the development of bioeconomic models of PMRs is a good biological description of the relationship between the protected and the unprotected areas. We need to know, for instance, the nature of fish movement between the protected and the unprotected areas. Clearly, this will depend on the type of fish we are trying to protect: movement of pelagic fish would certainly be different from that of demersal species.

Generally, to operationalize the principle of PMRs, a number of things need to be done.

1. Detailed multidisciplinary case-by-case empirical studies to determine the best way to design PMRs for each major fishery need to be carried out. This is because theory alone is insufficient to justify a PMR; the recommendation can follow only from empirical investigation (Farrow, 1996).
2. The key concerns, that is, biological uncertainty, and socio-economic, should all be incorporated into such an analysis.
3. The question of trade-offs between size of PMRs and the socio-economic benefits that may be derived from the resource in the short, medium and long term must be addressed.

In this way, it is possible to make the benefits of PMRs transparent for all to see, and thus enhance the chances of their implementation. At the moment there are a few scattered claims in the literature that suggest that there may be no loss in yield with the introduction of PMRs. For example, Quinn *et al.* (1993) show that under certain circumstances, closures can exert little or no penalty in catch when harvest is already low. In addition, closures can permit sustained catches which would otherwise be impossible when harvesting effort is intense. Lauck *et al.* (1997) also make some claims in this direction. The kind of comprehensive case-by-case studies advocated in this chapter will serve to bring out these claims more forcefully for all parties concerned, thus brightening the chances for the implementation of PMRs.

23.5 CONCLUSIONS

The point of view in this chapter is that the use of protected marine reserves appears to be one of the means by which we may be able to reinvent fisheries management, especially because they may be used to mitigate the impact of second-degree uncertainty. But, for the protected marine reserves to be operationalized, socio-economic variables need to enter the equation for the determination of their optimal size. In particular, the size or design of protected marine reserves need to pass a bioeconomic criteria test. The trade-offs between size of protected marine reserves and the socio-economic benefits from the fishery in the short, medium and long term, need to be explicitly evaluated. In this way the benefits of implementing reserves will be revealed to all parties concerned, thereby improving the chances for their implementation.

The need to subject the design or size of protected marine reserves to a bioeconomic criteria test cannot be overemphasized. To do so will automatically take into account (1) the nutritional needs of the population through the demand for fish, (2) the cost of implementation and enforcement of reserves, as well as all its benefits, and (3) the determination of the optimal size of fishing effort to be used in exploiting fish in the non-protected area of the habitat, which is arguably crucial for the success of the whole idea of PMRs.

On a more general note, the beauty of the principle of PMRs, in my opinion, lies in the willingness of scientists to say: 'Look here, no matter how hard we try, there is still something out there about fisheries that we will probably never be able to understand.' The new image that this will give fisheries scientists should help improve their credibility in the eyes of fishers, fisheries policy makers and the public as a whole.

ACKNOWLEDGEMENTS

I thank two anonymous referees and the editors for useful comments on an earlier draft. The work is supported by the Research Council of Norway, through the Norwegian Man and the Biosphere Research Programme.

REFERENCES

Andersen, P. and Sutinen, J.G. (1984) A survey of stochastic bioeconomics: methods and results. *Mar. Resour. Econ.* **1**(2), 117–136.

Bohnsack, J.A. (1990) The potential of marine fishery reserves for reef fish management in the U.S. Southern Atlantic. *NOAA Tech. Memo* NMFS-SEFC-261, 40 pp.

Clark, C.W. (1990) *Mathematical Bioeconomics: The Optimal Management of Renewable Resoures*, Wiley Intersciences, New York.

Davis, G.E. (1981) On the role of underwater parks and sanctuaries in the management of coastal resources in the southeastern United States. *Env. Conserv.* **8**, 67–70.

Farrow, S. (1996) Marine protected areas: emerging economics. *Mar. Policy* **20**, 439–446.

Gordon, H.S (1954) The economic theory of common property resource: the fishery. *J. Polit. Econ.* **62**, 124–142.

Holland, D.S. and Brazee, R.J. (1996) Marine reserves for fisheries management. *Mar. Resour. Econ.* **3**(11), 157–171.

Lauck, T., Clark, C.W., Mangel, M. and Munro, G. (1997) Implementing the precautionary approach in fisheries management through marine reserves. *Ecol. Appl.* (in press).

Ludwig, D., Hilborn, R. and Walters, C.J. (1993) Uncertainty, resource exploitation, and conservation: lessons from history. *Science* **260**, 17, 36.

Polacheck, T. (1990) Year round closed areas as a management tool. *Nat. Resour. Modeling* **4**, 327–354.

Man, A., Law, R. and Polunin, N.V.C. (1995) Role of marine reserves in recruitment to reef fisheries: a metapopulation model. *Biol. Conserv.* **71**, 197–204.

Munro, G. and Scott, A.D. (1985) The economics of fisheries management, in *Handbook of Natural Resource and Energy Economics* Vol. 2 (eds A.V. Kneese and J.L. Sweeney), North-Holland, Amsterdam, pp. 623–676.

Quinn, J.F., Wing, S.R. and Botsford, L.W. (1993) Harvest refugia in marine vertebrate fisheries: models and applications to the Red Sea urchin (*Strongylocentrotus franciscanus*). *Am. Zool.* **33**, 537–550.

Roberts, C.M. and Polunin, N.V.C. (1991) Are marine reserves effective in management of reef fisheries? *Rev. Fish Biol. Fish.* **1**, 65–91.

Rowley, R.J. (1994) Case studies and reviews, marine reserves in fisheries management. *Aquat. Conserv.: Mar. Freshwat. Ecosyst.* **4**, 233–254.

Wallis, O.L. (1971) Establishing underwater parks worldwide. *Trans. N. Am. Wildl. Nat. Resour. Conf.* **36**, 97–115.

Walters, C.J. and Hilborn, R. (1978) Adaptive control of fishing systems. *J. Fish. Res. Bd Can.* **33**, 145–159.

Wilson, J.A., Acheson, J.M., Metcalfe M. and Kleban, P. (1994) Chaos, complexity and community management of fisheries. *Mar. Policy* **18**(4), 291–305.

Chapter twenty-four

Rebuilding ecosystems, not sustainability, as the proper goal of fishery management

Tony J. Pitcher and Daniel Pauly

ABSTRACT

We propose that rebuilding ecosystems, and not sustainability *per se*, should be the goal of fishery management. Sustainability is a deceptive goal because human harvesting of fish leads to a progressive simplification of ecosystems in favour of smaller, high-turnover, lower-trophic-level fish species that are adapted to withstand disturbance and habitat degradation. Present fisheries management seems unable to reverse this trend for several reasons. Because of this effect on the ecosystem, sustainable harvests are generally incapable of ever being defined using single species-population dynamics, yet almost all fishery science has been long engaged in trying to do this. Even if our science and management were capable of sustaining exploitation at a defined ecosystem structure, we argue that this is the wrong goal. Aquatic systems are likely capable of producing large harvests of high-production, low-trophic-level species, perhaps much in excess of current global fishery yields of around 100 million tonnes per year, yet such exploitation would shift their structure and nature in a way, and lead to products, that would be unacceptable to many. Primal

Reinventing Fisheries Management. Edited by Tony J. Pitcher, Paul J.B. Hart and Daniel Pauly. Published in 1998 by Kluwer Academic Publishers, London. ISBN 0 412 83410 3.

systems, defined as those existing before humans used large-scale harvesting, are generally characterized by an abundance of large top-predator species. An approach to the primal abundance of such systems may have an increasingly higher economic value than present systems in an era where demand is outstripping supply. Therefore, we argue that management that moves aquatic systems in the direction of their primal states and abundance should be rewarded, and that this rebuilding and restoration of ecosystems should be the overarching goal of the new fisheries management.

24.1 INTRODUCTION

Sustainability is widely regarded as a desirable and proper goal of fishery management. The concept is unquestioned and deeply embedded within many disciplines that address fisheries issues (for example, Richards and Schnute, Chapter 30, Rice, Chapter 14, Charles, Chapter 6, and Sutton, Chapter 8, all in this volume, and Botsford *et al.*, 1997), even though it may be relatively new for some areas (McGuire, Chapter 18, and Jentoft, Chapter 12, both this volume). In this chapter we propose that, whatever its present degree of acceptance, sustainability is the wrong goal for fishery management. Moreover, we argue that success in implementing this goal, as currently understood in our discipline, is not only likely to reduce the biodiversity of the oceans, but will also seriously compromise future economic options for human wealth based on the harvest of aquatic resources. Our analysis suggests that the most effective goal for fisheries management is the rebuilding of natural ecosystems.

Much quantitative fishery research is judged by a sustainability criterion, assessment methods are defined and tested in terms of how well they address long-term sustainability, and the success of management measures is judged by reference to sustainability. For a long time, maximum sustainable yield (MSY) ruled supreme as the objective of fishery management (see contributions in Roedel, 1975) and despite its often-heralded demise (Larkin, 1977; Barber, 1988), sustainability is universally employed to delineate the many population criteria that have proliferated to complement or replace it (Smith, 1993, Smith *et al.*, 1993; Drummond and Symes, 1996). So deeply embedded is the concept of sustainability that when Ludwig *et al.* (1993) claimed that fishery management has failed, annoying many of their colleagues (the late Ray Beverton told us that the paper had 'done a great deal of harm'), they defined failure in terms of sustainability, in that no fishery, according to their analysis, has ever been sustainable.

Several mechanisms, some biological, the others socio-economic, jointly contribute to push exploited marine ecosystems toward states dominated by high-turnover, low-trophic-level species. To halt and reverse this trend, an understanding and resolution of both ecological and human factors is critical to success. Hence the proposed goal of rebuilding aquatic ecosystems is inherently interdisciplinary. Although the ecosystem concept was integral to the ideas of pioneers of fisheries science (Hardy, 1956; Skjoldal *et al.*, 1993) it was forgotten early in the era of domination of single-species population dynamics. Moreover, attempts to reinstate ecosystem management have lacked clear quantifiable goals (Larkin, 1996) and so a practical rebuilding goal would form part of a truly 'reinvented' fisheries management. Furthermore, if we look at the management practices of indigenous and aboriginal peoples (Johannes, 1987; Johannes *et al.*, 1993), we see that this goal is being revisited rather than reinvented.

24.2 ECOLOGICAL FACTORS

Harvesting alters ecosystem structure in ways that are only just beginning to be understood. We argue that long-term heavy commercial harvesting, that may well meet the most rigorous sustainability criteria, is likely to shift the ecosystem to high-turnover species with low trophic levels. Ludwig *et al.*'s (1993) argument that no fishery management has ever worked has the corollary that heavy commercial fishing (in many large fisheries over 80% of the fish are caught each year) alters systems irreversibly.

The dominant biological mechanism underlying species shifts is relatively well understood, and results from large, long-lived fishes, which also have low natural mortality rates (Pauly, 1980), being affected more strongly by a given rate of fishing mortality than smaller fishes that are part of the same community.

A second shift-inducing biological mechanism is habitat degradation caused by various fishing gears, especially bottom trawls. Here, the effect is through the elimination of the bottom structure, such as 'oyster reefs' and sponge beds (Young and Glaister, 1993; Pauly *et al.*, 1996) created, often over hundreds of years, by the sessile planktivores and detritivores upon which long-lived large benthic fish rely to complement their diet of small pelagic fishes. The destruction of these structures by trawls, besides depriving benthic fishes of habitats and prey, also has the effect of redirecting carbon flows toward the pelagic system, thereby adding to the effects of anthropogenic nutrient inputs.

Thirdly, these fluxes, and the fishery-induced reduction of predatory pressure by benthic fish, may then lead to an increase of small pelagic fish and squids which become available for exploitation, thus in part masking the decline of the demersal group (see contributions in Pauly and Murphy, 1982, and in Daan and Sissenwine, 1991). In the Gulf of Thailand, off Hong Kong, and other areas of the South China Sea, extremely heavy trawl pressure has resulted in a shift from valuable demersal table fish such as croakers, groupers and snappers (Sciaenidae, Epinephilinae, Lutjanidae) to a fishery dominated by small pelagics used for animal food, and invertebrates such as jellyfish and squids (Beddington and May, 1982; Boonyubol and Pramokchutina, 1984; Pitcher *et al.*, 1998).

These mechanisms may lead, through a positive feedback loop, to a fourth biological mechanism: harvesting small pelagic fish species at lower trophic levels reduces the availability of food for higher trophic levels, which then decline further, releasing more prey for capture by a fishery that finds it targets even lower down the food web, a process now occurring throughout the world (Fig. 24.1). The overall effect of harvesting aquatic ecosystems is summarized in Fig. 24.2, which plots a suitable index of biodiversity (such as abundance, biomass, species richness) against time, which may be extended to the left to the first known human fisheries – deer antlers carved into fish spears used to harvest giant river catfish some 85 000 years ago in what is now the Republic of Congo

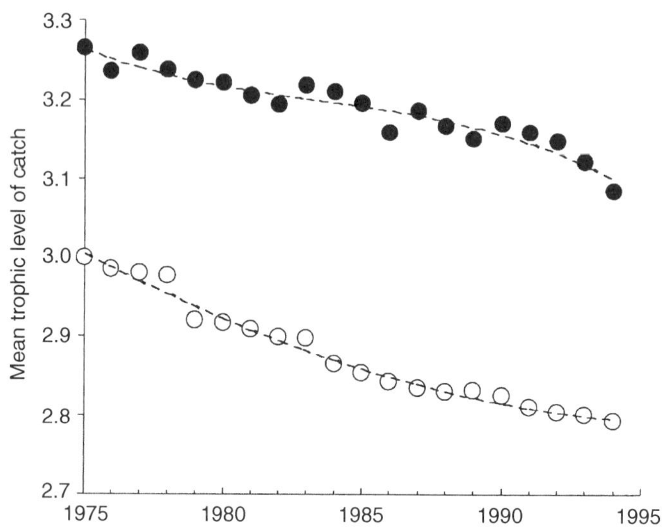

Fig. 24.1 Decline of the mean trophic level of fish and invertebrates caught by global marine (solid circles) and inland (open circles) fisheries, 1975–1994 (Pauly *et al.*, 1998). Cubic trend lines have been added to the two series. Figure is based on the FAO catch statistics and trophic level estimates included in FishBase 97 (Froese and Pauly, 1997).

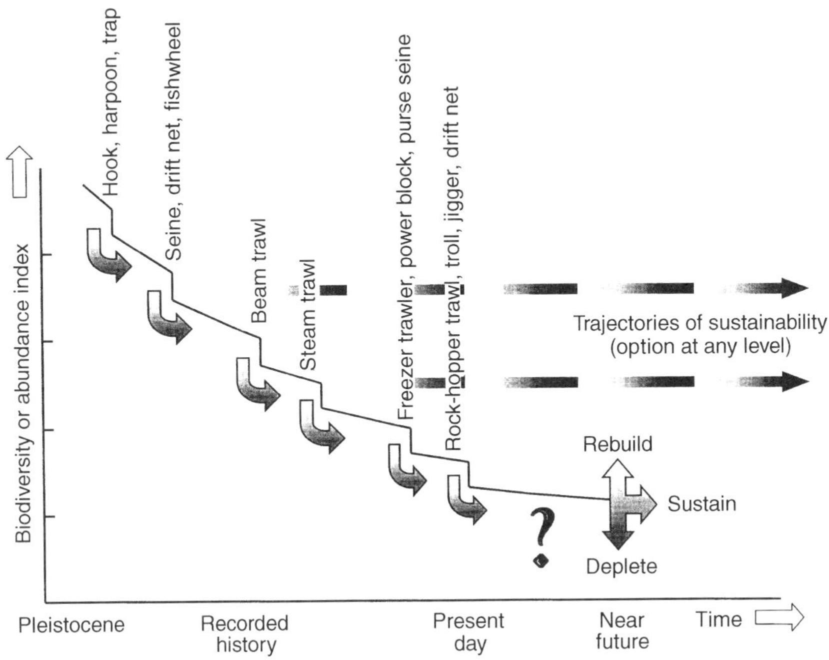

Fig. 24.2. Schematic representation of the loss of biodiversity in aquatic ecosystems since prehistory. Vertical axis represents abundance, biomass of top predators, a suitable diversity index or an ecosystem breadth measure. Horizontal axis is time. Stepped downward line represents serial depletion of the resources as new fishing technologies (some of the principal fishing gear inventions are indicated) are invented by humans. Horizontal arrows show sustainability which, in theory, may be pursued at any level if management instruments are capable. Future options are indicated by the three-way arrow.

(Yellen *et al.*, 1995). Successive stages in the development of human fishing are shown from left to right; historically the stages are congruent with major enhancements to fishing technology such as the development of the seine and cotton drift nets, beam trawl, steam trawls, freezer trawlers, power-block driven and drum purse seines, rock-hopper trawls, high-technology trollers for Pacific salmon, squid and tuna hook-and-line jiggers and nylon drift nets. Downward arrows indicate major depletion events as new technologies are rapidly deployed. A fascinating historical example of the spread of driftnet technology for catching herring in Scotland in the early 1800s is graphically described in the novel *The Silver Darlings* by Neil Gunn (1941).

At first sight, sustainability appears to be an option if management techniques can achieve this goal. Some fishery scientists think that this can be done (Beverton, 1998; many chapters in this volume), while others, notably Ludwig *et al.* (1993) and Parrish (1995), are pessimistic.

There is some evidence supporting the optimists and some supporting the pessimists. The latter might argue that fishing selects species with life history parameters that allow population persistence in the face of high mortality; this ecological argument is hard to counter. But optimists may contend that technical measures to allow for uncertainty provide risk-averse, fully precautionary management, and could allow us to maintain populations in the long term. Pessimists point to the growing series of collapses, which does not inspire great confidence. Moreover, fresh factors that should have been taken into account are revealed after each new collapse. The most recent example is the critical need for spatial analysis of abundance during depletion (Hutchings, 1996; Walters and Maguire, 1996; Pitcher, 1997), where only a short while ago, the majority in fishery science regarded such information as of only academic interest with no place in a stock assessment that pooled all other information. Again, contrary to the assumptions built into the majority of current stock assessments, there is now evidence of risk of recruitment failure at low-to-moderate biomass levels (Mertz and Myers, 1996; Myers and Barrowman, 1996) and analyses suggesting that truly sustainable rates cannot be anywhere near the 70–80% exploitation rates often encountered today.

The optimists can point to some recent papers (Mace, 1997) that, while deploring collapses, have cheerfully reported increases in some fished populations, and have hailed this as encouraging evidence of the potential success of present management science. Moreover, there is a current myth, based on pelagic fish such as herring, that, with the exception of skate in the Irish Sea (Brander, 1981), marine fisheries by humans have never brought about species extinction and are never likely to do so (Mace, 1997). But as exploited systems shift down Fig. 24.2, increases in species with life histories favoured by harvest are inevitable – Table 24.1 lists several such examples. For instance, it is likely that enormous populations of pollock, *Theragra chalcogramma*, in the North Pacific may be a direct result of exploitation and extinction of marine mammals since the 1700s. The huge expansion of trawl fishery in the 1980s, aimed at harvesting this resource, may have exacerbated the situation. Moreover, in the Bering Sea there are worrying reports of a vast increase in numbers of jellyfish (Brodeur *et al.*, 1996), a symptom of a shifting food web and decreasing trophic level known also from the South China Sea (Saeger, 1993), amplified to disaster level in the Black Sea (Shushkina and Vinogradov, 1991). Finally, supporting the pessimists, it often comes as a surprise to fishery scientists that living organisms, including fish, evolve in response to

Table 24.1 Some examples of documented shifts toward smaller, higher-turnover species in exploited multispecies fish communities

Fishing grounds/stock (period)	Remarks	Source(s)
Gulf of Thailand demersal stocks, 1960 to 1980s	Overall biomass reduced by 90%; residual biomass dominated by 'trash fish'	Boonyubol and Pramokchutina (1984)
Philippine Shelf small pelagics, 1950 to 1980s	Gradual replacement of sardine-like fishes by anchovies	Dalzell and Pauly (1990)
Carigara Bay, Samar, Philippines, all fish, 1970 to 1990	Fish replaced by jellyfish, now an export item	Saeger (1993)
Black Sea	Small pelagics and jellyfish replace large table fish	Shushkina and Vinogradov (1991)
North Sea	Halibut and small sharks extinct; cod and haddock threatened; demersal omnivores and small pelagics favoured	Rijnsdorp *et al.* (1996)
Humboldt, Current, Chile	Large hake depleted, small pelagics favoured	Aguayo (1995 and pers. comm.)
North Pacific	First marine mammal depletions, followed by huge trawl fisheries; pollock favoured	Trites (pers. comm.)*
South China Sea, Hong Kong	Croakers and groupers almost extinct; small pelagics bulk of fishery	Pitcher *et al.* (1998)

*Marine Mammal Research Unit, Fisheries Centre, UBC, Vancouver, Canada V6T 1Z4.

selective mortality, a process recently shown to be remarkably fast, even in vertebrates (Houde and Endler, 1990; Wiener, 1994). Astonishingly little effort has been made to trace such effects in fisheries (Policansky, 1993), but they may be much more widespread than previously thought. An example may be the halving of the size of mature chinook and pink salmon (*Oncorhynchus tshawytscha* and *O. gorbuscha*: McAllister *et al.*, 1992). In the semelparous salmon, early maturity means less time at risk of being caught and therefore higher fitness. These species have been intensively managed for over 80 years using the best that single-species quantitative science can offer, and yet chinook are in decline, something that has, not surprisingly, been blamed on environmental change (Beamish *et al.* 1995).

These considerations suggest to the pessimists, notably Parrish (1995), that the ultimate fate of all exploited aquatic ecosystems is analogous to the heavily depleted and unregulated fisheries of the South China Sea – dependent on jellyfish (which is now harvested), prawns, squids, small fast-growing pelagics, and little else. But whether or not sustainability is truly achievable using management based on single-species stock assessment actually proves to be irrelevant to our main argument.

24.3 SOCIO-ECONOMIC FACTORS

Two main socio-economic mechanisms contribute to species shifts. The first of these is increasing prices, both for traditional high-value species and for former 'trash' species. Astoundingly, bluefin tuna, *Thynnus thunnus* were discarded as trash fish in Danish waters until the 1930s (Pauly, 1995), as were Atlantic halibut, *Hippoglossus hippoglossus*, in the New England fishery until 1825 (Cushing, 1988). Such price increases are effective in masking the economic consequences of fishing at lower trophic levels, i.e. of catching what were previously low-value fish e.g. mackerel in Britain, (Lockwood, 1988), or jellyfish in the Gulf of Thailand (Soonthonvipat, 1976) or in the Philippines (Saeger, 1993).

The second mechanism, embodied in the scenarios represented in Fig. 24.2, is psychological, and more insidious. This is the 'shifting baseline syndrome of fisheries' (Pauly, 1995), wherein successive generations of fisheries scientists lose sight of the state of the resource during periods preceding their careers. Earlier accounts of high abundance (e.g. in Mowat, 1984, or MacIntyre *et al.*, 1995) are perceived as irrelevant anecdotes, though they pertain to a sensitive phase of the early fisheries, when a relatively small level of effort sufficed for a major reduction of resource biomass.

In fact, we show below how both of these mechanisms offer a means for resolving the problem.

24.4 METHODOLOGY

Is it practical to try to manage systems to rebuild? Do we have tools that enable us to assess the status of exploited aquatic ecosystems and to evaluate options for management? We think that the answers are 'Yes' and accordingly this section reviews five sets of techniques that may allow a rebuilding goal and determine how it might be evaluated.

1. Role of population dynamics

Reviewing what we can do with single-species population dynamics and what we cannot illustrates the limitations of even the best quantitative science. Paradoxically, the insufficiency of population dynamics is an important part of our argument, but the tools developed for population dynamics must be an essential part of any new methodology. Stated explicitly, detailed information on growth, mortality and recruitment schedules and their associated errors and uncertainties are all essential, i.e. the 'stuff' we teach, learn and elaborate in today's fisheries science is all necessary, but not sufficient for the ecosystem approach we advocate.

Likewise, management targets for particular species can as easily be defined as part of a system-rebuilding strategy as they are for single-species management. Such targets, fully bounded by risk evaluations, are structured in terms of spawning stock biomass, age structure, recruitment risks, effort or fishing mortality, although targets expressed directly in terms of catch quotas are probably not so wise (Walters and Pearse, 1996).

2. Role of ecosystem modelling

There are many recent developments in trophic models of aquatic ecosystems. Such modelling can now be performed more rapidly and rigorously than ever before (see contributions in Christensen and Pauly, 1993; Pauly and Christensen, 1995), providing a basis for viable and practical simulation models that have real predictive power (Walters *et al.*, 1997).

This was made possible by the development of Ecopath (Polovina, 1984; Christensen and Pauly, 1992), for construction of mass-balance models of ecosystems, based mainly on diet composition, food consumption rates, biomass, and mortality estimates, i.e. on information available for most commercial species, and easy to generalize to other species. The flux estimates obtained through this approach, though typically representative of an 'average' situation, are extremely useful for stock assessment purposes, e.g. to assess the relative impact of predation vs. fisheries, or to estimate the biomass of groups that cannot be sampled straightforwardly, such as oceanic squids (see contributions in Pauly and Christensen, 1996).

Such biomass values may sometimes be more reliable than estimates from independent surveys, although survey data can be used to tune biomass values in the model. One problem is that Ecopath estimates represent 'averages', inappropriate for ecosystems that are used as 'way stations' for migratory resources such as herring, salmon or tuna. Doubtless we may be accused here of setting up a new goal without the quantitative tools being yet in place. We suggest that more research effort should be put into honing tools for ecosystem management, such as Ecopath, rather than making old single-species tools ever more elaborate.

This new emphasis is particularly justified in that once an ecosystem has been described using Ecopath, its behaviour can be studied using Ecosim, the dynamic version of Ecopath (Walters *et al.*, 1997). Although still under development and far from perfected, this generic modelling software allows the validation and evaluation of alternative harvesting strategies on different parts of the system. It has recently been used, for the first time, for comparing the behaviour of different upwelling ecosystems (Mackinson *et al.* 1997), and the relation between stability and detritus recycling in marine ecosystems (Vasconcellos *et al.* 1998), while various applications to specific ecosystems are forthcoming.

In these applications, Ecosim is used to show the trajectories of various sectors of the resource when one of them is subjected to greater or lesser levels of harvest. A set of scenarios can be used to plot likely rebuilding strategies. At present, single-species dynamics would then be employed to investigate recommended harvest levels in detail; eventually, Ecosim, or some of its likely successors, should be used instead.

3. Role of fisheries management measures

Standard fishery management devices such as minimum size regulations, restriction of gear types, restriction of effort and catch (Pitcher and Hart, 1982 for a list) have an important role. But these devices have been termed 'band-aids' to be applied after the damage is done (Walters, 1995). Unless accompanied by draconian fishing regulations (such as the notorious '1-hour opening' of some Pacific fisheries), these devices are unlikely to contribute toward a rebuilding goal.

Marine reserves are critical (Bohnsack, 1993, 1996). This is not only because of their value in countering uncertainty (Clark, 1996), but also because they provide resilience against overexploitation (Guenette *et al.*, 1998) by allowing natural systems to regenerate (Ballentine, 1991) and act as a reservoir of large high-value species, precisely the objectives of the new goal put forward here.

4. Role of traditional environmental knowledge (TEK)

Historical records and indigenous knowledge can help tune the rebuilding models and define management objectives. The value of historical records of abundance and species composition is high. For instance, in the North Sea, there has been a call for 'data archeology' (Daan *et al.*, 1996) with which to try to evaluate major changes in the ecosystem. For example, a comparison of data from the 1990s with those from the early years of the

Table 24.2 Some costs and benefits in fish resources of depleting, sustaining and rebuilding ecosystem management strategies

Strategy	Product value	Total resource value	Harvest cost	Enforcement cost	Monitoring cost	Processing cost	Employment in sector
Depletion	Low	Stable	Low	Low	High	High	High
Sustaining	Medium	Rising	High	High	High	Medium	Low
Rebuilding	High	Fast rising	Low	Low	Low	Low	Low

20th century revealed shifts to lower species diversity, smaller fish and higher evenness in the demersal community (Rijnsdorp *et al.*, 1996). Finding ways to harness even older data from the 1800s and earlier (Cushing, 1988) may be valuable in evaluating the societal options for the nations around the North Sea that have been foregone by an unthinking industrialization of fisheries.

If we look at the management goals of indigenous and aboriginal peoples, we see that a goal related directly to primal abundance strikes an immediate chord (Haggan, Chapter 2, and Harris, Chapter 20, both this volume). Moreover, traditional environmental knowledge (TEK) can, through the cultural shadow cast by past abundances, help with the construction, tuning and validation of our scientific models of natural ecosystems. Present work is seeking ways of combining quantitative and qualitative information so that, together, the models and traditional knowledge can be used to describe the state of past natural systems

5. Role of socio-economics

In this section we compare the relative economic and social values of three scenarios: depletion, sustaining and rebuilding (Table 24.2).

The depletion scenario

If the pessimists are correct, the likely outcome of present-day management is depletion and ratcheting down the trophic levels, i.e. a continuation of the trends in Figs 24.1 and 24.2. In this scenario, the unit value of seafood products declines, although larger harvests, lower costs and higher demand may enable profits to be maintained. In the medium to long term, processing costs will rise as products for human consumption are created out of inauspicious species (such as myctophids or krill) and this may create industrial employment and market opportunities in the food industry. This scenario has low enforcement costs, but high monitoring costs as small pelagic and midwater organisms that are difficult to sample quantitatively would come to dominate the system. Fishing vessels and gear would have to change as their target size and location shifts, but the catching power of the world's fleet will likely stay the same, i.e. remain excessive. In many ways this scenario is the logical, and some will say, most likely, extension of the recent history of the world's fisheries.

The sustainability scenario

If optimists are correct and we are able to perfect management for sustainability in the face of the ecological and evolutionary mechanisms outlined above, then the present harvest regime will continue. In many ways, this scenario would be a dream come true for most fisheries scientists. But what is likely to happen? For the next century, product and resource values will likely rise with the inevitable further growth in the human population and, it is to be hoped, increasing living standards. Profits would be less than they might be, however, because of high harvest, enforcement and monitoring costs in a sustained fishery. There will likely be a panoply of strictly enforced management regulations: bycatch and discard regulations and limitations on fishing gear, closed areas and seasons, catch monitoring devices such as port inspection and onboard observers. Direct employment will likely be reduced as high technology is harnessed to reduce costs. The catching power of vessels will likely increase from this pressure, although strict regulation (or the market power of transferable harvest rights) would ensure that the total number fell. Given a stable catch, the economic efficiencies of large scale would ensure that the processing and harvesting industry became concentrated in fewer and fewer hands. Although they will try to appear to do so, governments will hence be hard pressed to reduce subsidies to fishing and to coastal communities. Clearly this scenario has major economic benefits that are much preferable to depletion. It also holds out the prospect of rewarding employment for the many fisheries scientists who will be needed to keep the adaptive management engines turning (Bundy, Chapter 26, this volume).

The rebuilding scenario

In what ways is a rebuilding strategy better than the sustainability scenario? The principal benefit in a world of increasing demand will be high prices for traditional large table fish. Indirect benefits will be increased opportunities for recreational fishing as top predators recover. Hence the total value of the resource will rise quickly for at least the medium term. Harvest costs will be low, as allowable catches will be easily made from an abundant resource. In most cases there will be no need for new technology to increase catching power, but technological solutions to provide highly selective harvest will be fostered. Enforcement costs, at least for a management system that includes large no-take reserves (Sumaila, Chapter 23, this volume), and uses the social power of local communities, should be relatively low. Monitoring costs should also be low as they can concentrate on simple indicators of rising abundance, for example using acoustics and range extension records. Because of the high product value, processing will be largely confined to traditional freezing, but distribution costs could be high for fresh fish (e.g. air freight). The only down side might be the employment, which would likely be relatively low in a small but valuable seafood industry.

24.5 CONCLUSIONS

Clearly, some of this crystal ball gazing may be flawed. But we suggest that rebuilding is likely to have the highest economic value of these three scenarios, and because of the high product values, it is the least likely to be prejudiced by high discount rates or inelastic seafood demand. An additional benefit is that the countries with 'rebuilt' fisheries resources would be able to meet the commitments implied in the various biodiversity conventions they have signed. Extrapolation of present practices will not allow for this.

In a major review of the issues facing world fisheries, Mace (1997) considers that sustainability has not been achieved in the past because we have adopted an inappropriate set of ground rules. She argues that once we successfully abandon these, true sustainability will be achievable using the instruments of modern fisheries management, such as the development of effective institutions to implement risk-averse management and the precautionary principle, set up workable access rights, a reduction of fleet capacity, and an environmentally friendly expansion of aquaculture. She presents a cogent analysis, with which we do not disagree, of the continued legacy of the inexhaustibility paradigm, a belief in the status quo, fishing as a birthright, and blaming the wrong scapegoats (such as MSY or the environment) for failure. But we disagree when Mace represents the maxim that 'recognizing the problem is half the solution' as not applying to fisheries. She considers that we have enough science, tools and insight to know what needs to be done. She comes closer to our vision when she argues that we need true environmental sustainability rather than single-species sustainability, but we contend that the outcomes she would like to see would become more probable with the rebuilding of fish stocks as the management goal.

Is the rebuilding of functioning, diverse systems, a concept valued in the culture of many aboriginal peoples but virtually absent from contemporary resource management, a realistic goal? Well, it's not so strange. We humans have just done it with large whale populations (Swartz, pers. comm., quoted in Mace, 1997) and, in the terrestrial environment, schemes are being attempted with prairie grassland and grazing bison (Callenbach, 1995). For terrestrial habitats, it has been argued that even to sustain natural habitats, some rebuilding must be done to offset natural destruction caused by wind, fire, and damage owing to the vicinity of unnatural ecosystems (Sinclair *et al.*, 1995; Arcese and Sinclair, 1997).

In summary, the suggested procedure is first to build an ecosystem model of a present ecosystem and explore the likely depletion pathway for the near future, using simulation software such as Ecosim and ecological and life history information on species likely to go extinct (Parrish, 1995). The second step is to construct mass-balance models of past systems using historical records and TEK. Both the inclusion of species and past abundance can be validated by TEK and historical documents. For example, inputting the likely number of humpback whales known to have occurred 100 years ago in the Strait of Georgia, British Columbia, into an Ecopath model also including the groups still occurring in that system will, upon rerunning the model, force a solution with much larger biomass of herring, krill and other humpback prey species. Thirdly, a socio-economic evaluation of an alternative rebuilt system can be compared with the value of the status quo and with the value of the depletion strategy. Note that in nearshore tropical habitats, an economic balance between rebuilding and sustaining will likely be found, because some species – prawn, squids – that are favoured as depletion occurs have high economic value. An important advantage is that an economic evaluation based on an ecosystem model can evaluate these options objectively, prior to their entering the political 'allocation' arena.

Some forms of habitat destruction other than trawling mentioned above, such as accumulation of debris and changes in flow reducing oxygen in river gravels for spawning salmon, or the loss of mangroves that provide nursery areas for many tropical species, may not result directly from fishing gear, but from other human activities, especially logging (Marchak, 1995). Conventional goals for fisheries management treat these influences at best as separate issues, to be solved separately from the standard population management. At worst, such issues form no part of any quantitative evaluation. In contrast, the rebuilding goal and its supporting methodology would have to explicitly take these factors into account.

There are some other advantages that come along with the adoption of an ecosystem view. For example, trade-offs between species are seen as a political issue in single-species management, and are often regarded in terms of 'allocation' between competing parts of the fishery sector. In contrast, all interacting species are included in a multispecies resource evaluation which is part of ecosystem-based management. Bioeconomic evaluation automatically includes harvest options for fish at different trophic levels. Indeed, maximizing economic and social returns is integral to the whole procedure.

Rebuilding implies some form of pre-existing structure. Although ecosystems are always in a state of flux, especially over geological time, what we refer to here is the state that obtained before the gross and rapid depletion caused by human fishing, as illustrated in Fig. 24.2. So the former state we rebuild towards is the species mix and levels of abundance of the system before significant human fisheries began. In some parts of the world this would be thousands of years BP, in others as little as 50 years BP. Of course, we would not necessarily wish to, nor be able to, restore things to their Pleistocene levels. We easily forget that we are living through one of the largest rapid mass extinctions in the history of the planet: all of them caused by ourselves. The normal fluxes of ecology have been so totally overwhelmed by human intervention, and in aquatic ecology our responses have been so ineffectual, that remedial action requires a drastic shift in attitude and concepts. How far we should go in this direction, and how much we should rebuild to counter continuing losses, are issues of detail that will be discussed in a future publication.

Unlike some sectors of the 'green' movement, we are not advocating a return to the Pleistocene (Fig. 24.2), which is clearly unrealistic. However, given the trend in Fig. 24.1, the more fisheries move down the slope on Fig. 24.2, the more valuable the seafood harvest in those jurisdictions that have the good sense and the courage to adopt rebuilding as a goal. If adopted, a rebuilding goal for resource management will encourage consultation and consent and will directly harness traditional knowledge and history to a societal goal.

REFERENCES

Aguayo, M. (1995) Biology and fisheries of Chilean hakes (*M. gayi* and *M. australis*), in *Hake: Fisheries, Ecology and Markets* (eds J. Alheit and T.J. Pitcher), Chapman & Hall, London, pp. 305–337.

Arcese, P. and Sinclair, A.R.E. (1997) The role of protected areas as ecological baselines. *J. Wildl. Manage.* **61**, 587–602.

Ballentine, W.J. (1991) *Marine Reserves for New Zealand, Leigh Lab. Bull.* no. 25, Univ. Aukland, 196 pp.

Barber, W.E. (1988) Maximum sustainable yield lives on. *N. Am. J. Fish. Manage.* **8**, 153–157.

Beamish, R.J., Riddell, B.E., Neville, C.E.M., Thomson, B.L. and Zhang, Z. (1995) Declines in chinook salmon catches in the Strait of Georgia in relation to shifts in the marine environment. *Fish. Oceanogr.* **4**, 243–256.

Beddington, J.R. and May, R.M. (1982) The harvesting of interacting species in a natural system. *Scient. Am.* **247**(5), 42–49.

Beverton, R.J.H. (1998) Fish, fact and fantasy; a long view. *Rev. Fish Biol. Fish.* (in press).

Bohnsack, J.A. (1993) Marine reserves. They enhance fisheries, reduce conflicts, and protect resources. *Oceanus* **36**, 63–71.

Bohnsack, J.A. (1996) Marine reserves, zoning, and the future of fishery management. *Fisheries* **21**(9), 14–16.

Boonyubol, M. and Pramokchutina, S. (1984) Trawl fisheries in the Gulf of Thailand. *ICLARM Trans.* **4**, 12 pp.

Botsford, L.W., Castilla, J.C. and Peterson, C.H. (1997) The management of fisheries and marine ecosystems. *Science* **277**, 509–515.

Brander, K. (1981) The disappearance of the common skate, *Raia batis*, in the Irish Sea. *Nature* **290**, 48–49.

Brodeur, R.D., Frost, B.W., Hare, S.R., Francis, R.C. and Ingraham, J. (1996) Interannual variations in zooplankton biomass in the Gulf of Alaska, and covariation with California Current zooplankton biomass. *CALCOFI Rep.* **37**, 80–99.

Callenbach, E. (1995) *Bring Back the Buffalo!: A Sustainable Future for America's Great Plains*. Island Press, 279 pp.

Christensen, V. and Pauly, D. (1992) ECOPATH II – a system for balancing steady-state ecosystem models and calculating network characteristics. *Ecol. Modelling* **61**, 169–185.

Christensen, V. and Pauly, D. (eds) (1993) *Trophic models of aquatic ecosystems*. ICLARM Conf. Proc. **26**, ICLARM, Manila, 390 pp.

Clark, C.W. (1996) Marine reserves and the precautionary management of fisheries. *Ecol. Appl.* **6**, 369–370.

Cushing, D.H. (1988) *The Provident Sea*, Cambridge, Cambridge Univ. Press, 329 pp.

Daan, N. and Sissenwine, M.P. (eds) (1991) Multispecies models relevant to management of living resources. *ICES Mar. Symp.* **193**, 358 pp.

Daan, N., Richardson, K. and Pope, J.G. (1996) Changes in the North Sea ecosystem and their causes: Århus 1975 revisited. *ICES J. Mar. Sci.* **53**, 879–883.

Dalzell, P. and Pauly, D. (1990) Assessment of the fish resources of Southeast Asia, with emphasis on the Banda and Arafura Seas. *Neth. J. Sea Res.* **25**, 641–650.

Drummond, I. and Symes, D. (1996) Rethinking sustainable development: the realist paradigm, in *Sustainable Fisheries?* (eds O. Otterstad and I. Drummond), (Special issue) *Sociol. Ruralis* **36**(2), 152–162.

Froese, R. and Pauly, D. (1997) *FishBase 97: Concepts, Design and Data Sources*, ICLARM, Manila, 257 pp. [incl. 2 CD-ROMs].

Guenette, S., Lauck, T. and Clark, C. (1998) Marine reserves: from Beverton and Holt to the present. *Rev. Fish Biol. Fish.* (in press).

Gunn, N.M. (1941) *The Silver Darlings*, Faber & Faber, London, 584 pp.

Hardy, A. (1956) *The Open Sea*, Collins, London, 335 pp.

Houde, A.E. and Endler, J.A. (1990) Correlated evolution of female mating preferences and male color patterns in the guppy *Poecilia reticulata*. *Science* **248**, 1405–1408.

Hutchings, J.A. (1996) Spatial and temporal variation in the density of northern cod and a review of hypotheses for the stock's collapse. *Can. J. Fish. Aquat. Sci.* **53**, 943–962.

Johannes, R.E. (1987) The value today of traditional management and knowledge of coastal marine resources in Oceania. *NAGA, ICLARM Q.* **10** (4), 5.

Johannes, R.E., Ruddle, K. and Hviding, E. (1993) People, society and Pacific islands fisheries development and management. South Pacific Comm., Nouméa, New Caledonia, *Inshore Fish. Res.* **5**, 1–7.

Larkin, P.A. (1977) An epitaph for the concept of Maximum Sustainable Yield. *Trans. Am. Fish. Soc.* **106**, 1–11.

Larkin, P.A. (1996) Concepts and issues in marine ecosystem management. *Rev. Fish Biol. Fish.* **6**, 139–164.

Lockwood, S.J. (1988) *The Mackerel: its Biology, Assessment and the Management of a Fishery*, Fishing News Books, Farnham, UK, 181 pp.

Ludwig, D., Hilborn, R. and Walters, C. (1993) Uncertainty, resource exploitation, and conservation: lessons from history. *Science* **260**, 17–18.

McAllister, M.K., Peterman, R.M. and Gillis, D.M. (1992) Statistical evaluation of a large-scale fishing experiment designed to test for a genetic effect of size-selective fishing on British Columbia pink salmon *(Oncorhynchus gorbuscha)*. *Can. J. Fish. Aquat. Sci.* **49**, 1294–1304.

Mace, P.M. (1997) Developing and sustaining world fisheries resources: the state of science and management, in *Developing and Sustaining World Fisheries Resources: the State of Science and Management* (eds D.A. Hancock, D.C. Smith, A. Grant and J.P. Beumer), CSIRO, Collingwood, Australia, pp. 1–20

MacIntyre, F., Estep, K.W. and Noji, T.T. (1995) Is it deforestation or desertification when we do it to the oceans? *Naga, ICLARM Q.* **18**(3), 4–6.

Mackinson, S., Vasconcellos, M., Sloman, K., Pitcher, T.J. and Walters, C.J. (1997) Ecosystem impacts of harvesting small pelagic fish using a dynamic mass-balance model, in *Forage Fishes in Marine Ecosystems* (ed. B. Baxter), Alaska Sea Grant Symp. pp. 731–749.

Marchak, P. (1995) *Logging the Globe*, McGill–Queen's Univ. Press, Montreal, 440 pp.

Mertz, G. and Myers, R.A. (1996) Influence of fecundity on recruitment variability of marine fish. *Can. J. Fish. Aquat. Sci.* **53**, 1618–1625.

Mowat, F. (1984) *Sea of Slaughter*, Atlantic Monthly Press, Boston, New York, 438 pp.

Myers, R.A. and Barrowman, N.J. (1996) Is fish recruitment related to spawner abundance? *Fish. Bull.* **94**, 707–724.

Parrish, R.H. (1995) Lanternfish heaven: the future of world fisheries? *Naga, ICLARM Q.* **18**(3), 7–9.

Pauly, D. (1980) On the interrelationships between natural mortality, growth parameters, and mean environmental temperature in 175 fish stocks. *J. Cons. Int. Explor. Mer* **39**(3), 175–192.

Pauly, D. (1995) Anecdotes and the shifting baseline syndrome of fisheries. *Trends Ecol. Evol.* **10**, 430.

Pauly, D. and Christensen, V. (1995) Primary production required to sustain global fisheries. *Nature* **374**, 255–257.

Pauly, D. and Christensen, V. (eds) (1996) Mass-balance models of North-eastern Pacific ecosystems: proceedings of a workshop held at the Fisheries Centre, University of British Columbia, Vancouver, B.C., Canada, November 6–10, 1995. *Fish. Centre Res. Rep.* **4**(1), 130 pp.

Pauly, D. and Murphy, G.I. (eds) (1982) *Theory and Management of Tropical Fisheries* (*ICLARM Conf. Proc.* **9**), ICLARM, Manila, 360 pp.

Pauly, D., Martosubroto, P. and Saeger, J. (1996) The *Mutiara 4* surveys in the Java and Soutern South China Seas, November 1974 to July 1976, in *Baseline Studies in Biodiversity: the Fish Resources of Western Indonesia* (*ICLARM Stud. Rev.* **23**), (eds D. Pauly and P. Martosubroto), ICLARM, Manila, pp. 47–54.

Pauly, D., Christensen, V., Dalsgaard, J., Froese, R. and Torres, F. Jr (1998) Fishing down marine food webs, *Science* **279**, 860–863.

Pitcher, T.J. (1997) Fish shoaling behaviour as a key factor in the resilience of fisheries: shoaling behaviour alone can generate range collapse in fisheries, in *Developing and Sustaining World Fisheries Resources: the State of Science and Management* (eds D.A. Hancock, D.C. Smith, A. Grant and J.P. Beumer), CSIRO, Collingwood, Australia, pp. 143–148.

Pitcher, T.J. and Hart, P.J.B. (1982) *Fisheries Ecology*. Chapman & Hall, London, 414 pp.

Pitcher, T.J., Courtney, A., Watson, R. and Pauly, D. (1998) Assessment of Hong Kong's inshore fishery resources. Fisheries Centre Research Reports, Vancouver, **6**(1), 149 pp.

Policansky, D. (1993) Evolution and management of exploited fish populations, in *Management Strategies for Exploited Fish* (eds G. Kruse, D.M. Eggers, R.J. Marasco, C. Pautzke and T.J. Quinn), Alaska Sea Grant, Anchorage, AK, pp. 651–664.

Polovina, J.J. (1984) Model of a coral reef ecosystem. Part I: the ECOPATH model and its application to French Frigate Schoal. *Coral Reefs* **3**, 1–11.

Rijnsdorp, A.D., van Leeuwen, P.I., Daan, N. and Heessen, H.J.L. (1996) Changes in abundance of demersal fish species in the North Sea between 1906–1909 and 1990–1995. *ICES J. Mar. Sci.* **53**, 1054–1062.

Roedel, P. (ed.) (1975) *Optimum Sustainable Yield as a Concept in Fisheries Management* (*Am. Fish. Soc. Spec. Publ.* **9**), American Fisheries Society, Bethesda, MD, 89 pp.

Saeger, J. (1993) The Samar Sea, Philippines: a decade of devastation. *Naga, ICLARM Q.* **16**(4), 4–6.

Shushkina, Eh.A. and Vinogradov, M.E. (1991) Long-term changes of plankton biomass in open areas of the Black Sea. *Oceanology* **31**, 973–980. (In Russian.)

Sinclair, A.R.E., Hik, D.S., Schmitz, O.J., Scudder, G.G.E., Turpin, D.H. and Larter, N.C. (1995) Biodiversity and the need for habitat renewal *Ecol. Appl.* **5**, 579–587.

Skjoldal, H.R., Noji, T.T., Giske, J., Fossaa, J.H., Blindheim, J. and Sundby, S. (1993) *Mare Cognitum. Science Plan for Research on Marine Ecology of the Nordic Seas*, Inst. Mar. Res., Bergen, 166 pp.

Soonthonvipat, U. (1976) Dried jelly fish, in *Fisheries Resources and their Management in Southeast Asia* (ed. K. Tiews), FAO, Rome, pp. 149–151.

Smith, S.J. (1993) Risk evaluation and biological reference points for fisheries management: a review, in *Management Strategies for Exploited Fish Populations* (eds G. Kruse, D.M. Eggers, R.J. Marasco, C. Pautzke and T.J. Quinn), Alaska Sea Grant, Anchorage, AK pp. 339–353.

Smith, S.J., Hunt, J.J. and Rivard, D. (eds) (1993) Risk Evaluation and Biological Reference Points for Fisheries Management. *Can. Spec. Publ. Fish. Aquat. Sci.* **120**, 450 pp.

Vasconcellos, M., Mackinson, S., Sloman, K. and Pauly, D. (1998) The stability of trophic mass-balance models of marine ecosystems: a comparative analysis. *Ecol. Modelling* (in press).

Walters, C.J. (1995) Fish on the line: the future of Pacific fisheries. Suzuki Foundation Report, Vancouver, 87 pp.

Walters, C.J. and Maguire, J.J. (1996) Lessons for stock assessment from the northern cod collapse. *Rev. Fish Biol. Fish.* **6**, 125–137.

Walters, C.J. and Pearse, P. (1996) Stock information requirements for quota management systems in commercial fisheries. *Rev. Fish Biol. Fish.* **6**, 21–42.

Walters, C.J., Christensen, V. and Pauly, D. (1997) Structuring dynamic models of exploited ecosystems from trophic mass-balance assessments. *Rev. Fish Biol. Fish.* **7**, 139–172.

Wiener, J. (1994) *The Beak of the Finch: Evolution in Real Time*, Jonathan Cape, London, 332 pp.

Yellen, J.E., Brooks, A.A., Cornelissen, E., Mehlman, M.J. and Stewart, K. (1995) A middle stone-age worked bone industry from Katanda, Upper Semliki Valley, Zaire. *Science* **268**, 553–556.

Young, P.C. and Glaister, J.P. (1993) Defining key factors relating marine fishes and their habitats, in *Sustainable Fisheries Through Sustaining Fish Habitat* (ed. D.A. Hancock), CSIRO, Canberra, Australia, pp. 84–94.

Chapter twenty-five

Ocean triads and radical interdecadal variation: bane and boon to scientific fisheries management

Andrew Bakun

ABSTRACT

Recent large-amplitude interdecadal-scale population variations in some of the world's largest fishery stocks have exhibited a remarkable degree of global synchrony. This leads to several conclusions: (1) on these scales, the individual regional fish communities are not driven entirely by their own autonomous dynamics; (2) the biological linkages must therefore be quite simple; (3) the problem of radical interdecadal fish stock variability would consequently appear to be eminently solvable. A 'fundamental triad' of enrichment, concentration and retention processes, augments the trophic energy in supporting biological processes with the mechanical energy of the earth's ocean–atmosphere system. Examples of such triad configurations in coastal upwelling regions may help scientific understanding of some crucial fisheries issues. The idea that experimental adaptive management could proceed effectively without a well-founded system of accounting for natural environmental variability is challenged.

Reinventing Fisheries Management. Edited by Tony J. Pitcher, Paul J.B. Hart and Daniel Pauly. Published in 1998 by Kluwer Academic Publishers, London. ISBN 0 412 83410 3.

25.1 INTRODUCTION

Traditionally, fisheries scientists have viewed marine fish communities as possessing the sort of intrinsic stability that tends to be a familiar aspect of our experience of populations of large terrestrial animals. Thus a prevailing conceptual framework has been that by properly culling the excess reproductive potential of a fish population, as one might do for populations of buffalo on a prairie grassland or guppies in a glass aquarium, one might expect to be able to maintain abundance at some high level that may be a major fraction of the 'carrying capacity' of the habitat. However, as the record of information and experience builds up, that preconception is becoming more and more difficult to maintain in the face of events. Apparently, 'different worlds have different rules' (Bakun, 1996).

On the other hand, there is certainly a limit beyond which removing additional quantities of fish will tend to destroy the productivity and sustainability of a fish stock. The trophic ecology of the ocean has evolved under, and has been maintained by, the imperative that a predator must maintain a positive near-term balance between its trophic energy return and its energy expenditures in locating and capturing food. Thus, as the density of a population of favoured prey is lowered, there is a point where the predator population, in order to avoid starvation, must either switch attention to alternative prey, migrate to richer feeding areas, or enter an energy-preserving inactive state. The resulting interruption of directed predation may constitute the ultimate refuge of a prey population not otherwise able to counter the distinct hydrodynamic advantages enjoyed by larger predators in a liquid environment (Bakun, 1989, 1996).

Modern industrial fisheries are able to evade this imperative by using fossil fuels (i.e. by accessing and rapidly utilizing the stored excess energy built up by past oceans' trophic systems over millions of years). Hence they are well capable of short-circuiting the mechanisms that act in the longer term to preserve diverse ocean ecology. Thus this chapter is in no way intended to offer ecosystem variability as an excuse for irresponsible exploitation. Rather it is intended as an exposition of facts and rationale that could perhaps be helpful in moving our science forward toward a soundly based capability for managing exploitation within the reality of continual environmental variation.*

*Because the path from scepticism to conviction is essentially a personal process, I beg the reader's indulgence for presenting some of the following argument from a personal perspective. Also, certain elements of the argument are extracted rather directly from Bakun (1996).

25.2 RADICAL DECADAL-SCALE VARIABILITY

About two decades ago, marine scientists were bemused and perplexed by the evidence from the continuous plankton recorder survey that the zooplankton biomass over a very large region of the north-eastern Atlantic had apparently undergone a drastic multidecadal decline (Glover *et al.*, 1972). At that time, it seemed to many of us that this was not the way the ocean should work, and we grasped for various explanations (potential problems with the instrument, the sampling programme, the analysis procedures, etc.) that might ease the shock to our preconceptions.

However, as more multidecadal time series have been accumulating, the realization is emerging that such radical, long-time-scale variability may not be at all extraordinary, but rather a quite normal condition for marine ecosystems.† For example, Roemich and McGowan (1995) report a 70% decrease in zooplankton biomass in the California Current since the 1950s, along with corresponding drastic declines in certain sea bird species. At the same time, there seems to have been a doubling of biomass of zooplankton (Brodeur and Ware, 1992) and of pelagic fish and squid in the subarctic North Pacific (Brodeur and Ware, 1995) in the 1980s compared with the late 1950s to early 1960s period. Notably, in several parts of the Pacific, there are indications of important changes beginning in the early to mid-1970s. For example, Venrick *et al.* (1987) report a doubling of depth-integrated chlorophyll in the subtropical North Pacific starting in the mid-1970s. South of the Equator, there seems to have been a very sharp 60–70% decline in zooplankton biomass off Peru in the mid 1970s, following the collapse of the anchoveta (Loeb and Rojas, 1988; Corrasco and Lozano, 1989; Alheit and Bernal, 1993)‡.

As regards fish, the populations of small coastal pelagic fishes such as anchoveta (anchovies, sardines, sardinellas, etc.), which support many of the largest fisheries in the world, have seemed to be particularly prone to wide population swings and fishery collapse (Lasker, 1989). Indeed, evidence derived from deposits of fish scales in sediments (Soutar and Isaacs, 1974; DeVries and Pearcy, 1982; Baumgartner *et al.*, 1992; Holmgren-

† Actually, in view of the peculiar 'high-risk' nature of reproductive strategies of modern marine teleost fishes (Bakun, 1996), one might have expected major reasons for, and repercussions of, the profound differences from the much more conservative reproductive strategies that are most familiar to us in the terrestrial world.

‡ One would assume that, after the collapse of the population that for years supported the largest fishery that has ever existed in the world, the enormous reduction in anchovy biomass would have reduced the predation pressure on the zooplankton populations off Peru and reduced any competition between zooplankton and anchovy for available phytoplankton. From the usual trophic carrying capacity arguments, therefore, one would expect the zooplankton standing stocks to increase rather than decrease. However, much the opposite evidently occurred.

Urba and Baumgartner, 1993) has indicated extreme fluctuations in population sizes of this type of fishes long before the advent of large-scale fishing. Thus, even in the absence of fishing, various environmental or interspecies effects must have acted to cause large perturbations in the abundance of these stocks.

The evidence suggesting extreme population variability even in the absence of fishing has led some to suggest that fish communities may represent intrinsically chaotic systems. Indeed, mathematical models of feedback interactions within simple, tightly linked predator–prey systems have provided some of the classical demonstrations of mathematical chaos (May, 1979). The idea that we might be trying to manage chaotic systems is certainly a disconcerting and even a defeating one, because it is characteristic of such systems that infinitesimal changes in antecedent conditions may lead to unexpected, drastically differing outcomes.

25.3 APPARENT GLOBAL SYNCHRONY

Worries such as this had already, a decade and a half ago, led to a major FAO consultation on the status of coastal fish populations, held in Costa Rica (Sharp and Csirke, 1983). There we reviewed example after example of the initiation and growth of a fishing industry followed by a precipitous decline or total collapse of the stock. We were busily involved in generalizing these results from what we considered a reasonable sampling of independent systems. In fact, our discussions were following similar lines to those along which the debate is being currently posed: Was overexploitation the culprit? Or was the cause environmental? Which was it? etc.

Then an eminent Japanese scientist, Professor Kawasaki, took the floor to present a very different view (Kawasaki, 1983). He showed that when one looks very broadly at the sardine landings from the three regions of the Pacific Ocean that have supported very large sardine fisheries, a pattern emerges (Fig. 25.1) that is suggestive of oceanwide synchrony in population fluctuations. This was a remarkable proposal, because the populations are certainly far too widely separated to interact in any direct way. In fact, the populations of sardines in the different 'corners' of the Pacific were considered until very recently (Parrish *et al.*, 1989) to be separate, distinct species. Moreover, the California sardine landings after 1970 had come mainly from inside the Gulf of California, while those from the earlier period were from the Pacific Coast.

My own first instinct was to discount the idea. What could be the mechanism? Perhaps market factors or technological innovations could have been involved in initiating events somewhat simultaneously in differ-

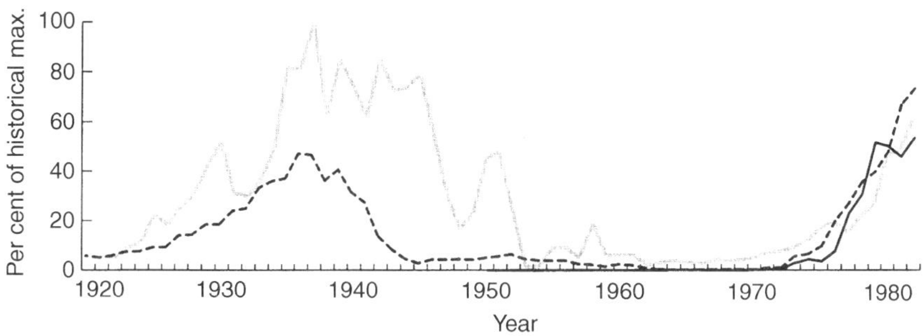

Fig. 25.1 Annual catches of sardines in the north-western, north-eastern, and south-eastern portions of the Pacific Ocean (modified from Kawasaki, 1983). Broken curve, Japanese sardine; light curve, California sardine; dark curve, South American sardine.

ent regions? Perhaps the apparent synchrony was mere happenstance (i.e. if you look long enough in a complex system you are bound to find some things that seem to form similar patterns)? And in any case, one must always be wary in trying to directly interpret fluctuations in fisheries landings as variations in abundance. I thought, 'Let's see if it *predicts* beyond the data set that established the correlation. Let him come back when he gets an additional *future* match in inflection points – *then* I'll be interested'.

Well, now I, together with many colleagues, am indeed interested. In the period since Kawasaki originally called attention to the synchronous pattern, an additional simultaneous reversal in trend has occurred (Fig. 25.2(a)). Moreover, two of the other largest coastal pelagic fish populations of the world, the Peruvian anchovy and the South African pilchard (Fig. 25.2(b)), appear to rise and fall in the 'gaps' of the same pattern (Lluch–Belda *et al.*, 1989, 1992), i.e. directly out of phase with the Pacific sardines§.

It was noted in the previous section that a recent major change in biological regime in several Pacific Ocean ecosystems appears to have

§Again, one must always be quite careful in attempting to directly link catch fluctuations to variations in abundance. Obviously, one cannot generally expect the increases in landings during the initial growth phases of these fisheries to reflect anything more than the growth in fishing effort, although in the South American case at least, ichthyoplankton data have clearly indicated a rapid rise from very low levels of sardine abundance following the collapse of the Peruvian anchovy (J. Csirke, pers. comm.). But after the initial growth phases, it seems rational at least to assume that the multi-year periods of rapid rises in landings to very high values must correspond to phases of relatively high abundance and that periods of minimal landings immediately following a fishery collapse must reflect low-abundance phases.

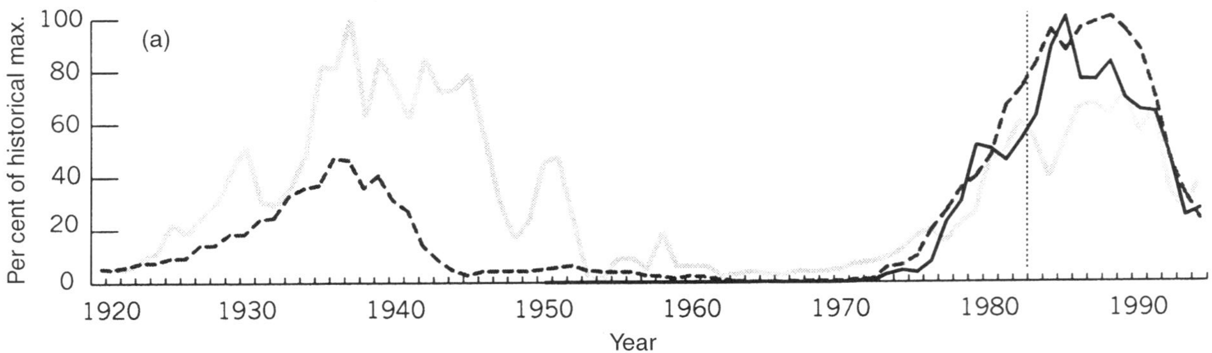
(a)
Per cent of historical max.
100
80
60
40
20
0
1920
1930
1940
1950
1960
1970
1980
1990
Year

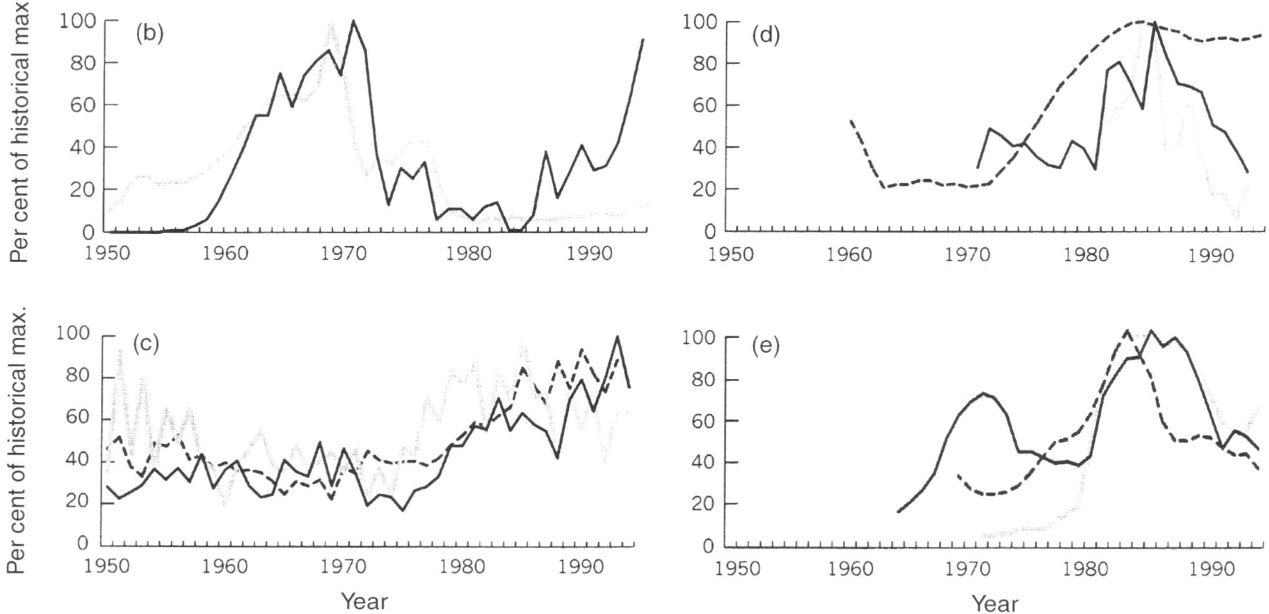

Fig. 25.2 Variations in some of the worlds largest fish stocks plotted as percentages of maximum historical annual values. Panels (a), (b), and (c) are based on landings data taken from the FAO files and from Csirke (1995), which, following the initial growth phase of the fisheries, are considered by workers in the field to validly reflect long-term variations in population abundance for these fisheries (Lluch-Belda *et al.*, 1989, 1992; Pearcy, 1992; Beamish and Boullion, 1993; Csirke, 1995; Csirke *et al.*, 1996). Panels (d) and (e) are based on actual abundance (biomass) estimates from Wespestad (1996) and from Anon. (1996). (a) Pacific Ocean sardines: broken curve, Japanese sardine; light curve, California sardine; dark curve, South American sardine. Vertical line at 1983 denotes end of Fig. 25.1. (b) Light curve, South African pilchard (sardine); dark curve, Peruvian anchoveta. (c) North Pacific salmon: broken curve, chum; light curve, pink; dark curve, sockeye. (d) Broken curve, yellowfin sole; light curve, sablefish; dark curve, Pacific hake. (e) Broken curve, Gulf of Alaska pollock; light curve, Pacific cod; dark curve, Bering Sea pollock.

commenced in the early to mid 1970s, corresponding to the time when the sardine populations of the Pacific began their most recent spectacular rises. One finds that the Pacific salmon species that attain large population sizes have followed a similar general pattern of increasing abundance starting in the early to mid 1970s and lasting to the mid 1980s (Fig. 25.2(c)). (Because the major portion of mature salmon returning to their native rivers are landed by fisheries, salmon landings also tend to be valid indicators of major long-period changes in abundance.) Moreover, the assessed abundance of many of the very large groundfish populations of the subarctic North Pacific (Figs. 25.2(d,e)) has also been varying substantially in phase with the sardine pattern.

This pattern of increases in large important fish stocks, followed by declines after the mid 1980s, seems to have been remarkably widespread and consistent in a number of regional ecosystems of the world's oceans. Moreover, many of the stocks that have not been in phase with the pattern have been quite directly out of phase (i.e. mirror images), or at least have matched some of the main inflexion points (Fig. 25.3). As one views more and more time series of fish stock fluctuations covering the past two and one-half decades, and repeatedly notes aspects of the same pattern, the idea that something is acting in one way or another to synchronize these variations becomes really quite compelling. I have found the facts to be extremely difficult to discount or to evade.

25.4 HOPEFUL IMPLICATIONS

If radical interdecadal variability is indeed occurring synchronously in distantly separated fish populations distributed over the world's oceans, there would appear to be some very hopeful implications for fisheries science. The separation between the populations must in most cases be far too great for significant population exchanges or other purely biological interactions that might provide linkage mechanisms which could explain the synchrony. It would therefore seem that such synchronous behaviour would have to be driven by some type of very large-scale external forcing, most probably through global-scale climatic effects (Kawasaki and Omori, 1988). Moreover, it seems unlikely that a set of separate autonomous ecosystems, each dominated by its own internal chaotic dynamics, could somehow self-organize themselves to generate mutual synchronous variability on a global scale. Thus, if these populations are indeed varying synchronously, the conclusion seems to be that they must not be functioning primarily as independent chaotic systems, at least on the time scales of the synchronized variability (which are in these cases the important time scales for management).

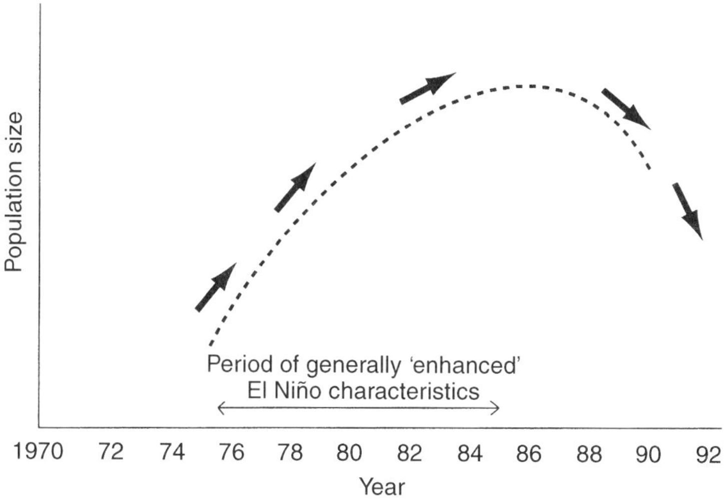

Fig. 25.3 Diagram characterizing the pattern of variation, observed in many marine fish populations, where a period of increased stock productivity and rapid population growth in the decade from the mid 1970s to the mid 1980s is followed by stock declines after the mid 1980s (figure modified from Bakun, 1996). Examples of various important fish populations having stock fluctuations in phase with (or in opposite phase to) the pattern include the following. (1) 'Dome-shaped' population curve, rising from the mid 1970s to the mid 1980s, peaking in the mid 1980s, falling thereafter: sardines (Japan, Peru–Chile, California), Benguela anchovy, North Pacific groundfish (Alaskan pollock and other stocks), and in the tropical North Pacific, lobsters, sea birds, seals and reef fishes. (2) In opposite phase: anchovies (Japan, Peru–Chile, California), Benguela sardine, North Pacific albacore tuna. (3) Population expansion beginning in mid 1970s: many stocks of Pacific salmon, and *Sardinella aurita* in the Gulf of Guinea. (4) Population declining following mid 1980s: Brazilian sardine (*Sardinella*), northern cod stocks, and *Balistes* in West Africa.

Moreover, because biological models representing anything but the simplest of marine trophic interactions are characterized by chaotic behaviour, the implication of global synchrony would seem to be that the biological dynamics involved must be very simple: i.e. a rather direct effect of the external physical forcing acting either on the fish themselves at some sensitive life stage in their complex life cycles, or directly on a primary food source. It must not be, for example, an effect working its way through a complex planktonic food web (the unravelling of which might keep biological oceanographers and trophic modellers busily employed for many decades). Thus, global synchrony would seem to be an unexpected and marvellous gift. That is, if the evident synchrony is not merely fortuitous, the environment–fisheries problem that we face would appear to be much simpler than it might very well have been. In such a case, there would appear to be a realistic hope of success in gaining a real scientific understanding of, and resulting predictive capability for, the most prominent factors determining the population dynamics of many of the most important fishery resources of the world.

25.5 THE GLOBAL SIGNAL

The details of the mechanisms linking climatic variations on decadal time scales, in widely separated marine ecosystems spread over the world's oceans, are not currently well defined. Much scientific activity is currently directed toward working out just how the ocean and atmosphere interact to produce climate variability on scales of decades to centuries (Gu and Philander, 1997; Stevens, 1997).

It has appeared to me (Bakun, 1996, 1997) that the El Niño–Southern Oscillation (ENSO) system of the Pacific must be in one way or the other involved, either as the driving engine or at least as an integral component of the mechanism. Of all the sources of short-term global climatic variability, ENSO is by far the strongest, completely dominating any other mode of variability in the Pacific Ocean (Glantz, 1996). Moreover, because the Pacific is so large, its effect on the earth's climate system is global. El Niño events have been associated with drought and catastrophic wildfires in Indonesia and Australia, with both drought and flooding in different parts of Brazil, and with anomalies in monsoon rainfall and drought in India. Correlations with climate anomalies in the central and eastern United States, western Europe, Africa, central Russia, China, and Japan have also been indicated.

A precipitous drop in the Southern Oscillation Index, a measure of the atmospheric pressure difference between the eastern and western sides of the tropical Pacific, is the much-watched-for signal of the onset of an El Niño event. When this occurs, the dynamic system of the equatorial Pacific relaxes in intensity: the trade winds slacken or reverse direction, the area of rising air and heavy rainfall usually situated in the far west (near Indonesia) expands far eastward into normally arid parts of the ocean, and large, intensely warm sea temperature anomalies cover much of the eastern tropical Pacific.

El Niño episodes generally last only about a year, but there are decadal-scale modulations in intensity and frequency of the individual events. In particular, the period from the early to mid 1970s to the mid 1980s which has seemed (Fig. 25.3) to synchronize long-period fish population variations occurring in many different regions, was a period of prolonged long-term drop in the low-passed Southern Oscillation Index, i.e. a sort of decadal-scale analogue of a standard El Niño episode. Thus, the coupled ocean–atmosphere system of the equatorial Pacific appears to have undergone a long-term relaxation during this particular period. Correspondingly, judging from time series of indicators of atmosphere–ocean mechanical energy and momentum fluxes (Bakun, 1990, 1992, 1996; Roy, 1990), many peripheral regions of the world's oceans, including the eastern boundary upwelling systems as well as the subarctic Pacific, appear to have undergone long-term intensification during the same period.

Temperature itself does not appear to be the causative agent in the apparent biological synchrony. Temperature is an easily accessible signal and although the mid 1970s to mid 1980s was a period of long-term rise in global mean temperature, it is clear that the various different regions involved did not covary in temperature in any direct or consistent manner. That leaves the aforementioned transfers of mechanical energy and momentum as the most likely candidates.

25.6 OCEAN TRIADS

The question then arises as to exactly how mechanical energy and momentum transferred from the atmosphere to the ocean through the sea surface skin might affect fish population dynamics so dramatically as to produce apparent global decadal scale synchrony in abundance fluctuations. Comparative studies of fish habitat climatology have identified three major classes of physical processes that combine to yield favourable reproductive habitat for many types of fishes:

1. enrichment processes (upwelling, mixing, etc.);
2. concentration processes (convergence, frontal formation, water column stability); and
3. retention processes favouring retention within (or drift toward) appropriate habitat.

I have called these (in a recent book in which I was drawing a number of musical analogies) the 'fundamental triad' underlying reproductive habitat suitability (Bakun, 1996).

The importance of *enrichment* processes is widely recognized and appreciated. Perhaps less widely appreciated is the importance of *concentration* processes. For small organisms, such as fish larvae, sea water represents quite a viscous fluid: major energy expenditures may be necessary just to move to the next food particle. Rapid growth from high energy gain enables a quick passage through size-related intense predation for fish larvae. Consequently, processes whereby food particles are concentrated (e.g. Fig. 25.4) may be critical.

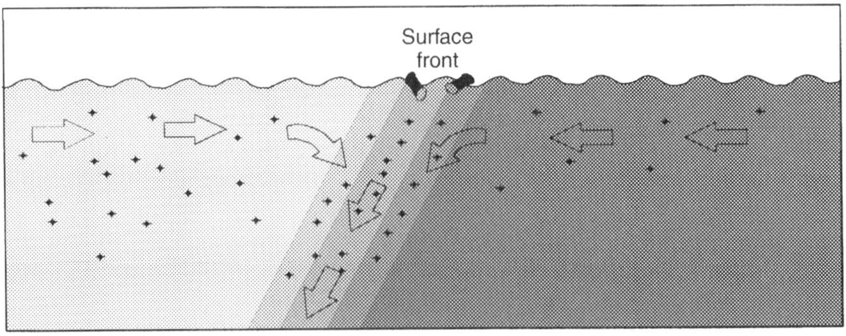

Fig. 25.4 Schematic diagram of a front between waters of differing density (lightest shading, lower-density water, fresher and/or warmer; darkest shading, higher-density water, more saline and/or cooler). Arrows indicate density-driven transverse flows associated with the front. Particle symbols indicate planktonic organisms capable of resisting vertical dispacement. (Scales are distorted: vertical scale greatly expanded relative to horizontal; particles greatly magnified; surface waves not to scale, etc.) (Redrawn from Bakun, 1993.)

This is probably a major reason why various types of interfaces, or *ergoclines* (Legendre and Demers, 1985), tend to be sites of enhanced biological activity in the ocean. Such interfaces tend either to be maintained by, and/or to maintain, mechanisms of concentration (Bakun, 1996). Ocean fronts (Fig. 25.4) are obvious examples. The importance of processes occurring in or near ocean fronts is suggested by the widespread attraction of fish and other marine animals to drifting objects. The actual convergent water motions associated with a front may be too subtle to be directly sensed by pelagic organisms operating in an environment devoid of fixed reference points. However, drifting objects tend to be carried into and to accumulate within frontal structures. An innate attraction to drifting objects serves to position the fish within the zones of enhanced biological activity and correspondingly improved feeding conditions. Conversely, turbulence is a dispersive process and so tends to act counter to concentration processes. Thus intense turbulent mixing events have appeared to be detrimental to larval survival (Lasker, 1978, 1981a,b). On the other hand, extremely small-scale turbulence might actually act like a concentration mechanism by increasing the encounter rate of small organisms with food particles (Rothschild and Osborn, 1988).

The third element in the triad is *retention*. Life cycles of marine organisms tend to include at least one stage of passive larval drift. Thus, in a dispersive fluid medium, loss of early life stages from the population habitat may represent serious wastage of resources. Consequently, fish populations tend to spawn in locations and seasons that minimize such losses (Parrish *et al.*, 1981; Sinclair, 1988).

Example: ocean triad configurations in upwelling regions

The spawning activity of most successful fish populations appears generally to be tuned geographically and seasonally to occur within habitat configurations where the three elements of the triad occur in mutually supportive combinations. In such cases, intensification of the dynamic aspects of the physical system would appear to yield even more of what may generally be good things for reproductive success and resulting population growth. And indeed, during the mid 1970s to mid 1980s period, the subarctic Pacific ocean–atmosphere system seems to have experienced a general intensification (enhanced Aleutian Low pressure system, etc.) and was favourable for growth of many salmon and groundfish populations of the region. Conversely, the relaxation of the near-equatorial system during this same period corresponded to an apparent decline in the abundance of North Pacific albacore tuna (Fig. 25.3).

However, in the environmental configurations of coastal upwelling regions, the three elements of the triad tend not to be mutually supportive. Coastal upwelling occurs because wind-driven (Ekman) transport of surface water is directed offshore, causing a situation of flow divergence near the coast. Accordingly, wind-induced coastal upwelling is directly linked to loss of larvae from the coastal habitat. The strong winds which drive the upwelling also impart turbulent mixing energy to the water column. Such mixing may abet enrichment, but destroy vital small-scale structure in food particle distributions (Lasker, 1978). Thus, upwelling ecosystems appear to be difficult places for fish to reproduce successfully (Parrish *et al.*, 1981). Nevertheless, the intensity of enrichment that occurs makes them so productive that the fishes that use them as feeding grounds have supported many of the world's largest fisheries.

It is notable that nearly all fishes inhabiting these regions avoid spawning in the core centres of maximum upwelling intensity. Rather, spawning habits appear to be tuned seasonally and geographically to enable injection of larvae into appropriately ordered sequences of the triad elements, such that initial enrichment is followed at some temporal–spatial lag by processes promoting concentration and retention.

Coastal bight downstream from an upwelling centre

For example, a favourite type of spawning habitat for fishes inhabiting upwelling regions appears to be a coastal bight that is situated downstream from an upwelling centre (Fig. 25.5). Here the coastal topography tends to shelter the bight interior from the full force of the alongshore wind, with the result that both wind-driven offshore transport and input of turbulent mixing energy drop dramatically. Correspondingly, the continental shelf edge topography may steer the large-scale ocean boundary current flow so that it skirts the open outer edge of the bight rather than flowing through the bight interior. This acts to leave a somewhat isolated water mass over the continental shelf which may be distributed in one or more rather enclosed anticyclonic circulation cells, which are 'spun' by the lateral friction of the deep ocean boundary flow at their outer edge and by strong cyclonic wind stress curl (Bakun and Nelson, 1991) produced by the sheltering effect of the coastal topography.

Thus the spatial arrangement converts to a temporal sequence. While waters enriched at the upstream upwelling centre are being advected in the alongshore flow toward the bight, entrained planktonic communities may already be undergoing trophic succession from microscopic primary producers to organisms representing suitable food for larval and juvenile fishes. Some of the enriched waters with their entrained planktonic organisms may then be diffused into the bight interior as a result of turbulent transfers between the oceanic boundary flow and the enclosed gyral circulations within the coastal bight. Finally, within the sheltered, less turbulent bight interior, fish larvae and larval food may be retained and concentrated together in frontal structures (Fig. 25.4) produced at the interfaces of the ambient bight shelf waters and the denser upwelling-conditioned water, which having entered the bight will tend to be overridden by the less dense shelf waters and thus to slowly sink and disappear from evidence at the sea surface.

Reproductive habitat configurations of this general type appear to include the coastal bights of north-western Africa located near Casablanca and near Sidi Ifni, the large bight in the South American coast near the border between Peru and Chile, the south-eastern Brazilian bight off eastern South America, and to some degree the slightly concave stretch of the coastline of south-western Africa near Walvis Bay (Bakun, 1996).

Two-celled upwelling circulation

A second type of triad habitat configuration that can occur in upwelling regions has been identified by Roy (1998). The well-known, classical wind-driven coastal upwelling circulation structure consists of general offshore transport of surface waters which induces upwelling at the coast and perhaps, in situations where the upwelling-favourable wind may be particularly strong, also near the continental shelf break (Fig. 25.6(a)). This appears to describe, for example, the upwelling situation off the

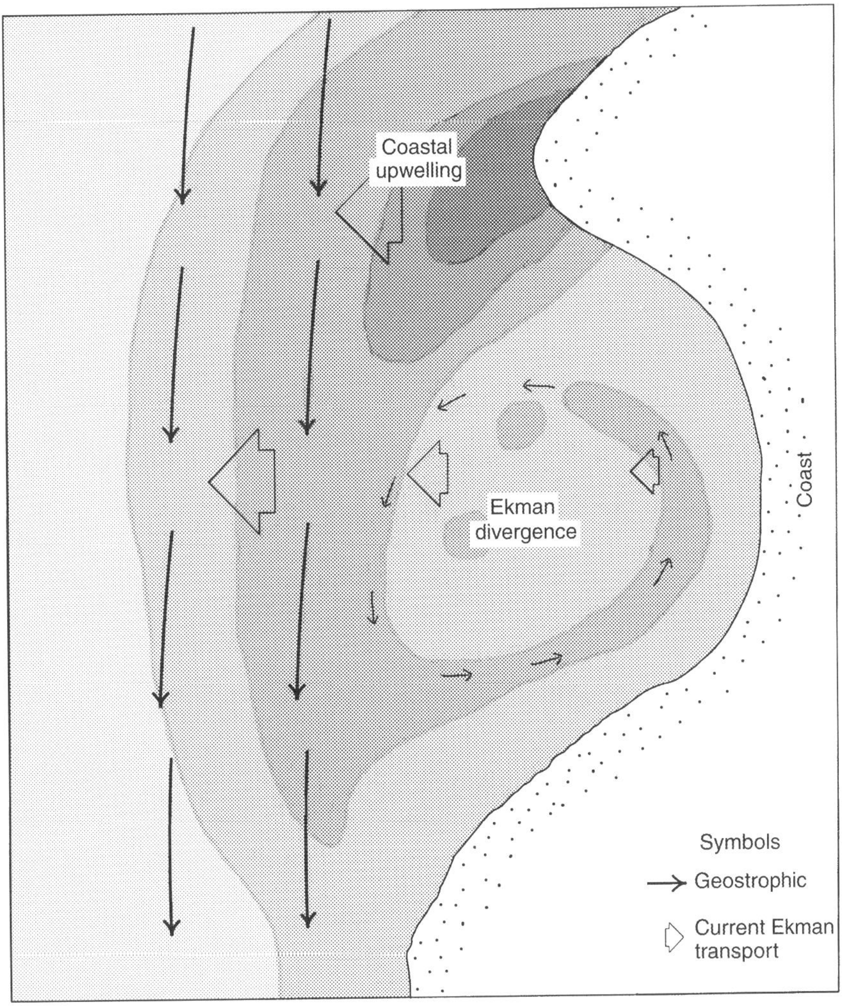

Fig. 25.5 Schematic diagram of a spawning habitat within a coastal bight downstream of an upwelling centre. Darker shading indicates higher-density (upwelled) water. Upwelled water advects southward and diffuses into the sheltered coastal bight, where wind-driven offshore transport and turbulent mixing drop sharply. Within the bight, the shoreward decline of wind stress intensity produces divergent Ekman transport, corresponding upward Ekman pumping (leading to additional local enrichment), and closed gyral circulation (which may be broken into multiple cells) in the bight interior. (Redrawn from Bakun, 1993.)

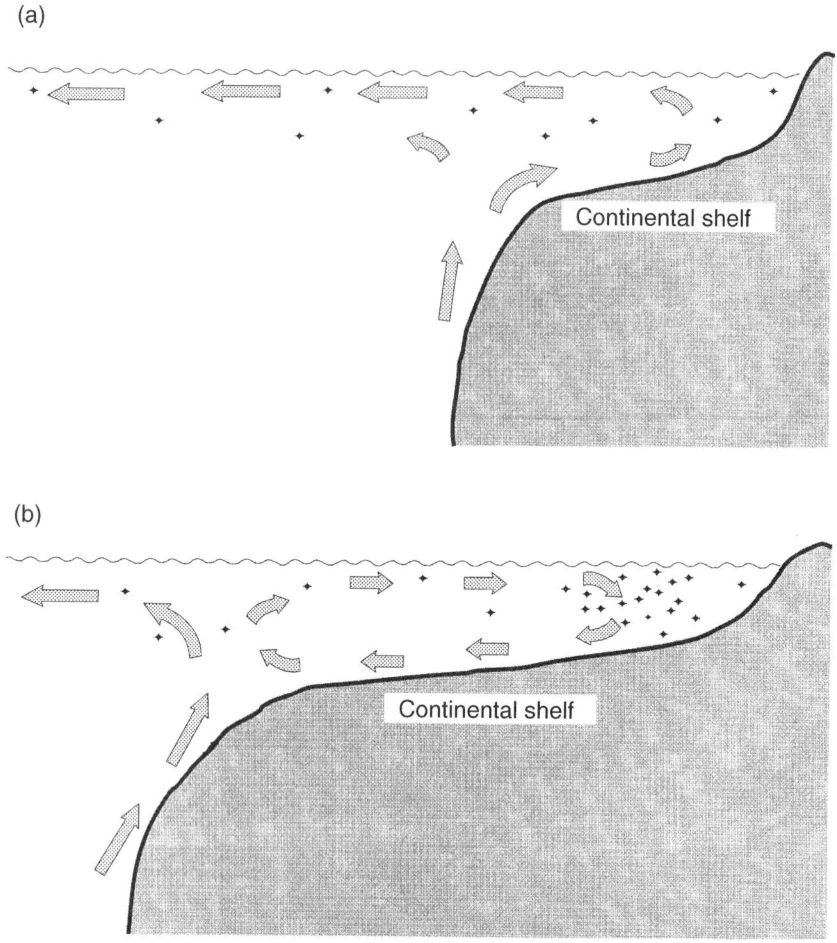

Fig. 25.6 Schematic profiles of upwelling circulation patterns (modified from Roy, 1998) and associated distributions of planktonic organisms. (a) Normal upwelling circulation. (b) 'Double-celled' upwelling circulation, where enrichment occurs at the shelf break. Enriched surface water is carried shoreward to a zone of sinking water next to the coast, which produces concentration and retention.

section of the Senegalese coast of western Africa situated to the north of Cap Vert.

However, Roy notes that nearly all the fishes that feed in this richly productive area appear to undertake a migration to spawn to the south of

Cap Vert, where the upwelling circulation structure has a different form. Here the continental shelf becomes wider and the cape itself provides some shelter from the large-scale northerly (southward-blowing) wind. The upwelling occurs primarily at the continental shelf break, tending to divide the upwelling circulation structure into two separate cells (Fig. 25.6(b)). Because of the low latitude, reduced Coriolis effect produces strong upwelling without correspondingly strong input of turbulent mixing energy.

Clearly, this represents a favourable triad configuration, with the enriched upwelled waters and their entrained biological system being carried toward the coast, where at some point they must flow against and sink beneath lighter coastal surface waters. Organisms with the ability to maintain a depth level near the sea surface thus tend to be concentrated in frontal structures (Fig. 25.4) and retained over the inner part of the continental shelf.

This triad configuration is so newly recognized that it is unclear how widely distributed over the coastal upwelling regions of the world's oceans it might be. Roy's opinion, which I share, is that it will probably turn out to be quite important in a large number of places. For example, examination of satellite imagery off Morocco reveals the identifying signature (i.e. a temperature minimum toward the outer edge of the continental shelf dividing warmer sea surface temperatures both coastward and seaward) in the important sardine reproductive habitat situated near the port of Dakhla (Do Chi and Kiefer, 1996).

Topographically trapped meso-scale eddies

Most readers with experience in upwelling regions will be familiar with the characteristic patterns of offshore squirts and jets visible in satellite-sensed infrared imagery. These give the impression of being mechanisms for rapid offshore transport, dispersion, and loss of fish eggs and larvae and of other small organisms from the near-coastal zone. But, on the contrary, it appears that at least some of these may actually be components of favourable triad configurations.

Navarro-Perez and Barton (1998) describe such a squirt structure that characteristically appears off the coast of western Africa south of the Canary Islands. Apparently, in this case, the cold squirt signature is merely a filament of recently upwelled surface water that becomes entrained near the coast and carried seaward on a stationary meso-scale eddy (Fig. 25.7). It has a tendency to be trapped at this location by the dynamical constraints imposed by the local sea floor topography. Because the entrained upwelling filament has a greater density than the surrounding offshore surface waters, it tends to sink, and thus lose visibility at the sea surface, as it circuits the eddy. This imparts the visual impression of a

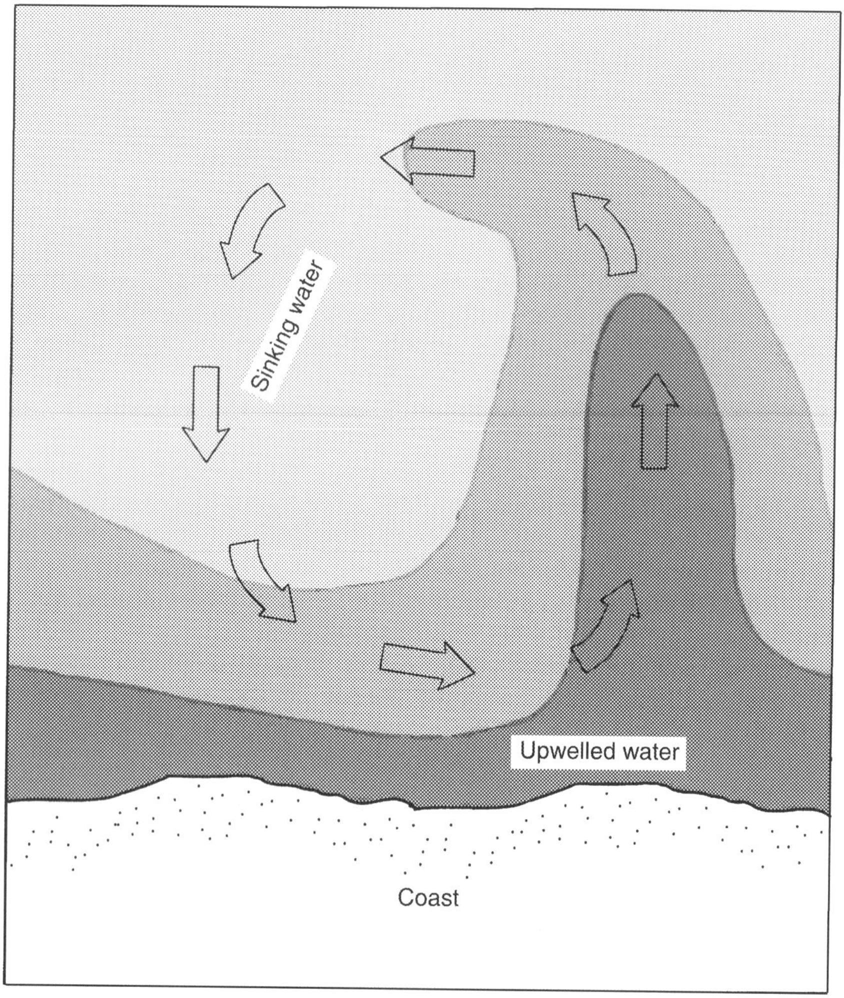

Fig. 25.7 Schematic diagram of the effect of a topographical-trapped meso-scale eddy in entraining upwelling conditioned surface water which, being of greater density, slowly sinks from the surface layer as it circuits the eddy, producing a visual impression of an offshore 'squirt' of coastal water. The sinking motion concentrates fish larvae and food organisms along the axis of the eddy circulation; they are returned to the vicinity of the coast after circuiting the eddy.

one-way offshore-flowing jet, whereas in reality the cold filament represents a slowly disappearing surface tracer circuiting around the eddy.

This evidently constitutes a favourable triad sequence. Enriched coastal surface waters and their planktonic contents are carried around the eddy circuit in a pattern of generally sinking confluence with offshore surface waters, where organisms that can maintain their depth level tend to be concentrated (Fig. 25.4). Adult fishes experiencing favourable feeding conditions in the frontal structures may spawn eggs and larvae into the circulating filament, where the larvae in turn may find correspondingly enriched and concentrated feeding conditions while being removed offshore from the intense predation that pervades the nearshore habitat. (Indeed, fish larvae have been observed to be concentrated in this particular filament structure; D. Barton, pers. comm.) Later, the surviving older, more competent juvenile fishes may be returned in the eddy circulation and deposited in their resident coastal habitat.

Other triad configurations in upwelling regions

In certain locations a major break in coastline may offer the opportunity to circumvent the characteristic opposition of triad elements that seems to hamper reproductive success in upwelling regions. For example, in the southern part of the Benguela upwelling regions, the truncation of the African continent allows adult anchovies and sardines to migrate out of the upwelling zone, around the Cape of Good Hope, to spawn over the Agulhas Bank where they may encounter a triad with mutually supportive elements (Bakun, 1993, 1996). A similar situation appears to exist off north-western Spain, where adult sardines migrate around Cape Finisterre to spawn along the northern coast of Spain. In these situations, the poleward propagation of the upwelling 'signal' and its entrained equatorward jet current 'around the corner' to the reproductive grounds via coastal trapped waves (Yoshida, 1967; Gill and Clarke, 1974; Allen, 1975) may offer enrichment from local upwelling without corresponding intensities of wind-driven offshore transport and turbulent mixing. The entrained upwelling jet current then serves to carry the larvae at a somewhat more advanced age and competency back around the corner to inject them into the rich feeding area of the upwelling zone proper. Clearly, the combination of circumstances must be sufficiently favourable to have led to the habit of expending, for the adult migrations, energy that might otherwise have been used in elaborating reproductive products. However, there may be an associated hazard of catastrophic loss of larvae; satellite views of these currents indicate that, as they intensify and make the equatorward turn, they may be subject to amplifying meanders, instabilities, and shedding of current fragments and eddies that lose contact with the coastal boundary regime and drift away to the west.

A situation with similar aspects may be found off south-western India, where poleward propagation of the signal of intense wind-driven offshore surface transport off the southern tip of India may afford a triad-favourable spawning habitat for small pelagic fishes along the Malabar coast (Bakun *et al.*, 1997).

Finally, although large river plumes are not very common in upwelling regions, where they do occur they may form a suitable habitat for small pelagic fish reproduction. For example, the Columbia River Plume environment provides a spawning habitat for a northern subpopulation of anchovy, *Engraulis mordax*, of the California Current (Richardson, 1973). Here, well to the north of the zones of most intense upwelling, a generally relaxed summer wind situation, as well as the near-surface density gradient owing to the lighter, fresher plume waters overlying the oceanic waters, provide a rather stable, non-turbulent environment for formation of food particle strata. The interfaces between the surface waters of contrasting densities are zones of frontal formation and resultant concentration of biological activity (Fig. 25.4). In addition, the lens of lighter water may induce a superficial local anticyclonic circulation around its periphery, which may aid in maintaining the integrity of the plume and in retaining larvae and other organisms within the plume area. Both the Columbia River outflow and the intermittent mild upwelling provide sources of nutrients to sustain a locally enriched trophic system. Except directly in front of the river mouth, the plume is pushed offshore during periods of coastal upwelling. However, during the upwelling season in this region there are intermittent relaxations or reversals in upwelling-producing wind stress. At these times, the plume tends to collapse coastward (Smith, 1974), providing a mechanism for returning juvenile fishes and other organisms entrained within the plume waters to the near-coastal zone.

25.7 MOVING THE SCIENCE FORWARD

River plumes, unstable current meanders, squirts and jets, eddies, temperature minima displaced offshore to the continental shelf break, current filaments either skirting or sweeping through coastal bights, warm patches indicating retained water parcels, etc., are all features that one can see in satellite images. Moreover, we expect soon to have several special ocean-orientated satellites in orbit simultaneously, where we have never before had more than one at any one time, and have had none at all for a number of years. Furthermore, the new satellites are designed with capabilities that we never have had before (direct estimates of primary production, etc.).

To reiterate the train of logic developed above, (1) the global climatic signal that may have acted to synchronize fish population variability in recent decades is most probably transferred locally through the sea surface skin; (2) the physical transfer probably consists of the mechanical factors, energy and momentum; (3) the transfer is likely to be expressed biologically through alterations of the triad in spawning habitat configurations. Following this train of logic, it would seem to make sense to try to make use of the opportunities suggested by this gift that we seem to have been granted. The resources required to investigate the effects on fish stocks of coastal ocean features observable by satellites may not need to be large – modest sampling with respect to satellite-observed triad features, to see what types of fish may be utilizing them for spawning, and which type of larvae appear to be surviving and growing well under given circumstances.

There are now available some very promising tools that might be conveniently and economically employed (Hunter and Alheit, 1995). Fish eggs may be sampled, identified, and counted in a recently developed 'egg pump' sampling system (J. Hunter, pers. comm.). Adult fish behaviour with respect to triad features can be readily studied during the sonic biomass surveys that are becoming a standard stock assessment methodology around the world. Recent nutrition and growth progress might be gauged by biochemical analysis (Buckley and Lough, 1987) of the tissues of a modest representative sample of larval or early juvenile fish. When satellite images may indicate short-term variability in triad features, one may later sample juvenile fish and count daily otolith increments to determine birthdate frequencies of surviving individuals (i.e. to develop a time history of variation in early life survival) and daily growth histories (Anon., 1994).

But satellite pictures and within-season birthdate and growth rate information constitute short-time-scale descriptions, while the time scales most pertinent for fishery management are annual (Cochrane and Starfield, 1992), and even more important multiyear and interdecadal (Figs. 25.2 and 25.3). The key in such a case is to manage to telescope the findings from the studies of short-time-scale variability that can be encompassed in research projects to the much longer time scales which may be of greatest management interest.

In this case the problem appears rather straightforward. The active agent in the atmosphere–ocean transfer of mechanical energy and momentum is the action of the wind on the sea surface. Moreover, wind intensity and direction is largely a linear function of the atmospheric pressure distribution which serves therefore as a 'stream function' for linearly averaging wind data to extend them in an undistorted manner from shorter to longer time scales. Consequently, using for example the satellite pictures and at-sea studies as suggested above, one can define on short operational time-scales (e.g.) the relative favourability for a given species of observed variations of a triad configuration that controls the suitability of a particular reproductive habitat and the corresponding variations in large-scale wind pattern. One can then use the much longer record of large-scale climatic data to hindcast the descriptions over previous decades for which historical stock information is available.

As a specific example, consider Roy's (1998) two-celled upwelling circulation (Fig. 25.6(b)). If this configuration may be controlling stock variability in a given case, there must be some change in wind character (for instance an increase in wind intensity or a change in wind direction to remove some of the sheltering effect of a cape) that will cause the upwelling circulation to shift from the favourable two-celled configuration (Fig. 25.6(b)) to the unfavourable single-celled configuration (Fig. 25.6a). Once the relationship to the wind is defined on short time scales, then one could readily go to the historical climatic data archives to define multiyear periods in which the large-scale pattern of atmospheric pressure highs and lows may be such as to favour either the single-cell or two-celled circulation.

25.8 ANALOGIES, POLEMICS, AND DISCUSSION

Recently, important fish stocks around the world have been collapsing, causing economic dislocations and personal suffering to people whose livelihood depends on fishing or on related commercial activities (Buckworth, Chapter 1, this volume). The realization is spreading that current theory and practice may be insufficient for management (Apollonio, 1993). This growing realization has, in fact, led to the assembly of this volume.

Many fisheries scientists are saying that it is not the conventional methodologies that are at fault, but that the problem lies in imperfect application; i.e. we should do the same things, but do them better next time. Indeed it seems that, retrospectively, one can nearly always find a point where something different could have been done, as well as vocal advocates for such an action. But it seems fair to say that the sudden onset of most of the collapses comes as a relative surprise to the mainstream of fishery scientists involved.

Others have been writing that the problem is simply too difficult and that we should give up trying to understand the causes of natural fish stock variability and perhaps transfer the effort (and the research funding, of course) to the development of robust strategies that may at least take a bit of the sting from the consequences of having, instead of a true prognostic science, a sort of operational art designed to make the best of a situation of 'fishing in the dark'. To take an analogy from financial stock markets, dollar cost averaging is a robust strategy, that may or may not work in a given situation. But everyone knows that dollar cost averaging can never be a match for insider information. Contrary to the situation in the stock market, development of insider information for use in fishery management is not against the law. It is clear that we need it. Ought we give up trying to produce it? Can we afford to surrender the prospect that fishery science could become a soundly based, truly prognostic science?

Some who appear to advocate giving up on the scientific aspects of natural stock variability are also vocal advocates of experimental adaptive fishery management (Bundy, Chapter 26, this volume). This seems to me a contradiction, at least as regards the very large regional fish populations examined herein. The apparent importance of small-spatial-scale pattern in the habitat (as exemplified by the triad processes, etc.) would seem to invalidate any arbitrary division of the habitat into separate equivalent subareas that could serve as effective experimental controls on multiyear and interdecadal scales (i.e. on time scales where recruitment becomes the major factor determining population dynamics). If so, given the evident radical decadal-scale natural stock variability, how can the results of any experimental manipulation of such a system be rationally interpreted without our having developed a means to filter the natural trends? With apologies for drawing yet another analogy, this strikes me as being something akin to trying to perform agricultural experiments from a moving train.

Our business is changing. With the advent of efforts such as the Marine Stewardship Council (Sutton, Chapter 8, this volume; Sutton, 1997) which propose to use eco-labelling in the retail marketplace to enforce sustainable fishing, the consumer may be entering the fishery management business as never before. We may soon need to be much more precise as to our definition of sustainable fishing. Are we to consider any fish population experiencing a declining trend in biomass to be unsustainably managed and therefore a pariah in the marketplace? Consider, for example, the very large groundfish stocks of the US zones of the Bering Sea and Gulf of Alaska. These stocks fall within the purview of the Magnuson Fishery Conservation and Management Act of 1976, which requires formal fishery management plans which are designed to achieve sustainable yield. Regular resource surveys are undertaken for purposes of fishery-independent stock assessment. All of the major fishery stocks of the region are managed by catch quotas or harvest guidelines. However, in recent years pollock and other important stocks continue to decline (FAO, 1992). Sustainable yield may be the law in this case, but it is a law fashioned by human legislators. Evidently, it is not a law of nature.

With respect to stock manageability in the conventional sense, there probably exists a spectrum of types of species–environment configurations. At one end of the spectrum may be situations like flatfish in the North Sea or Pacific salmon. Flatfish in a largely enclosed shelf sea, where there is a strictly limited number of square metres of suitable bottom habitat available and where the enclosed nature of the circulation may essentially ensure an excess of larvae to fill that habitat, may in fact almost manage themselves (i.e. as fishing pressure grows, the individuals in the catch merely get smaller until reduced marketability may act to intervene). Likewise, salmon compete for a strictly limited resource of suitable gravel beds to hold the spawn, and so compensatory density dependence is strong. Interestingly, North Sea flatfish and Pacific salmon are the fish types upon which much of the development of classical fishery science has been based.

At the other end of the spectrum may be situations like the triggerfish outbreak off tropical western Africa of a couple of decades ago (Caverivière, 1991), the similar population explosion of trumpetfish off Morocco (Brêthes, 1979), or the more recent great abundance of filefish in Hawaii (Hobson and Chess, 1996). It is possible that trying to manage stocks like these may be similar to trying to sustainably manage locust plagues.

One might think that the Japanese sardine, for example, and many of the small pelagic fish populations of intense upwelling systems, may be much closer to this latter end of the spectrum, whereas some Atlantic cod stocks may be closer to the former. Many others (Alaskan pollock, etc.) may lie somewhere between. Perhaps, in view of its new challenges, fisheries science could benefit from some well-posed efforts toward classifying the various exploited fish stocks of the world in terms of their probable position along such a spectrum (i.e. in terms of amenability to, and suitability for, attempts at management for sustained relatively constant yield).

In any case, it seems that we may have been given a gift in the evident synchronized variability of the recent decades. Certainly, the pattern of global synchronization may not continue to hold in the future. The situation of the decade from the mid 1970s to mid 1980s in which the equatorial Pacific evidently experienced decadal-scale relaxation of its physical dynamics while many of the peripheral boundary regions of the world's oceans apparently intensified, may not be as clear cut in the present and future decades. Time will tell, of course.

The important thing is that the global synchronization that we seem to have experienced tells us much about the nature of our problem. For one thing, it must be simpler than it could well have been. To reiterate some earlier points, logic tells us that radical fish population variability on decadal scales cannot be a complex food-web issue. It seems that it must be quite a direct effect of physical forcing on the life cycle processes of the fish themselves (or directly on a dominant food source). And if so, it is probably through processes such as the hypothesized triad configurations described above. There seem to be no better alternatives to account for the patterns that occurred. Even if global synchrony may not occur in the future, this gift we have experienced seems to tell us that the problem in each local system, with respect to each separate local set of physical driving forces, must be eminently solvable.

So we have something to work on, and we have tools with which to work. And we have hope of, eventually, a real prognostic science. We do need one, now more than ever.

REFERENCES

Alheit, J. and Bernal, P. (1993) Effects of physical and biological changes on the biomass yield of the Humboldt Current ecosystem, in *Large Marine Ecosystems – Stress, Mitigation, and Sustainability* (eds K. Sherman, L.M. Alexander, and B. Gold), Am. Ass. Adv. Sci., Washington, DC, pp. 53–68.

Allen, J.S. (1975) Coastal trapped waves in a stratified ocean. *J. Phys. Oceanogr.* **5**, 300–325.

Anonymous (1994) Report of the Working Group on Recruitment Processes, Lysekil, Sweden, June 14–17, 1994. *ICES C.M.* 1994/L:12, 53 pp.

Anonymous (1996) Stock Assessment Example Documents: summaries of other species in the eastern Bering Sea. NOAA–NMFS Alaska Fish. Sci. Center REFM Stock Assess. Home Page (www.wrc.noaa.gov/~jianelli/home/htm.).

Apollonio, S. (1993) Introduction, in *The System in the Sea, Applying Ecosystem Principles to Marine Fisheries, Vol. 2: Conf. Proc.* (ed. D.D. Platt), The Island Inst., Rockland, ME, pp. 10–13.

Bakun, A. (1989) L'océan et la variabilité des populations marines, in *L'homme et les Ressources Halieutiques* (ed. J.-P. Troadec), IFREMER, Brest, France, pp. 155–187.

Bakun, A. (1990) Global climate change and intensification of coastal ocean upwelling. *Science* **247**, 198–201.

Bakun, A. (1992) Global greenhouse effects, multi-decadal wind trends, and potential impacts on coastal pelagic fish populations. *ICES Mar. Sci. Symp.* **195**, 316–325.

Bakun, A. (1993) The California Current, Benguela Current, and South-western Atlantic Shelf ecosystems: a comparative approach to identifying factors regulating biomass yields, in *Large Marine Ecosystems – Stress, Mitigation, and Sustainability* (eds K. Sherman, L.M. Alexander and B. Gold), Am. Ass. Adv. Sci. Washington, DC, pp. 199–224.

Bakun, A. (1996) *Patterns in the Ocean: Ocean Processes and Marine Population Dynamics*, Univ. California Sea Grant, San Diego, CA (in cooperation with Centro de Investigaciones Biológicas de Noroeste, La Paz, BCS, Mexico), 323 pp.

Bakun, A. (1997) A dynamical scenario for occurrence of simultaneous 'regime-scale' marine population shifts in widely separated LMEs of the Pacific, in *The Large Marine Ecosystems of the Pacific Rim* (eds K. Sherman and Q. Tang), Blackwell Science Inc., Malden, MA.

Bakun, A. and Nelson, C.S. (1991) Wind stress curl in subtropical eastern boundary current regions. *J. Phys. Oceanogr.* **21**, 1815–1834.

Bakun, A., Roy, C. and Lluch-Cota, S. (1997) Coastal upwelling and other processes regulating ecosystem productivity and fish production in the western Indian Ocean, in *Large Marine Ecosystems of the Indian Ocean. Assessment, Sustainability, and Management* (eds E. Okemwa, M. Ntiba and K. Sherman), Blackwell Science Inc., Malden, MA.

Baumgartner, T.R., Soutar, A. and Ferreira-Bartrina, V. (1992) Reconstruction of the past history of Pacific sardine and northern anchovy populations over the past two millenia from sediments of the Santa Barbara Basin, California. *CalCOFI Rep.* **33**, 24–40.

Beamish, R.J. and Bouillon, D.R. (1993) Pacific salmon production trends in relation to climate. *Can. J. Fish. Aquat. Sci.* **50**, 1002–1016.

Brêthes, J.C. (1979) Contribution à l'étude des populations de *Macrorhamphosus scoplopax* (L., 1758) des côtes atlantiques marocaines. *Bull. Inst. Pêches-Marit., Casablanca* **24**, 3–62.

Brodeur, R.D. and Ware, D.M. (1992) Long-term variability in zooplankton biomass in the subarctic Pacific Ocean. *Fish. Oceanogr.* **1**, 32–38.

Brodeur, R.D. and Ware, D.M. (1995) Interdecadal variability in distribution and catch rates of epipelagic necton in the North Pacific Ocean, in *Climate Change and Northern Fish Populations.* (ed. R.J. Beamish), *Can. Spec. Publ. Fish. Aquat. Sci.* **121**, pp. 329–356.

Buckley, L.J. and Lough, R.G. (1987) Recent growth, biochemical composition, and prey field of larval haddock (*Melanogrammus aeglefinus*) and Atlantic cod (*Gadus morhua*) on Georges Bank. *Can. J. Fish. Aquat. Sci.* **44**, 14–25.

Caverivière, A. (1991) L'explosion démographique du baliste (*Balistes carolinensis*) en Afrique de l'Ouest et son évolution en relation avec les tendances climatique, in *Pêcheries Ouest-Africaines Variabilité, Instabilité et Changement* (eds P. Cury and C. Roy), Editions de l'ORSTOM, Paris, pp. 354–367.

Cochrane, K.L. and Starfield, A.M. (1992) The potential use of predictions of recruitment success in the management of the South African anchovy resource, in *Benguela Trophic Functioning* (eds A.I.L. Payne, K.H. Brink, K.H. Mann and R. Hilborn), *S. Afr. J. Mar. Sci.* **12**, 891–902.

Corrasco, S. and Lozano, O. (1989) Seasonal and long-term variations of zooplankton volumes in the Peruvian Sea, 1964–1987, in *The Peruvian Upwelling Ecosystem: Dynamics and Interactions* (*ICLARM Conf. Proc.* **18**) (eds D. Pauly and I. Tsukayama), Instituto del Mar del Peru (IMARPE), Callao, Peru; Deutsche Gesellschaft für Technische Zusammenarbeit (GTZ), GmbH, Eschborn, Germany; and International Center for Living Aquatic Resources Management (ICLARM), Manila, Philippines, pp. 82–85.

Csirke, J. (1995) Fluctuations in abundance of small and mid-size pelagics. *Sci. Mar.* **59**, 481–490.

Csirke, J., Guevara-Carrasco, R., Cárdenas, G., Ñiquen, M. and Chipollini, A. (1996) Situación de los recursos pesqueros anchoveta (*Engraulis ringens*) y sardina (*Sardinops sagax*) a principios de 1994 y perspectivoas para la pesca en el Perú, con particular referencia a la region norte y centro de la costa Peruana. *Bol. Inst. Mar. Callao, Perú* **15**(1), 23 pp.

DeVries, T.J. and Pearcy, W.G. (1982) Fish debris in sediments of the upwelling zone off central Peru: a late Quarternary record. *Deep-Sea Res.* **28**, 87–109.

Do Chi, T. and Kiefer, D.A. (eds) (1996) *Workshop on the Coastal Pelagic Resources of the Upwelling Ecosystem of Northwest Africa: Research and Predictions*, FAO, Rome, 194 pp.

FAO (1992) Review of the state of world fishery resources. *FAO Fish. Circ.* **710**, Rev. 8, 114 pp.

Gill, A.E. and Clarke, A.J. (1974) Wind-induced upwelling, coastal currents, and sea level changes. *Deep-Sea Res.* **21**, 325–345.

Glantz, M.H. (1996) *Currents of Change: El Niño's Impact on Climate and Society*, Cambridge, UK, Cambridge Univ. Press, 194 pp.
Glover, R.S., Robinson, G.A. and Colebrook, J.M. (1972) Plankton in the North Atlantic – an example of problems of analyzing variability in the environment, in *FAO: Marine Pollution and Sea Life* (ed. M. Ruivo), Fishing News Books, London, pp. 439–445.
Gu, D. and Philander, S.G.H. (1997) Interdecadal climate fluctuations that depend on exchanges between the tropics and extratropics. *Science* **275**, 805–807.
Hobson, E.S. and Chess, J.R. (1996) Examination of a great abundance of filefish, *Pervagor spilosoma*, in Hawaii. *Env. Biol. Fish.* **47**, 269–278.
Holmgren-Urba, D. and Baumgartner, T.R. (1993) A 250-year history of pelagic fish abundances from the anaerobic sediments of the central Gulf of California. *CalCOFI Rep.* **34**, 60–68.
Hunter, J.R. and Alheit, J. (eds) (1995) International GLOBEC Small Pelagic Fishes and Climate Change Program. Report of the First Planning meeting, La Paz, Mexico, June 20–24, 1994. *Globec Rep.* **8**, 72 pp.
Kawasaki, T. (1983) Why do some pelagic fishes have wide fluctuations in their numbers? – biological basis of fluctuation from the viewpoint of evolutionary ecology, in *Reports of the Expert Consultation to Examine Changes in Abundance and Species Composition of Neritic Fish Resources* (eds G.D. Sharp and J. Csirke), *FAO Fish. Rep.* **291**(3), 1065–1080.
Kawasaki, T. and Omori, M. (1988) Fluctuations in the three major sardine stocks in the Pacific and the global trend in mean temperature, in *Int. Symp. on Long Term Changes in Marine Fish Populations* (eds T. Wyatt and M.G. Larrañeta), Instituto de Investigaciones Marinas, Vigo, Spain, pp. 273–290.
Lasker, R. (1978) The relation between oceanographic conditions and larval anchovy food in the California Current: identification of the factors leading to recruitment failure. *Rapp. P.-v. Réun. Cons. Int. Explor. Mer* **173**, 212–230.
Lasker, R. (1981a) Factors contributing to variable recruitment of the northern anchovy (*Engraulis mordax*) in the California Current: contrasting years 1975 through 1978. *Rapp. P.-v. Réun. Cons. Int. Explor. Mer* **178**, 375–388.
Lasker, R. (1981b) The role of a stable ocean in larval fish survival and subsequent recruitment, in *Marine Fish Larvae: Morphology, Ecology and Relation to Fisheries* (ed. R. Lasker), Univ. Washington Press, Seattle, pp. 80–87.
Lasker, R. (1989) Les déterminants du recrutement, in *L'homme et les Ressources Halieutiques* (ed. J.-P. Troadec), IFREMER, Brest, France, pp. 189–222.
Legendre, L. and Demers, S. (1985) Auxiliary energy, ergoclines and aquatic biological production. *Nat. Can. (Rev. Écol. Syst.)* **112**, 5–14.
Lluch-Belda, D., Crawford, R.J.M., Kawasaki, T., MacCall, A.D., Parrish, R.H., Schwartzlose, R.A. and Smith, P.E. (1989) World-wide fluctuations of sardine and anchovy stocks: the regime problem. *S. Afr. J. mar. Sci.* **8**, 195–205.
Lluch-Belda, D., Schwartzlose, R.A., Serra, R., Parrish, R.H., Kawasaki, T., Hedgecock, D. and Crawford, R.J.M. (1992) Sardine and anchovy regime fluctuations of abundance in four regions of the world oceans: a workshop report. *Fish. Oceanogr.* **1**, 339–347.
Loeb, V.J. and Rojas, O. (1988) Interannual variation of ichthyoplankton composition and abundance relations off northern Chile, 1964–83. *Fish. Bull., U.S.* **86**, 1–14.
May, R.M. (1979) Bifurcations and dynamic complexity in ecological systems. *Ann. NY Acad. Sci.* **316**, 517–529.
Navarro-Perez, E. and Barton, D. (1998) The structure of an upwelling filament off

the North-West African coast during August 1993, in *Benguela Dynamics. Impact of Variability on Shelf–sea Environments and their Living Resources* (eds S.C. Pillar, C.L. Moloney, A.I.L. Payne and F.A. Shillington), *S. Afr. J. Mar. Sci.* **19**, (in press).

Parrish, R.H., Nelson, C.S. and Bakun, A. (1981) Transport mechanisms and reproductive success of fishes in the California Current. *Biol. Oceanogr.* **1**, 175–203.

Parrish, R.H., Serra, R. and Grant, W.S. (1989) The monotypic sardines, *Sardina* and *Sardinops*: their taxonomy, distribution, stock structure, and zoogeography. *Can. J. Fish. Aquat. Sci.* **46**, 2019–2036.

Pearcy, W.G. (1992) *Ocean Ecology of North Pacific Salmonids* (Washington Sea Grant Program), Univ. Washington Press, Seattle and London, 179 pp.

Richardson, S.L. (1973) Abundance and distribution of larval fishes in waters off Oregon, May–October 1969, with special emphasis on the northern anchovy, *Engraulis mordax*. *Fish. Bull. U.S.* **71**, 697–711.

Roemich, D. and McGowan, J. (1995) Climatic warming and the decline of zooplankton in the California Current. *Science* **267**, 1324–1326.

Rothschild, B.J. and Osborn, T.R. (1988) The effects of turbulence on planktonic contact rates. *J. Plankton Res.* **10**, 465–474.

Roy, C. (1990) Réponses des stocks de poissons pélagiques à la dynamique des upwellings en Afrique de l'Ouest: analyse et modélisation, thèse de doctorat, Université de Bretagne Occidental, Brest, France, 147 pp.

Roy, C. (1998) An upwelling-induced retention area off Senegal: a mechanism to link upwelling and retention processes, in *Benguela Dynamics. Impact of Variability on Shelf–sea Environments and their Living Resources* (eds S.C. Pillar, C.L. Moloney, A.I.L. Payne and F.A. Shillington), *S. Afr. J. Mar. Sci.* **19**. (in press).

Sharp, G.D. and Csirke, J. (eds) (1983) Proceedings of the Expert Consultation to Examine Changes in Abundance and Species Composition of Neritic Fish Resources, *FAO Fish. Rep.* **291**(2,3), 1224 pp.

Sinclair, M. (1988) *Marine Populations. An Essay on Population Regulation and Speciation*, Washington Sea Grant Program, Seattle, 252 pp.

Smith, R.L. (1974) A description of current, wind, and sea level variations during coastal upwelling off the Oregon Coast, July–August 1972. *J. Geophys. Res.* **79**, 435–443.

Soutar, A. and Isaacs, J.D. (1974) Abundance of pelagic fish during the 19th and 20th centuries as recorded in anaerobic sediment off the Californias. *Fish. Bull. U.S.* **72**, 257–273.

Stevens, W.K. (1997) Study of ocean currents offers clues to global climate shifts. *The New York Times* (March 18, 1997), pp. C1 and C6.

Sutton, M. (1997) The Marine Stewardship Council: new hope for marine fisheries. *Naga ICLARM Q.* **19**(3), 10–12.

Venrick, E.L., McGowan, J.A., Cayan, D.R. and Hayward, T.L. (1987) Climate and chlorophyl a: long-term trends in the Central North Pacific Ocean. *Science* **238**, 70–72.

Westpestad, V. (1996) Trends in North Pacific pollock and cod fisheries (presentation to the NORFISH Symposium). NOAA–NMFS Alaska Fish. Sci. Center, Stock Assess. Home Page at www.wrc.noaa.gov/~jianelli.

Yoshida, K. (1967) Circulation in the eastern tropical oceans with special reference to upwelling and undercurrents. *Japan. J. Geophys.* **4**, 1–75.

Part Five

Modelling Through in a Reinvented Fisheries Management

Chapter twenty-six

The red light and adaptive management

Alida Bundy

ABSTRACT

Adaptive management directly addresses uncertainty and risk. It implicitly and explicitly aims to increase knowledge. However, its practical, as opposed to theoretical, application in fisheries has been limited. Some reasons for this are discussed. They include a lack of understanding of the methodology, unwillingness to risk livelihoods, unwillingness to be in a non-optimal fishing situation, and the high economic and social costs of adaptive management. Adaptive management policies may range from a passive approach where a system's response to a policy is monitored, to an actively experimental approach where policies are designed to probe and produce definitive feedback information. It is the latter which is commonly associated with adaptive management. Adaptive management is a very valuable and useful methodological approach as it can provide a framework for all other fisheries methods. Attempts should be made to make it more accessible.

26.1 INTRODUCTION

The concepts of uncertainty and feedback are gaining acceptance as valid and necessary approaches to fisheries management (Walters and Hilborn, 1976; Walters, 1985, Hilborn and Walters, 1992, Ludwig *et al.*, 1993).

Reinventing Fisheries Management. Edited by Tony J. Pitcher, Paul J.B. Hart and Daniel Pauly. Published in 1998 by Kluwer Academic Publishers, London. ISBN 0 412 83410 3.

Because there is more that we do not know than we do know about fisheries and their interactions with natural aquatic ecosystems (Pitcher, 1994), it makes sense to use an approach that increases our knowledge. Adaptive management is a method that explicitly addresses our uncertainty about resource dynamics and uses feedback information to improve our knowledge. However, adaptive management has not been widely adopted in fisheries management. In fact, actual examples of adaptive management in action are very few; the best known is perhaps that of Sainsbury (1989, 1991) from the north-west shelf of Australia. Others include Johnson *et al.* (1992), who used an adaptive management framework to evaluate the effect of a rehabilitation plan on the yellow perch fishery in Lake Michigan; Ward and Jacoby (1992), who reported on their preliminary success using an adaptive management approach in Jervis Bay, Australia; Alcala and Russ (1990), who used adaptive management to test the effectiveness of no-take zoning, where 25% of a subtidal coral reef was protected from all forms of exploitation, on the abundance and yield of marine resources in the Philippines; and Milliman *et al.* (1987), who proposed adaptive management as a means to evaluate the success of rehabilitation of the lake trout fishery in the Laurentian Great Lakes and to make policy choices.

In adaptive management, hypotheses or models are made about key uncertainties in the resource system (Walters, 1986). Short-term management policies are then developed to discriminate between the alternative models. After an experimental period where the fishery is carefully monitored for response to the chosen policy, information from this period is fed into the decision-making process for the next management step. If all goes well, it might be possible to place more certainty on one model than the other, or it may be necessary to undergo another experimental or short-term management period before a longer-term management strategy is set. Adaptive management should be a continuous feedback system where the responses of the resource to management are used constructively to inform future management decisions.

26.2 ON THE LIMITED APPLICATION OF ADAPTIVE MANAGEMENT

Given such a well-defined role for adaptive management in fisheries, the question is, why has the approach not been more widely adopted? There are several, interrelated reasons that can be discerned. The first is that although in concept, adaptive management is quite straightforward, in detail it is complex and involves sophisticated numerical modelling (Walters, 1986). Such modelling is not readily accessible to many (McAllister and Peterman, 1992). Briefly, Bayesian statistics are employed to place initial probabilities on the uncertainties of different resource models or hypotheses (Lindley, 1983; Walters, 1986; Hilborn *et al.*, 1994; Walters and Ludwig, 1994). Simulations under the different short-term and long-term policies then produce updated probabilities for the models. Decision theory is subsequently used to work out which is the best overall strategy to follow, using criteria such as 'expected value'. For many involved in resource use and management, it is difficult to conceive how the above process relates to actual conditions, yet alone to understand the various steps and their complexities.

The second problem with adaptive management is that it requires cooperation from those involved in the fishery. This is not always easy to obtain, owing in part, to a lack of understanding or to disbelief in the process described above. Once all the different options are spelled out, there is an essential trade-off between the risk to fishers, to the resource, to the managers and the value of knowledge gained about the resource, or, more simply, the trade-off between risk to livelihoods and knowledge. It is, of course, the fishers who are most at risk in this process. It has been argued that dissent within the fishing sector (Jake Rice, pers. comm.) and the fundamentally different time horizons of fishers and scientists (Paul Hart, pers. comm.) exacerbate the challenges of cooperation.

At a more fundamental level, problems arise when adaptive management policies involve reduction in fishing effort: fishers do not want to be in the experimental area where fishing intensity is low (McAllister and Peterman, 1992). Experimental design may circumvent this problem, or extensive discussion may be required. In either case, additional time and economic resources are required.

Associated with the above are the economic and social costs of adaptive management. Adaptive management costs are high. For example, the adaptive management experiment on the north-west shelf of Australia cost, in personnel, three full-time people over 5 years, and this is regarded as high (Keith Sainsbury, pers. comm.). However, in addition to the scientific costs, there is also the very real issue of the costs to displaced fishers and their families. These social and economic costs can be extremely high, and far outweigh the scientific costs of the experiment. They are included in the 'expected value' as, for example, the value of lost catches, or unemployment costs. However, the calculation of such may not be readily apparent to the fisher or manager. The north-west shelf of Australia adaptive management experiment, for example, did include the costs of failure, but it is not clear how this was broken down into economic and social costs (Sainsbury, 1991).

Such costs are incurred with any management decision. The difference with adaptive management is that they are made more explicit. But, what about the biological, economic and social costs of *not* doing the experiment? According to Walters (Chapter 21, this volume), the costs of doing experiments are trivial compared with the costs of not doing them. Walters and Staley (1987) estimated that if an experimental approach had been used in the Fraser River sockeye salmon fishery in the 1930s and 1940s, the fishery would be about $1 billion ahead now. However, it is much easier in hindsight to ask questions about the costs of not doing an experiment than it is to assess the future benefits of doing an experiment. In the political lifespan, current costs (let alone votes) take precedence over future savings or profits. This also holds true for fishers.

In addition to the sophistication of the method, the need for cooperation of fishers and the cumulative costs of adaptive management, there is a further reason why adaptive management has not been applied more widely. Adaptive management has come to be equated with active or experimental adaptive management. But this is said not to be the case (K. Sainsbury, pers. comm.). Adaptive management is, at its core, a management system which responds to the feedback that results from passive management policies. It can include manipulative experimental policies which deliberately probe the resource, often at extremes to get a definite response. In some instances this will be the most useful method. However, this is not by any means always the case. Simulation modelling often shows that there is no real value to be gained from deliberate experimentation and that passive policies may give as much information. The difference between passive adaptive management and 'normal' management is that in the former, the system is actively monitored and the response used to adapt theories and policy formation. The key to adaptive management is the feedback response.

26.3 THE ROLE OF ADAPTIVE MANAGEMENT

Having listed several reasons why adaptive management has been limited in its application, I now return to the role of adaptive management in fisheries. Adaptive management can be used with several other key fisheries methods. For example, it provides a framework for the precautionary approach, now formally adopted by the UN Law of the Sea (FAO, 1995a). How precautionary should the precautionary approach be? (FAO, 1995b)? Decision theory is already used within the adaptive management methodology (Peterman *et al.*, Chapter 29, this volume). Where decision theory has been used in multi-objective decision making to resolve conflict and choose between management strategies, the problems of the cooperation and involvement of resource users are also found. There may be lessons to be learned for both methods. Management methods as diverse as multi-species management and single-species constant harvest quota management would benefit from using adaptive management to reduce uncertainty. Adaptive management is essentially a universal approach which could be used in all fisheries management situations. Hilborn and Sibert (1988), for example, suggest that adaptive management is the optimum way to manage developing fisheries.

Spatially based management is another approach in vogue where adaptive management is useful. Adaptive management techniques can aid in the determination of harvest rates, size of areas to be open/closed to fishing, timing of closures etc. Walters and Parma (1996) proposed that harvest rates should be constant, and adaptive management techniques can be used to set the value. For example, if the rate was set at say 10%, then only 10% of the stock should be vulnerable to exploitation. That is, if the harvest rate is to be limited, then the easiest way to limit it is to expose only that percentage of the stock to fishing.

A radical proposal made by Walters (Chapter 21, this volume) suggests that we should turn the problem of fisheries management around and show most of the world the red light – i.e. a stop signal! Instead of trying to decide how much of an area should be closed to fishing, the whole area should be closed and then experimental areas opened to see how much should be opened to fishing. This is the extreme of precautionary approaches – if the uncertainty is too great, don't take any more risks. It could also be described as the extreme in adaptive management. Large spatial refugia used to exist offshore as a consequence of physical and economic conditions, that is, they were dangerous and expensive to exploit. But over time, as fishing power has increased, these refugia have been systematically plundered. There are no longer any large havens from exploitation. Hence the continuing demise of many of the world's major fish stocks (for example, the cod fishery off the east coast of Canada). So instead of experimentally trying to see how much we can get out of a fishery by continually pushing, Walters suggested that we experiment to see how much we can sustainably exploit, by closing everything down first and then probing.

The challenges of this 'red light' approach are enormous and it is tempting to discount it as extreme and impossible. Given the difficulties in reducing fishing effort anywhere in the world, how would it be possible to close down entire fisheries? The social and economic costs would be enormous. In western countries with welfare systems this might be tenable, but in less fortunate areas of the world, such draconian measures could only result in extreme hardship, poverty and suffering. However, the idea should not be dismissed entirely. To test the hypothesis that large offshore spatial refugia are critical for healthy fisheries, offshore fisheries could be closed whilst the inshore fisheries continue to operate. Ecologically, this could only have a positive outcome.

26.4 CONCLUSIONS

Adaptive management has a role in fisheries management. The problems outlined earlier with its application should be addressed so as to make the methodology more accessible and feasible. Many of the costs associated with adaptive management might be alleviated in a co-managed or community-managed fishery. Costs and fishing privileges could be shared and a community decision made on acceptable risk. There is still a steep learning curve on how best to convey the complexities of adaptive management to fishers, managers and stakeholders and this problem needs to be addressed. It might be a question of time and familiarity: however, simplified approaches should also be considered. One such approach may be qualitative adaptive management (Kendall, 1984; Lave and Dowlatabadi, 1993).

In summary, then, there are several undeniable facts associated with fisheries management:

- there is commonly a large degree of uncertainty;
- there are risks associated with this uncertainty;
- fisheries management means managing people, not fish;
- fisheries management costs money.

Many management approaches ignore one or more of these factors. Adaptive management takes all these factors into account and in the process checks out the possibility of actively experimenting to improve our knowledge about the resource. It could be argued that the biggest failing of adaptive management is that it is 'up front' about all these factors. In many cases, it would be easier to ignore them. All fisheries management is an experiment because we can never be certain of the outcome. The difference is that now we are experimenting in a disorganized manner and cannot associate effect with cause. Adaptive management offers a controlled experimental approach, with experimental design, monitoring and evaluation. In effect, adaptive management can provide a framework for other fisheries methods.

ACKNOWLEDGEMENT

I would like to thank Rögnvaldur Hannesson for comments on an earlier draft of this manuscript.

REFERENCES

Alcala, A.C. and Russ, G.R. (1990) A direct test of the effects of protective management on abundance and yield of tropical marine resources. *J. Cons. Int. Explor. Mer* **46**, 40–47.

FAO (1995a) *The Code of Conduct for Responsible Fisheries*, FAO, Rome, 41pp.

FAO (1995b) The Precautionary Approach to Fisheries. Part 1. Guidelines on the precautionary approach to capture fisheries and species introductions. FAO/ Swedish Board of Fisheries, *Fish. Tech. Pap.* **350**, Part 1, 47 pp.

Hilborn, R. and Walters, C.J. (1992) *Quantitative Fisheries Stock Assessment: Choice, Dynamics and Uncertainty*. Chapman & Hall, New York, 570 pp.

Hilborn, R. and Sibert, J. (1988) Adaptive management of developing fisheries. *Mar. Policy* **12**, 112–121.

Hilborn, R., Pikitch, E.K. and McAllister, M.K. (1994) A Bayesian estimation and decision analysis for an age-structured model using biomass survey data. *Fish. Res.* **19**, 17–30.

Johnson, B.L., Milliman, S.R., Bishop, R.C. and Kitchell, J.F. (1992) Evaluating fishery rehabilitation under uncertainty: a bioeconomic analysis of quota management for the Green Bay yellow perch fishery. *N. Am. J. Fish. Manage.* **12**, 703–720.

Kendall, S. (1984) Fisheries management planning: an operational approach to adaptive management of a renewable resource for multiple objectives using qualitative information. *Can. J. Reg. Sci.* **7**(2), 153–180.

Lave, L.B. and Dowlatabadi, H. (1993) Climate change. The effects of personal belief and scientific uncertainty. *Env. Sci. Technol.* **27**, 1962–1972.

Lindley, D.V. (1983) Theory and practise of Bayesian statistics. *The Statistician* **32**, 1–11.

Ludwig, D., Hilborn, R. and Walters, C.J. (1993) Uncertainty, resource exploitation and conservation: lessons from history. *Science* **260**, 17, 36.

McAllister, M. and Peterman, R. (1992) Experimental design for fisheries management. *N. Am. J. Fish. Manage.* **12**, 1–18.

Milliman, S.R., Grima, A.P. and Walters, C.J. (1987) Policy making within an adaptive management framework, with an application to lake trout (*Salvelinus namaycush*) management. *Can. J. Fish. Aquat. Sci.* **44**, 425–430.

Pitcher, T.J. (1994) Stewardship and sustainability of Pacific fishery resources: the need for critical insight and an encyclopaedia of ignorance, in *Our Common Shores and our Common Challenge* (ed. D.G. Shaw), Alaska Sea Grant, Fairbanks, AK, pp. 5–28.

Sainsbury, K.J. (1989) The ecological basis of multispecies fisheries, and management of a demersal fishery in tropical Australia, in *Fish Population Dynamics* (ed. J.A. Gulland), John Wiley and Sons Ltd, Chichester, pp. 349–382.

Sainsbury, K.J. (1991) Application of an experimental approach to management of a tropical multispecies fishery with highly uncertain dynamics. *ICES Mar. Sci. Symp.* **193**, 301–320.

Walters, C.J. (1985) Approaches to adaptive policy design for harvest management, in *Modelling and Management of Resources Under Uncertainty*, (*Lecture Notes in Biomathematics* Vol. 72) (eds T.L. Vincent, Y. When, W.J. Grantham and G.P. Kirkwood), Springer-Verlag, Berlin, pp. 114–124.

Walters, C.J. (1986) *Adaptive Management of Renewable Resources*, MacMillan, New York, 374 pp.

Walters, C.J. and Hilborn, R. (1976) Adaptive control of fishing systems. *J. Fish. Res. Bd Can.* **33**, 145–159.

Walters, C.J. and Ludwig, D. (1994) Calculation of Bayes posterior probability distributions for key population parameters. *Can. J. Fish. Aquat. Sci.* **51**, 713–722.

Walters, C.[J.] and Parma, A.M. (1996) Fixed exploitation rate strategies for coping with effects of climate change. *Can. J. Fish. Aquat. Sci.* **53**, 148–158.

Walters, C.J. and Staley, M.J. (1987) Evidence against the existence of cyclic dominance in Fraser River sockeye salmon (*Oncorhynchus nerka*), in *Sockeye Salmon*, Oncorhynchus nerka, *Population Biology and Fisheries Management* (eds H.D. Smith, L. Margoles and C.C. Wood), Dept of Fisheries and Oceans, Ottawa, pp. 375–384.

Ward, T.J. and Jacoby, C.A. (1992) A strategy for assessment and management of marine ecosystems: baseline and monitoring studies in Jervis Bay, a temperate Australian embayment. *Mar. Poll. Bull.* **25**, 163–171.

Chapter twenty-seven

Fundamental obstacles to the application of ecological science in fisheries management

James P. Scandol

ABSTRACT

In this chapter the difficulties in the application of ecological science to fisheries management are discussed. The simplistic, but commonly held, presupposition that more ecological science leads to better fisheries management is erroneous. An examination of fundamental obstacles to the application of science to public policy formulation is used to illuminate the difficulties in the application of ecology to fisheries management. Methods that are, or could be, used for the efficacious application of ecology to management are briefly overviewed.

27.1 INTRODUCTION

Hammond *et al.* (1983) suggested that fundamental obstacles exist to the application of scientific information to public policy formulation. The present point-of-view chapter examines those obstacles relevant to the use of ecology for fisheries management and briefly reviews options available to deal with these barriers.

Reinventing Fisheries Management. Edited by Tony J. Pitcher, Paul J.B. Hart and Daniel Pauly. Published in 1998 by Kluwer Academic Publishers, London. ISBN 0 412 83410 3.

Ecology is the scientific study of the interactions that determine the distribution and abundance of organisms (Krebs, 1972). Ecology is a vast subject and this chapter makes little specific reference to particular areas of ecology. Fisheries managers who are attempting to regulate fishing mortality will naturally support the development of a science that describes and forecasts the distribution and abundance of target species. Fisheries ecology applied to fisheries management can give a predictive understanding of the response of target and non-target species to fishing pressures. Some studies within fisheries ecology are exceedingly theoretical with no obvious application, whilst other studies are specific analyses to provide information for particular management decisions. Both types of studies within fisheries ecology applied to fisheries management are susceptible to the fundamental obstacles outlined in this chapter, but to differing degrees.

27.2 ANALYSIS

Three types of fundamental obstacle – situational, cognitive and scientific – are discussed in turn.

Situational context

Lindblom (1959, 1979), a policy theorist, suggested that the best descriptive model for policy formulation was one of incrementalism or muddling through. The prescriptive suggestion of identifying objectives, specifying constraints, articulating and evaluating policy alternatives to deduce an optimal policy (Chechile and Carlisle, 1991) simply does not occur in practice (Lindblom, 1959). Instead, policy makers make decisions that are small, or incremental variations to previous arrangements, because for individuals involved these are already the optimal decisions. The challenge set by the prescriptive model to identify common objectives for all players was particularly unrealistic. Hammond *et al.* (1983) reiterated this incremental model by stating that the process for formulating policy was not rational as a consequence of the ambiguity of scientific information and social values.

Facts are not value free

Policy analysts Morgan and Henrion (1990) noted that a particular class of uncertainty was disagreement between experts. Those with experience with fisheries management would rapidly concur that disagreements occur within and between groups of experts in social, technical and even logistical areas. The context of management, or situational context in which ecology (or any science for that matter) is to be applied, cannot readily be rationalized. Attempts to determine simply, and then to implement, the optimal policy are likely to be flawed, either in calculation or in execution. The difficulty of this situational context is particularly acute when dealing with uncertainty. Although fisheries science has seen the development of imaginative and robust quantitative methods developed to deal explicitly with uncertain data and processes (Walters, 1986), many policy makers have difficulty interpreting and utilizing probabilistic results. This is because many policy decisions are discrete choices (Hammond *et al.*, 1983). For example, the decision whether or not to close down a hatchery is not greatly aided by being informed that trophic arguments lead to the conclusion that there is a 30% chance that average total catch from the fishery will decrease. Several authors have suggested that rational methods such as statistical decision theory can and should play an important role in decision making under uncertainty (Ludwig *et al.*, 1993). I also support that approach. However, even the most optimistic of us must acknowledge that social and technical barriers will always make the interpretation of rationalized probabilistic information difficult.

Another facet to situational context is that fisheries management is a political process. Incremental decision making often involves individuals making decisions that are optimal for them, not for the fishery in the long term (Hart and Pitcher, Chapter 16, this volume). This individual may be an elected politician, a middle-level bureaucrat or a field officer. The public often sees individuals and the institutions they represent as more disciplined and effective if there exists a constancy or fixed attitude of purpose (Hammond *et al.*, 1983). Scientific information that is counter to current thinking may be ignored or downplayed because it is not in the best interest of the individuals involved in decision making to do otherwise. This was undoubtedly the case for the eastern Canadian cod fishery (Finlayson, 1994). The insidious nature of this problem has been highlighted with the benefit of hindsight (Fraser, 1995), but it is unlikely that any inquiry or review will eliminate it.

Cognitive obstacles

Hammond *et al.* (1983) also suggested that the integration of all types of information for policy formulation is not an analytical process, but involves mental processes that are probably impossible to completely understand. Balancing many facts, opinions and values for decisions that must be made under pressure can be done only in a quasi-rational fashion. This is particularly true for the judgement of social impacts.

The contribution of science to policy decisions is further complicated by the fact that, by definition, science is done by scientists. No scientist comes to a policy problem baggage free. Information from scientists is not treated equally because of the different individuals involved (for example historical credibility). In some cases scientists will argue to win the debate rather than to uncover the truth. Decision makers will add additional value judgements to scientific information, based simply upon which scientists were chosen to be involved in a study and how credibly their responses are treated. This problem is often complicated when some scientists present too much, or irrelevant, information to policy makers. This will involve an additional value-laden process of discarding excess information. These are some of the fundamental cognitive limitations of decision-makers (Hammond *et al.*, 1983).

Scientific obstacles

Ecologists are trained to adhere to the inferential algorithm of falsification (Peters, 1991). The cycle of hypothesis formulation, experimental design and hypothesis rejection or corroboration has proved exceedingly potent in many scientific disciplines (for example, molecular biology after Watson and Crick, 1953). Peters (1991) has critically analysed ecology and suggested that failure to adhere to strict falsification has caused an erosion of ecological credibility. The fundamental scientific obstacle identified is that inductive inference does not apply across large changes in physical scale (Levin, 1992) and that it is difficult and sometimes impossible to perform experiments relevant to policy choices (Hammond *et al.*, 1983). Fisheries scientists have long recognized this problem (Hilborn and Walters, 1992) and the explicitly experimental components of adaptive management address these issues (Walters, 1986). The adaptive management approach to this issue is correct and must continue to be researched and applied. Ecologists must be made more aware of the fundamental nature of the problem and must be prepared to engage in the difficult statistical inference exercise required for non-replicable systems.

Cross-scale or cross-system inference requires deduction (essentially conceptual extrapolation). Depending upon the system being evaluated and the experience of the scientists involved, this may or may not lead to a correct result. This uncertainty is another fundamental obstacle to the application of science to policy. The difficulty of cross-scale inference in ecology has been understood for some time (Allen and Hoekstra, 1992; Holling, 1995) but there appears to be no consensus on the best approach forward. I believe that such a development may require a re-evaluation of how best to model space and time in ecosystems (Sugihara and May, 1990). Unfortunately such a highly theoretical approach is unlikely to be received enthusiastically by a stakeholder whose livelihood is influenced by policies inferred from these models.

Contemporary large-scale experiments designed to generate informed fisheries policies still have significant difficulties, even in wealthy nations (Mapstone *et al.*, 1996). This is because of the difficulties and expense of effective stakeholder consultation. Rational theories are not necessarily accepted in quasi-rational political environments.

27.3 CONCLUSIONS

This point-of-view chapter has emphasized the obstacles to the application of ecological science to the development of fisheries policies. The unquestioned optimism exhibited by scientists (and those who sell science) that their profession is the panacea for environmental problems is misleading (Wakeford, 1995) and possibly regressive for environmental policy. This regressive nature of oversold science is described below.

Ecological science has significant opportunity costs. Science is expensive in terms of both time and financial resources. Although the delaying of decisions because of forthcoming scientific results is now officially discouraged by the latest definition of the precautionary principle (Gilpin, 1996), there is little doubt that it still occurs. The mutual benefits to the scientist and the politicians are simply too great for this funding and delaying tactic to be unused, but delayed decisions are often missed opportunities.

Financial opportunity costs are less subtle. The challenge of policy implementation was a repeated theme at the Reinventing Fisheries Management Conference and in the literature of contemporary applied ecology (Mangel *et al.*, 1996). A decision is futile if it cannot be implemented because of logistic, economic, social or political barriers. If resources are directed to science at the expense of research into policy implementation methods, then we may not be using the most cost-effective approach to solve problems.

The final consequence of oversold science in policy is more esoteric and is associated with the disruption of negotiation processes. Scientists are, or should be, becoming aware that stakeholders are treating them as simply another player within environmental negotiations. The author has been in negotiations where some players treat scientists with open hostility. Purporting to represent an objective or value-free scientific position within discussions is simply not possible because we do not know what impacts our values have on our interpretation of facts. Recent suggestions by applied ecologists (Peters, 1991, Mangel *et al.*, 1996) that scientists differentiate their political and scientific views are well intentioned but are somewhat naive from a historical or political perspective. Frank discussion about the baggage that stakeholders (including scientists) bring into negotiations may result in more effective and honest discussions. Scientists may need reminding that many lobbyists with either conservation or exploitation-based ideologies are highly skilled, intelligent individuals who probably enjoy disputing with scientists as much as they do with politicians and bureaucrats. After all, that is their career. Ecologists who do not understand the reasons for this behaviour from other players may become disillusioned and disengage from the application of their work. This would be a tragic loss to fisheries management and can be avoided by encouraging debate about the difficulties of applying science in policy settings.

Many of the points raised in this chapter are generic to the application of science to any environmental issue. Fisheries science is probably one of the more conceptually advanced and integrated areas of environmental science because evidence of serious problems has existed for so much longer. The arguments of Hammond *et al.* (1983), already discussed, identified fundamental obstacles to the application of science to public policy, showing that the application of ecology to fisheries management has sidestepped some of the obstacles. They have also made clear that many obstacles remain. Identification of these would be a productive step forward for any situation where science interacts with policy formulation.

This point of view is not anti-science. No credible person doubts the magnificent contribution that science has made to the human understanding of nature and the fantastic predictive capability exhibited by many scientific disciplines. However, successful science is fuelled by effective criticism. To downplay some glaring difficulties of integrating science into environmental decision making may offer some short-term benefits but would involve very large long-term costs.

ACKNOWLEDGEMENTS

The author would like to thank the two anonymous referees who made extensive and valuable comments on an earlier draft of this essay. Alistair Gilmour and Sandra Gauld also suggested important changes to the penultimate draft. Financial support for this work was provided by a Canadian NSERC strategic grant to M. Healey, P. LeBlond and C. Walters.

REFERENCES

Allen, T.F.H. and Hoekstra, T.W. (1992) *Towards a Unified Ecology*, Columbia Univ. Press, New York, 384 pp.

Chechile, R.A. and Carlisle, S. (1991) *Environmental Decision Making: A Multidisciplinary Perspective*, Van Nostrand Reinhold, New York, 296 pp.

Finlayson, A.C. (1994) *Fishing for Truth: A Sociological Analysis of the Northern Cod Stock Assessments from 1977 to 1990*, Inst. Social Econ. Res., Memorial Univ., St John's, Newfoundland, 176 pp.

Fraser, J.A. (1995) *Fraser River Sockeye 1994: Problems and Discrepancies*, Fraser River Sockeye Public Review Board, Vancouver.

Gilpin, A. (1996) *Dictionary of Environment and Sustainable Development*, John Wiley & Sons, Chichester, 247 pp.

Hammond, K.R., Mumpower, J., Dennis, R.L., Fitch, S. and Crumpacker, W. (1983) Fundamental obstacles to the use of scientific information in public policy making. *Technol. Forecast. Soc. Change* **24**, 287–298.

Hilborn, R. and Walters, C.J. (1992) *Quantitative Fisheries Stock Assessment: Choice, Dynamics and Uncertainty*, Chapman & Hall, London, 570 pp.

Holling, C.S. (1995) What barriers? What bridges? in *Barriers and Bridges to the Renewal of Ecosystems and Institutions* (eds L.H. Gunderson, C.S. Holling and S.S. Light), Columbia Univ. Press, New York, pp. 3–34.

Krebs, C.J. (1972) *Ecology: The Experimental Analysis of Distribution and Abundance*, Harper and Row, New York, 694 pp.

Levin, S.A. (1992) The problem of pattern and scale in ecology. *Ecology* **73**, 1943–1967.

Lindblom, C. (1959) The science of muddling through. *Pub. Admin. Rev.* **19**, 79–88.

Lindblom, C. (1979) Still muddling, not yet through. *Pub. Admin. Rev.* **39**, 517–526.

Ludwig, D., Hilborn, R. and Walters, C.J. (1993) Uncertainty, resource exploitation, and conservation: lessons from history. *Science* **260**, 17, 36.

Mangel, M. and 41 others (1996) Principles for the conservation of wild living resources. *Ecol. App.* **6**, 338–362.

Mapstone, B.D., Campbell, R.A. and Smith, A.D.M. (1996) Design of experimental investigations of the effects of line and spear fishing on the Great Barrier Reef. CRC Reef Research Centre, Townsville, 86 pp.

Morgan, M.G. and Henrion, M. (1990) *Uncertainty: A Guide to Dealing with Uncertainty in Quantitative Risk and Policy Analysis*, Cambridge Univ. Press, Cambridge, 332 pp.

Peters, R.H. (1991) *A Critique for Ecology*, Cambridge Univ. Press, Cambridge, 366 pp.

Sugihara, G. and May, R.M. (1990) Applications of fractals in ecology. *Trends Ecol. Evol.* **5**, 79–86.

Wakeford, T. (1995) Bringing science down to earth, in *Science for the Earth* (eds T. Wakeford and M. Walters), John Wiley & Sons, Chichester, pp. 349–363.

Walters, C.J. (1986) *Adaptive Management of Renewable Resources*, MacMillan, New York, 374 pp.

Watson, J.D. and Crick, F.H.C. (1953) Molecular structure of nucleic acid. A structure for deoxyribose nucleic acid. *Nature* **171**, 737–738.

Chapter twenty-eight

Evolutionary models for fisheries management

Jarl Giske

ABSTRACT

The potential use of functional evolutionary models and mechanistic ecological models as predictive tools in fisheries ecology is discussed. Evolution by natural selection is a force that leads to ecological adaptations of the individuals in the populations. By mechanistic modelling of the sense organs, we may model how individuals perceive their environment, and by life history theory we may predict optimal responses to environmental variation. Modelling tools such as dynamic programming and individual-based modelling are discussed, and the coupling to physical models of the sea allows analyses that hindcast with explanatory power. Currently, our ability to forecast is limited by weather models. However, the quasi-periodic oscillations of the hemisphere-level ocean–atmosphere interactions bring hope for increased ability to make predictions in the future.

28.1 INTRODUCTION

In the first five years of the 20th century, the Norwegian Vöringen expedition investigated the physical and biological conditions of the Norwegian Sea, an area that was almost unknown beforehand. Helland-Hansen and Nansen discovered that the Barents Sea was fed by irregular transport of warmer Atlantic water along the coast of Norway. As there

Reinventing Fisheries Management. Edited by Tony J. Pitcher, Paul J.B. Hart and Daniel Pauly. Published in 1998 by Kluwer Academic Publishers, London. ISBN 0 412 83410 3.

was a two-year delay in the pulse of water from western Norway to the Barents Sea, they anticipated that ocean temperatures might be monitored and predicted. Large-scale biological events such as invasions of beluga whales and Greenland seals to Norwegian coastal waters, as well as poor growth and low condition of Atlantic cod, *Gadus morhua*, indicated food shortage in these waters at the time, and Helland-Hansen and Nansen linked this to water transport: 'We think that these discoveries give us the right to hope that by continued investigations it will be possible to predict the character of climate, fisheries, and harvests, months or even years in advance' (Helland-Hansen and Nansen, 1909, pp. iv–v).

While Helland-Hansen and Nansen cruised the unknown Norwegian Sea, the Wright Brothers were constructing the first motorized aeroplane. As this chapter is written, the Hubble telescope seeks out distant stars and planets. There has been tremendous scientific achievement in the 20th century, but the hopes of the early oceanographers have not been fulfilled. Will we ever be able to predict the future of fish stocks?

During the past 30 years, fisheries assessment has been based on catch-at-age analyses of various sorts, including virtual population analysis (VPA) and elaborations of it (Sampson, 1988). By measures of recruitment and assessment of fishing mortality, the year class strengths may be calculated. Hence, the number and biomass of fish aged $n + 1$ in year $y + 1$ may be predicted from numbers of fish aged n in year y. This method works well, in many cases. However, there are two cases where this is not a good tool: for short-lived fish and for variable environments.

Short-lived fish exist, but the stable ecosystem does not. Because the classical mathematical assessment tools assume stability, new methods are needed. In a world with stable ecosystems from which high-quality empirical relationships exist, theoretical approaches are of only academic interest. However, such models are valuable for management in fluctuating environments, i.e. where it is hardest to make predictions. And even if future events could not be predicted with much certainty, adaptive management would benefit from revealing causal relationships, as this understanding is needed to evaluate possible effects of a proposed management.

28.2 EVOLUTION AS A FORCE

The theory of evolution by natural selection gives a frame for understanding the processes of development, allocation patterns, life cycles and habitat choice. Individuals with a genetic composition that makes them more successful in reproduction, leave on average more offspring than other members of the population. By this process, the gene pool of the population is under continuous selection for alleles that are suited for coping with the physical and biological environment. After 3.5 billion years of mutations, recombinations and natural selection, we are left with the evolutionary winners (Dawkins, 1976, 1996). This implies both: (1) that all species on Earth are reasonably well adapted to the environment in which their ancestors lived; and (2) that individuals are likely to have a genome that allows them to live almost optimally in order to spread their genes.

Therefore, there exists a predictive evolutionary force that operates on units exposed to natural selection: the behaviour of individuals will be aimed at maximizing their potential reproductive rate (Sibly, 1989, McNamara and Houston, 1996). This force can be utilized in environmental modelling to yield predictions of individual behaviour and growth patterns as well as life cycles and population dynamics (Fiksen *et al.*, 1995; Rosland and Giske, 1997; Giske *et al.*, 1998a). The objective for predictive ecological and fisheries modelling is to provide functional models, relating the activities of the individuals to fitness (and thereby to understand their ultimate motivations for behaviour), and mechanistic models relating the potentials for feeding, growth, reproduction and survival to the state of the environment. The mechanistic models will estimate the immediate gains and losses associated with, for example, a series of habitats, while the functional models will trade off these forces so that the action taken is the one that is most likely to maximize fitness. Mechanistic models will focus on sensory systems and physical, chemical and bioenergetical laws, while functional models must address which traits will spread in a population under natural selection.

28.3 THE STATUS OF COMBINING MECHANISTIC AND FUNCTIONAL MODELS

For a long period, mechanistic models of aerodynamics and hydrodynamics have been used to predict water movements, salinities, and temperatures in the sea. More recently, dynamics of nutrients and primary production have also been modelled mechanistically and have been coupled to the physical models (Aksnes, 1993; Aksnes *et al.*, 1995; Giske *et al.*, 1998b).

Far fewer models are available for the higher trophic levels: zooplankton, fish and sea mammals. The new challenge at these levels is that the organisms are (relative to phytoplankton and bacteria) long lived with complex life histories. They have, through evolution, been selected for performing different tasks through their lives, and they are to a high degree able to sense both their internal state and the external environment and behave optimally using evolved responses. Hence, modelling these higher trophic levels may require different, but perhaps not more complex, mechanistic models. Modelling populations of animals that are able to relocate requires functional models that properly describe the fitness values of the multiple choices these individuals continuously perform (Tyler and Rose, 1994; Giske *et al.*, 1998a). The meagre success in modelling higher trophic levels in classical bottom-up biogeochemical models is probably related to the lack of proper behavioural control that only fitness-maximization imposes.

Theories that combine optimal behaviour with optimal life cycles first appeared in the 1980s (Mangel and Clark, 1986; Aksnes and Giske, 1990). Models of the sensory fields of fish (Jumper and Baird, 1991; Schellart, 1992; Bleckmann, 1993; Aksnes and Utne, 1997), bioenergetics (Kitchell *et al.*, 1977, Hewett and Johnson, 1992; Salvanes *et al.*, 1995) and of the environmental impact on growth and survival (Clark and Mangel, 1986; Giske *et al.*, 1994) are available, although only developed for a limited number of species and situations. The relatively homogeneous and transparent pelagic environment should give the best opportunities for linking ecological variables like depth and diet selection, group size, and swimming speed to growth and survival, as required by managers.

Predictive models are in need of biological theories but also of numerical methods to find solutions. Over the past decade, a series of techniques have become available. Optimizations based directly on life history theory (by the Euler–Lotka equation or a derivative) are often called static optimization (Leonardsson, 1991), as the state of the individual is not included and motivation is not allowed to change over short time intervals. Static optimizations are therefore best suited for principal analyses or large-scale phenomena, and can often yield analytical solutions. Examples are the growth–mortality trade-off for juvenile fish (Werner and Gilliam, 1984), habitat choice as a function of life history (Aksnes and Giske, 1990), age-dependent difference in acceptance of mortality risk (Giske and Aksnes, 1992), optimal timing for settling of codlings (Salvanes *et al.*, 1994), diet versus depth selection (Giske and Salvanes, 1995) and density-dependent habitat profitability (Giske *et al.*, 1997). There now also exists an extensive theoretical treatment of demography (Tuljapurkar, 1990) and adaptations in stochastic environments (Yoshimura and Clark, 1993).

Optimizations are called dynamic if the outcome will depend on changes in state of the individual. The state may be internal factors such as stomach fullness, fat reserves, body mass, or external factors such as group size or the spatial position within the habitat, and the optimal policy of the individual will change over time according to these variables and their fitness values. The most widely used dynamic optimization technique is stochastic dynamic programming (SDP, Houston *et al.*, 1988; Mangel and Clark, 1988). This method has been used to study vertical distributions of fish populations (Clark and Levy, 1988; Rosland and Giske, 1994, 1997) as well as large-scale horizontal distributions (Fiksen *et al.*, 1995).

The use of individual-based modelling (IBM), whereby the population is represented by a large number of artificial individuals, has recently provided interesting results on how fish population processes are governed by the success of the few, rather than by the average success rate (Crowder *et al.*, 1992; DeAngelis and Rose, 1992; Rose *et al.*, 1993; Brandt and Hartman, 1993, and references therein). While SDP models are used to study individual variation in motivation for behavioural decisions, IBMs are good at resolving population consequences of variability in performance owing to stochastic processes and at interactive processes including density dependencies or trophic couplings. This approach has recently been much used in studying the ecology of fish larvae and in recruitment studies (Hinckley *et al.*, 1996; Letcher *et al.*, 1996; Werner *et al.*, 1996) and on growth rate estimates (Mason *et al.*, 1995; Mason and Brandt, 1996). The potential uses of IBMs in resource management have been discussed by Barnthouse (1992), Hansen *et al.* (1993) and Bart (1995).

Biological models of oceanic populations must interact with physical models of ocean dynamics. The forcing functions for the most-used oceanographic models are meteorological fields, such as air temperatures, air pressures, wind field, precipitation, solar radiation and cloud cover. Historical data for these forcing variables are stored in open archives, enabling models to mimic quite precisely the physical conditions that forced a particular historical biological event. For example Aksnes *et al.* (1989) calculated the spread of the toxic flagellate *Chrysochromulina polylepis* (killing salmon in aquaculture) in the North Sea in 1988, Hinckley *et al.* (1996) used meteorological forces to simulate drift of eggs and larvae in the Gulf of Alaska, and Fiksen *et al.* (1995) utilized meteorological data to calculate sea temperatures and zooplankton advection (to calculate habitat profitability for capelin) in the Barents Sea in 1979–1981. But, while this method gives eminent opportunity to recapitulate history, it also disables predictions of future events, as the horizon is limited to the weather forecast. Therefore, these models are currently used in hindcast analyses of past events, not in predictions.

To predict, a different type of physical forcing of hydrodynamical and biological models is needed. There is a strong chaotic element in the current weather forecasting models, but weather itself is not chaotic: seasonality itself shows that strong deterministic forces are operating. Also, on longer time scales, oceanographic and meteorological phenomena indicate strong influence from ordering forces. Examples are the El Niño–Southern Oscillation (ENSO: Rogers, 1984; Ropelewski and Halpert, 1987; Halpert and Ropelewski, 1992) in the Pacific and the North Atlantic Oscillation (NOA: Walker and Bliss, 1932; Van Loon and Rogers, 1978;

Wallace and Gutzler, 1981). These and other huge weather systems exhibit quasi-periodic oscillations (WMO and UNEP, 1995). Understanding these oscillations and how they are forced (Ikeda, 1990), will put us in a different situation with regard to predicting biological change as a function of climate oscillations months or years in advance.

28.4 PERSPECTIVES

The use of predictive models on historical data has already revealed causal relationships from physics and phytoplankton to fish. Theoretical models have also given insight into how the different external and internal forces interact to produce life histories and behaviour. With the continuous increase in computer capacity, in numerical methods, and in quantitative biological theories, a variety of theoretical models and predictive simulation models may soon play an increasing role in marine management.

More recently, complex ecological models (Aksnes *et al.*, 1995; Fiksen *et al.*, 1995) also indicate that the biological basis is improving for the creation of models predicting the future state of biological resources and environmental variables. However, these efforts are at present limited by the physical forcing, which so far hinges on meteorology. A different set of forcing functions, e.g. based on statistical analyses of climate variability or forcing by Earth and astronomical events, or on measured upstream changes in physical factors (Helland-Hansen and Nansen, 1909), will be needed for this purpose. Still, the uncertainties associated with these predictions will be formidable.

Meanwhile, we can do no better than to forecast by scenario modelling. From the current state of the environment, it may be possible to select a limited number of similar past situations and through them simulate possible outcomes. For capelin in the Barents Sea, the present ice and temperature conditions could be compared with those in previous years, and the meteorological conditions for similar situations could then be used to simulate some scenarios for the further development of temperatures and currents. In these physical scenarios, the biological models could be run, to simulate the probable development of the capelin stock. This may seem unsatisfactory, but nature is inherently stochastic, and the future in particular.

Scenario modelling is also a means of both quantifying and reducing the inherent uncertainties of nature as well as in process representations, parameter values and field data (Moxnes, 1996; Hagen *et al.*, 1998). The precautionary principle is already in use in the management of sea mammals (Gambell, 1993), and may also be adopted for fisheries (Garcia, 1994). By developing more advanced quantitative tools for understanding nature, and by using them in models accounting for the potential variabilities in the forcing functions, we may be able to narrow the security measures needed for the protection of stocks and the environment (Mangel *et al.*, 1996), so allowing a sound utilization of the resources.

As the different theoretical approaches and numerical methods are only capable of handling bits of the total complexity, managers should be trained at using a variety of tools. By comparing results from very different models, managers will be more able to assess possible outcomes of their recommendations and acquire an understanding of the functioning of the system and its interacting components. This skill is a prerequisite for successful adaptive management.

ACKNOWLEDGEMENTS

This study was supported by the Marine Resource Management programme under the Research Council of Norway. Svein Sundby, Øyvind Fiksen, Geir Huse, two anonymous referees and the editors are thanked for valuable comments.

REFERENCES

Aksnes, D.L. (1993) Mathematical models for aquatic populations, in *Human Impact on Self-recruiting Populations* (Royal Norw. Soc. Sci. Lett. Found., Kongsvoll Symp. 1993) (ed. G. Sundnes), Tapir Forlag, Trondheim, pp. 285–308.

Aksnes, D.L. and Giske, J. (1990) Habitat profitability in pelagic environments. *Mar. Ecol. Prog. Ser.* **64**, 209–215.

Aksnes, D.L. and Utne, A.C.W. (1997) A revised model of visual range in fish. *Sarsia* **82**, 137–147.

Aksnes, D.L., Aure, J., Furnes, G.K., Skjoldal, H.R. and Stœre, R. (1989) Analysis of the *Chrysochromulina polylepis* bloom in the Skagerrak, May 1988. Environmental conditions and possible causes. Bergen: Bergen Scientific Centre, BSC 89/1, 38 pp.

Aksnes, D.L., Ulvestad, K.B., Baliño, B.M., Berntsen, J., Egge, J.K. and Svendsen, E. (1995) Ecological modelling in coastal waters: towards predictive physical-chemical-biological simulation models. *Ophelia* **41**, 5–36.

Barnthouse, L.W. (1992) The role of models in ecological risk assessment – a 1990s perspective. *Env. Toxicol. Chem.* **11**, 1751–1760.

Bart, J. (1995) Acceptance criteria for using individual-based models to make management decisions. *Ecol. Appl.* **5**, 411–420.

Bleckmann, H. (1993) Role of the lateral line in fish behaviour, in *Behaviour of Teleost Fishes*, 2nd edn (ed. T.J. Pitcher), Chapman & Hall, London, pp. 201–246.

Brandt, S.B. and Hartman, K.J. (1993) Innovative approaches with bioenergetics models – future applications to fish ecology and management. *Trans. Am. Fish. Soc.* **122**, 731–735.

Clark, C.W. and Levy, D.A. (1988) Diel vertical migration by juvenile sockeye salmon and the antipredation window. *Am. Nat.* **131**, 271–290.

Clark, C.W. and Mangel, M. (1986) The evolutionary advantages of group foraging. *Theor. Pop. Biol.* **30**, 45–75.

Crowder, L.B., Rice, J.A., Miller, T.J. and Marschall, E.A. (1992) Empirical and theoretical approaches to size-based interactions and recruitment variability in fishes, in *Individual-based Models and Approaches in Ecology* (eds D.L. DeAngelis and L.J. Gross), Chapman & Hall, New York, pp. 237–255.

Dawkins, R. (1976) *The Selfish Gene*, Oxford, Oxford Univ. Press.

Dawkins, R. (1996) *Climbing Mount Improbable*, Penguin Books, London.

DeAngelis, D.L. and Rose, K.A. (1992) Which individual-based approach is most appropriate for a given problem? in *Individual-based Models and Approaches in Ecology* (eds D.L. DeAngelis and L.J. Gross), Chapman & Hall, New York, pp. 67–87.

Fiksen, Ø., Giske, J. and Slagstad, D. (1995) A spatially explicit fitness-based model of capelin migrations in the Barents Sea. *Fish. Oceanogr.* **4**, 193–208.

Gambell, R. (1993) International management of whales and whaling: an historical review of the regulation of commercial and aboriginal subsistence whaling. *Arctic* **46**, 97–107.

Garcia, S.M. (1994) The precautionary principle: its implications in capture fisheries management. *Ocean Coast. Manage.* **22**, 99–125

Giske, J. and Aksnes, D.L. (1992) Ontogeny, season and trade-offs: vertical distribution of the mesopelagic fish *Maurolicus muelleri*. *Sarsia* **77**, 253–261.

Giske, J. and Salvanes, A.G.V. (1995) Why pelagic planktivores should be unselective feeders. *J. Theor. Biol.* **173**, 41–50.

Giske, J., Aksnes, D.L. and Fiksen, Ø. (1994) Visual predators, environmental variables and zooplankton mortality risk. *Vie Milieu* **44**, 1–9.

Giske, J., Huse, G. and Fiksen, Ø. (1998a) Modelling spatial dynamics of fish. *Rev. Fish Biol. Fish.* **8**, 57–91.

Giske, J., Rosland, R., Berntsen, J. and Fiksen, Ø. (1997) Ideal free distribution of copepods under predation risk. *Ecol. Mod.* **95**, 45–59.

Giske, J., Skjoldal, H.R. and Slagstad, D. (1998b) Ecological modelling for fisheries, in *Models for Multispecies Management* (ed. T. Rødseth), Physica-Verlag, Heidelberg, pp. 11–68.

Hagen, G., Hatlebakk, E. and Schweder, T. (1998) Scenario Barents Sea, a tool for evaluating fisheries management regimes, in *Models for Multispecies Management* (ed. T. Rødseth), Physica-Verlag, Heidelberg, pp. 173–226.

Halpert, M.S. and Ropelewski, C.F. (1992) Surface temperature patterns associated with the Southern Oscillation. *J. Climate* **5**, 577–593.

Hansen, M.J., Boisclair, D., Brandt, S.B., Hewett, S.W., Kitchell, J.F., Lucas, M.C. and Ney, J.J. (1993) Applications of bioenergetics models to fish ecology and management – where do we go from here? *Trans. Am. Fish. Soc.* **122**, 1019–1030.

Helland-Hansen, B. and Nansen, F. (1909) The Norwegian Sea. *Rep. Norw. Fish. Mar. Invest.* **2**(2), 1–390.

Hewett, S.W. and Johnson, B.J. (1992) An upgrade of a generalized bioenergetics model of fish growth for microcomputers. Univ. Wisconsin, Madison, Wisconsin Sea Grant Program, Sea Grant Tech. Rep. WIS-SG-92-250, 79 pp.

Hinckley, S., Hermann, A.J. and Megrey, B.A. (1996) Development of a spatially explicit, individual-based model of marine fish early life history. *Mar. Ecol. Prog. Ser.* **139**, 47–68.

Houston, A., Clark, C., McNamara, J. and Mangel, M. (1988) Dynamic models in behavioural and evolutionary ecology. *Nature* **332**, 29–34.

Ikeda, M. (1990) Decadal oscillations of the air–ice–ocean system in the Northern Hemisphere. *Atmosphere–Ocean* **28**, 106–139.

Jumper, G.Y. Jr and Baird, R.C. (1991) Location by olfaction: a model and application to the mating problem in the deep-sea hatchetfish, *Argyropelecus hemigymnus*. *Am. Nat.* **138**, 1431–1458.

Kitchell, J.F., Stewart, D.J. and Weinigner, D. (1977) Application of a bioenergetics model to yellow perch (*Perca flavescens*) and walleye (*Stizostedion vitreum*). *J. Fish. Res. Bd Can.* **37**, 2159–2168.

Leonardsson, K. (1991) Predicting risk-taking behaviour from life-history theory using static optimization technique. *Oikos* **60**, 149–154.

Letcher, B.H., Rice, J.A. Crowder, L.B. and Rose, K.A. (1996) Variability in survival in larval fish: disentangling components with a generalized individual-based model. *Can. J. Fish. Aquat. Sci.* **53**, 787–801.

McNamara, J.M. and Houston, A.I. (1996) State-dependent life histories. *Nature* **380**, 215–221.

Mangel, M. and Clark, C.W. (1986) Towards a unified foraging theory. *Ecology* **67**, 1127–1138.

Mangel, M. and Clark, C.W. (1988) *Dynamic modeling in behavioral ecology*. Princeton Univ. Press, Princeton, 308 pp.

Mangel, M. and 41 others (1996) Principles for the conservation of wild living resources. *Ecol. Appl.* **6**, 338–362.

Mason, D.M. and Brandt, S.B. (1996) Effects of spatial scale and foraging efficiency on the predictions made by spatially-explicit models of fish growth rate potential. *Env. Biol. Fish.* **45**, 283–298.

Mason, D.M., Goyke, A. and Brandt, S.B. (1995) A spatially explicit bioenergetics measure of habitat quality for adult salmonines: comparison between Lakes Michigan and Ontario. *Can. J. Fish. Aquat. Sci.* **52**, 1572–1583.

Moxnes, E. (1996) Optimal quota policies for a stochastic predator–prey system. *Norw. School Econ. Business Admin.*, Bergen, SNF-report 76/96. 30 pp.

Rogers, J.C. (1984) The association between the North Atlantic Oscillation and the Southern Oscillation in the northern hemisphere. *Monthly Weather Review* **112**, 1999–2015.

Ropelewski, C.F. and Halpert, M.S. (1987) Global and regional scale precipitation patterns associated with the El Niño/Southern Oscillation (ENSO). *Monthly Weather Rev.* **115**, 1606–1626.

Rose, K.A., Christensen, S.W. and DeAngelis, D.L. (1993) Individual-based modeling of populations with high mortality: a new method based on following a fixed number of individuals. *Ecol. Mod.* **68**, 273–292.

Rosland, R. and Giske, J. (1994) A dynamic optimization model of the diel vertical distribution of a pelagic planktivorous fish. *Prog. Oceanog.* **34**, 1–43.

Rosland, R. and Giske, J. (1997) A dynamic model for the life history of the pelagic planktivore *Maurolicus muelleri*. *Fish. Oceanogr.* **6**, 19–34.

Salvanes, A.G.V., Giske, J. and Nordeide, J.T. (1994) A life-history approach to habitat shifts for coastal cod (*Gadus morhua* L.). *Aquac. Fish. Manage.* **25**, Suppl. 1, 215–228.

Salvanes, A.G.V., Aksnes, D.L. and Giske, J. (1995) A surface-dependent gastric evacuation model. *J. Fish Biol.* **47**, 679–695.

Sampson, D.B. (1988) The stability of virtual population analysis cohort size estimates. *J. Cons. Explor. Mer* **44**, 135–142.

Schellart, N.A.M. (1992) Interrelations between the auditory, the visual and the lateral line systems of teleosts; a mini-review of modelling sensory capabilities. *Neth. J. Zool.* **42**, 459–477.

Sibly, R.M. (1989) What evolution maximizes. *Funct. Ecol.* **3**, 129–135.

Tuljapurkar, S.D. (1990) *Population Dynamics in Variable Environments* (*Lecture Notes in Biomathematics* **85**), Springer Verlag, New York, 154 pp.

Tyler, J.A. and Rose, K.A. (1994) Individual variability and spatial heterogeneity in fish population models. *Rev. Fish Biol. Fish.* **4**, 91–123.

Van Loon, H. and Rogers, J.C. (1978) The seesaw in winter temperatures between Greenland and Northern Europe. *Monthly Weather Rev.* **106**, 296–310.

Walker, G.T. and Bliss, E.W. (1932) World weather V. *Mem. R. Meteor. Soc.* **4**, 53–84.

Wallace, J.M. and Gutzler, D.S. (1981) Teleconnections in the geopotential height field during the northern hemisphere winter. *Monthly Weather Rev.* **109**, 784–812.

Werner, E.E. and Gilliam, J.F. (1984) The ontogenetic niche and species interactions in size structured populations. *Ann. Rev. Ecol. Syst.* **15**, 393–425.

Werner, F.E., Perry, R.I. Lough, R.G. and Naimie, C.E. (1996) Trophodynamic and advective influences on Georges Bank larval cod and haddock. *Deep-sea Res. II – Topical Stud. Oceanogr.* **43**, 1793–1822.

WMO and UNEP (1995) *The Global Climate System Review – climate system monitoring* June 1991–November 1993. WMO no. 819, 150 pp.

Yoshimura, J. and Clark, C.W. (1993) *Adaptation in Stochastic Environments* (*Lecture Notes in Biomathematics* **98**), Springer Verlag, New York, 193 pp.

Chapter twenty-nine

Bayesian decision analysis and uncertainty in fisheries management

Randall M. Peterman, Calvin N. Peters, Christina A. Robb and Shane W. Frederick

ABSTRACT

Large variability and estimation errors in data create challenges for estimating risks and identifying appropriate fisheries management strategies. The formal quantitative method of decision analysis, sometimes referred to as statistical decision theory, can help deal with this challenge because it explicitly considers uncertainties in quantities such as parameters of dynamic processes in fish populations or fishing fleets. Field data can be used in conjunction with Bayesian statistical analysis to calculate probabilities associated with different estimates of the uncertain parameters. These probabilities can then be used as part of a decision analysis to identify the optimal management action for each specified management objective. We illustrate this approach of decision analysis with three examples. (1) The optimal decision for opening an in-river sockeye salmon fishery depended, among other things, on the assumed functional form (not just parameter values) of the stock–recruitment relationship, i.e. whether it was a Ricker model or a more flexible Shepherd model, which can take on various shapes, including a Ricker shape. (2) When uncertainties in density-dependent growth and in size-dependent vulnerability to fishing

Reinventing Fisheries Management. Edited by Tony J. Pitcher, Paul J.B. Hart and Daniel Pauly. Published in 1998 by Kluwer Academic Publishers, London. ISBN 0 412 83410 3.

gear were accounted for, the optimal stocking density for juvenile rainbow trout in British Columbia lakes increased considerably compared with the case where uncertainties were ignored. (3) Marine fish stocks such as Atlantic menhaden typically exhibit large uncertainty in estimates of current stock biomass and stock–recruitment relationships. We show that commonly used arbitrary reductions in harvest rate (such as a 20% 'safety margin') that qualitatively try to allow for such uncertainties are not necessarily optimal, and can lead to significant reduction in benefits. Instead, optimal safety margins should be estimated for each situation. These examples demonstrate that it is worth explicitly considering uncertainties in analyses of fisheries management options because they can potentially alter the optimal decision.

29.1 INTRODUCTION

With increasing fishing pressure around the world, fisheries management agencies need to design management strategies that sustain harvests and human fishing communities, without compromising fish stocks. However, complex and variable ecological processes, as well as estimation errors in fisheries data, introduce large uncertainties into forecasts of effects of management actions (Neill, Chapter 22, Walters, Chapter 21, Cochrane and Payne, Chapter 5, all in this volume). These uncertain forecasts make it difficult to estimate ecological risks and identify appropriate management strategies. However, explicitly recognizing such uncertainties and considering them in the policy formulation process by methods such as decision analysis can identify policies that are appropriate.

29.2 APPROACHES TO CONSIDERING UNCERTAINTIES IN DECISION MAKING

Historically, fisheries management agencies have addressed uncertainties in various ways. In some cases they have ignored uncertainties by relying solely on the best point estimates for parameters and state variables that describe the dynamics of fish populations and harvesters. However, if these point estimates are inaccurate, this procedure may lead to overfishing, to collapse of the stock, or to harvests that are smaller than those the stock is able to support. In other situations, uncertainties have been acknowledged and have been used to justify (1) maintaining the status quo (when agencies are reluctant to change their policies because outcomes are very uncertain), (2) aggressive harvest policies (when pressures are high for large, short-term economic yields and the potential negative consequences cannot be convincingly demonstrated), or (3) area closures, application of arbitrary safety margins, and other precautionary actions (when risks associated with continued harvest are assumed to be high). However, none of these approaches reflects a growing trend in fisheries agencies to explicitly and quantitatively consider the implications of uncertainties for decisions or their possible outcomes (Kruse *et al.*, 1993).

Decision analysis

Decision analysis is an approach to decision making that quantitatively considers uncertainties when evaluating alternative management actions (Walters, 1986; Morgan and Henrion, 1990), which helps management agencies make more informed choices. An important feature of decision analysis is that rather than using only point estimates of parameter values or state variables to simulate outcomes of alternative management options, several hypothesized values of the parameters or states of nature are used and, as a result, several possible outcomes for each management action are generated (Keeney, 1982; von Winterfeldt and Edwards, 1986). Each hypothesized value has an associated probability that reflects the relative degree of belief in that value. This information is typically displayed in a decision tree.

Depending on the management objective, the decision tree is used in various ways to determine the optimal management action. For instance, if performance is measured in terms of average annual tonnage of fish harvested, then the decision tree, which includes probabilities for the states of nature, can be used to calculate an expected value, or weighted average, of outcomes for each management action. In this case, the optimal management action is the one that maximizes the expected annual harvest. In contrast, if the management objective is to choose the option that has the best worst-case outcome, then this can be found simply by inspecting the decision tree and identifying which action has the worst-case outcome that is the best among all actions. To reflect risk preferences, decision makers may convert objective outcomes (such as monetary value of the harvest) to their subjective equivalent (utilities), using a utility function. We provide three applications of decision analysis to fisheries management decisions to illustrate its main features.

Opening a surplus fishery for Nass River sockeye salmon

Robb and Peterman (1998) applied decision analysis to the management of sockeye salmon, *Oncorhynchus nerka*, on the Nass River, British Columbia (Link and English, Chapter 10, this volume). The Canadian Department of Fisheries and Oceans opens a 'surplus fishery' for the First Nation community upstream from all conventional (commercial, sport, and native food) fisheries in years when escapement is predicted to exceed a target value. A major decision is thus when to open the surplus fishery during the season. If it is opened too early, too many fish could be harvested to reach the escapement goal. If it is opened too late, escapement targets may be exceeded, which could decrease future recruitment if high densities actually depress the total number of eggs, juveniles, or adults that survive. In either case, future harvests may be less than what is expected if the escapement target was met exactly.

Robb and Peterman (1998) evaluated several 'decision rules' for opening the surplus fishery, defined as the percentage of the target escapement estimated to have passed through the commercial fishing area before a surplus fishery was opened (alternative decision rules are shown in Fig. 29.1 as branches coming from the square 'decision' node at the left). The model simulated the escapement, catch in the surplus fishery (surplus catch), and total catch (conventional plus surplus) resulting from each decision rule. Uncertainties in run timing, recruitment variability, and catchability components were included via Monte Carlo simulation.

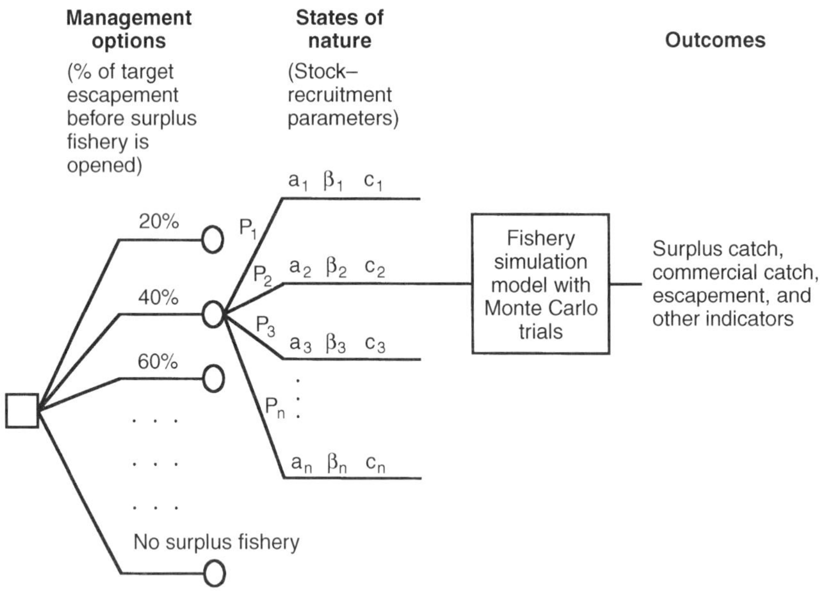

Fig. 29.1 An example portion of the decision tree for analysing when to open the in-river surplus fishery for sockeye salmon on the Nass River, British Columbia. For simplicity, only 4 of the 13 decision rules and 4 of the 3240 parameter sets are shown. Adapted from Robb and Peterman (1998).

Uncertainty in a Shepherd stock–recruitment model (Shepherd, 1982) was explicitly accounted for by simulating the outcome of each decision rule using each of 3240 different sets of Shepherd parameters (shown in Fig. 29.1 as branches coming from the circular 'chance' node). The different parameter sets generated stock–recruitment curves that included dome-shaped curves (similar to a Ricker curve), various Beverton–Holt curves, and Cushing curves (Shepherd, 1982). Each parameter set had an associated probability, calculated from stock–recruit data using Bayesian statistics. For each decision rule, the expected outcome was the weighted average of all the outcomes for that decision rule, where weightings were the probabilities for the sets of Shepherd-model parameters.

The optimal decision rule depended on the management objective (Robb and Peterman, 1998). For example, never opening a surplus fishery maximized expected conventional (commercial plus sport and native food) catches, despite the more frequent occurrence of overescapements than when the surplus fishery opened occasionally. However, if the objective was to maximize expected surplus catch without reducing the expected conventional catch by more than 5% from what its value would be if there were no surplus fishery, the optimal decision was to open the surplus fishery when 100% of the target escapement was estimated to have passed the commercial fishing area (which meant the surplus fishery was opened in 78% of years).

Furthermore, when only dome-shaped stock–recruitment curves (similar to the Ricker curve) were used, the optimal decision rule was to open the surplus fishery when fewer fish had escaped past conventional fisheries than when full uncertainty in the parameters of the Shepherd model was admitted (Robb and Peterman, 1998). This is because the strong density-dependent feedback implicit in a dome-shaped stock–recruitment curve meant that overescapement had a greater negative effect on future recruitment and catches than when the recruitment function was not constrained to a dome shape. Thus, management errors may result if scientists are incorrect in assuming that only a Ricker curve is an appropriate representation of Pacific salmon stock–recruitment relationships. The small variance accounted for in stock–recruitment data by any particular stock–recruitment model suggests that a decision analysis approach may be appropriate, because it can explicitly account for uncertainty in the functional form of the relationship.

Optimizing stocking densities for rainbow trout fisheries

In another example of decision analysis, Peters (1996) determined the optimal number of yearling rainbow trout, *Oncorhynchus mykiss*, to stock per area in Alleyne Lake, British Columbia. Yields are optimized at some intermediate stocking density. Stocking too few fish will not take full advantage of the lake's productivity, and stocking too many fish may lead to lower yields because of density-dependent reductions in survival and growth rates. The purpose of this analysis was to identify the optimal stocking density while explicitly considering uncertainties in density-dependent and size-dependent processes that affect yields.

The analysis evaluated stocking densities from 50 yearlings ha^{-1} to 800 yearlings ha^{-1} (Fig. 29.2). For each stocking density, four measures of quality of fishing were calculated: average catch per unit effort (CPUE), average length of fish caught, percentage of stocked fish captured in the fishery ('return rate'), and the number of successful anglers (those catching at least one fish per day) per dollar spent on stocking. Values of these outcomes were computed using an empirically based simulation model. Growth and vulnerability data were used in a Bayesian statistical analysis to calculate posterior probabilities for parameter values of the two key model relationships where uncertainty was included: density-dependent

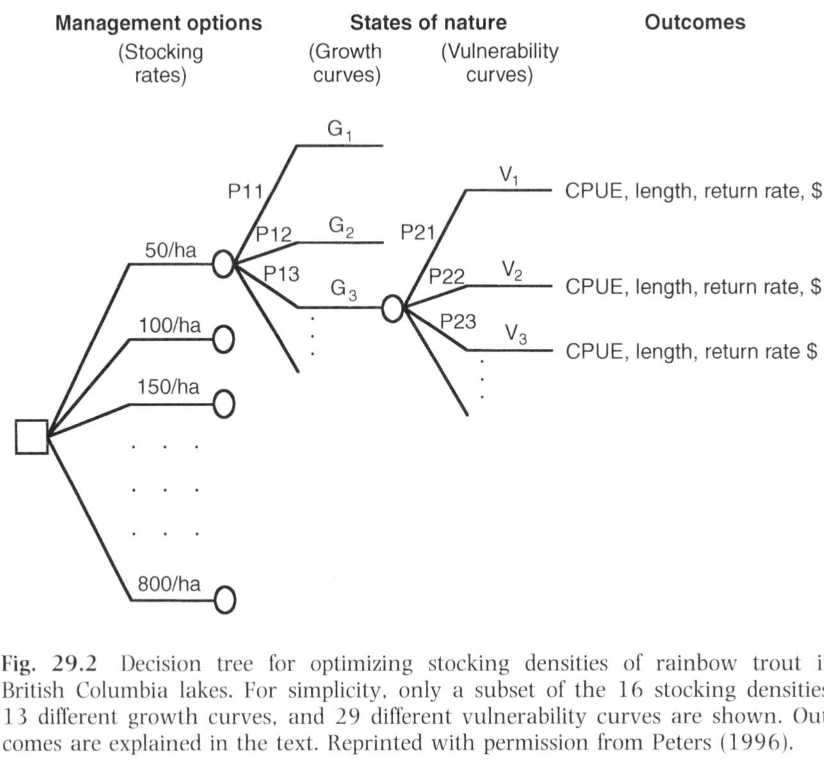

Fig. 29.2 Decision tree for optimizing stocking densities of rainbow trout in British Columbia lakes. For simplicity, only a subset of the 16 stocking densities, 13 different growth curves, and 29 different vulnerability curves are shown. Outcomes are explained in the text. Reprinted with permission from Peters (1996).

growth and vulnerability of fish to angling as a function of body size. The model also included density-dependent survival rates.

Optimal stocking densities for a given angling effort varied according to the management objective. For instance, to maximize the expected CPUE, the optimal stocking rate was 800 yearlings ha^{-1} (Fig. 29.3(a)). However, the expected number of 'successful' anglers per dollar of stocking cost was maximized at 200 yearlings ha^{-1} (Fig. 29.3(b)). Peters (1996) used a multiattribute utility function to combine these and the two other objectives into a single measure of quality of fishing. Maximizing this multiattribute utility function led to an optimum at some intermediate stocking rate, the exact value of which depended on the relative importance placed on each component of this function.

Taking uncertainties into account also changed the estimated optimal policy considerably. For example, when outcomes were measured as CPUE, ignoring uncertainty by using only point estimates in growth and vulnerability relationships resulted in an optimal stocking density of 550

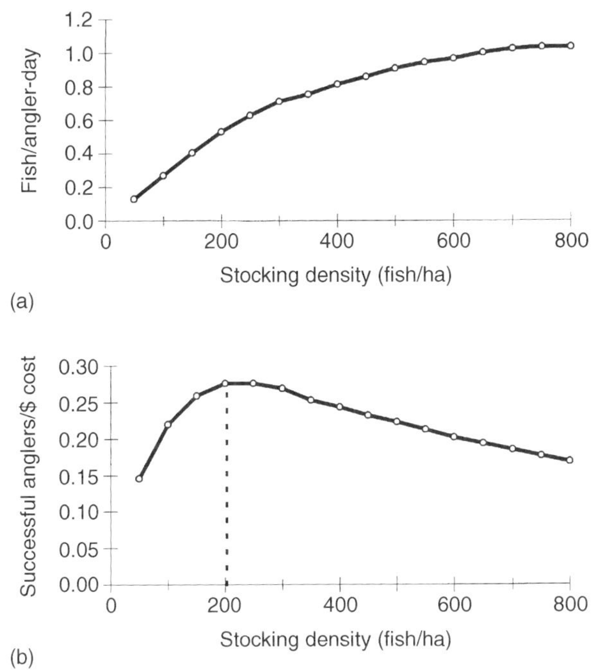

Fig. 29.3 Expected CPUE (a) and successful anglers per dollar (b) as functions of stocking density, as calculated by a decision analysis, which accounted for uncertainties. Reprinted with permission from Peters (1996).

yearlings ha^{-1} and an expected CPUE of 0.95 fish per angler per day. However, accounting for uncertainty in those relationships resulted in an optimal stocking density of 800 yearlings ha^{-1} with an expected CPUE of 1.05 (Fig. 29.3(a)).

Selecting appropriate uncertainty adjustments to harvest rates

Frederick and Peterman (1995) used decision analysis to determine the optimal adjustment in harvest rates for Atlantic menhaden, *Brevoortia tyrannus*. While adjustments are often made when uncertainties in abundance estimates or in the stock–recruitment relationship prevent simple determination of an optimal harvest rate, these adjustments are often arbitrary and are not based on quantitative analyses.

Frederick and Peterman (1995) used an empirically based, age-structured simulation model to calculate the cumulative catch over a 20-year time period for each of several possible uncertainty adjustments, which they defined as a percentage change from the deterministic optimal harvest rate (the one calculated using the best point estimates). The deterministic optimal harvest rate was 72%, and annual catches were determined for each uncertainty adjustment applied to that deterministic harvest rate (Fig. 29.4). This process was repeated for uncertainty adjustments ranging from −20% to +20%, which generated harvest rates on the vulnerable portion of the fish stock ranging from 52% to 92%.

The model explicitly considered two sources of uncertainty: (1) errors in estimates of stock abundance, and (2) uncertainty in the underlying stock–recruitment parameters. The model estimated stock abundance with a coefficient of variation (*CV*) to represent measurement error in obtaining these estimates. Uncertainty in the parameters of the stock–recruitment relationship was considered directly by calculating cumulative catches for each uncertainty adjustment using 1000 different bootstrapped sets of parameter values for a Shepherd stock–recruitment model, (i.e. each parameter set estimated by randomly sampling the historical stock–recruitment data points, with replacement, and then fitting the model). In this example, the frequency with which a certain parameter set appeared in the bootstrapping procedure reflected the probability that the parameter set described the 'true' stock–recruitment relationship. The optimal uncertainty adjustment was the one that maximized the expected present economic value of the average cumulative catch over the 1000 bootstrapped parameter sets. The analysis also considered interannual variability in recruitment using Monte Carlo simulation.

The optimal uncertainty adjustments for Atlantic menhaden were all negative, indicating that a 'safety margin' (reduction in the deterministic harvest rate) should be applied, given the uncertainties (Fig. 29.5). The

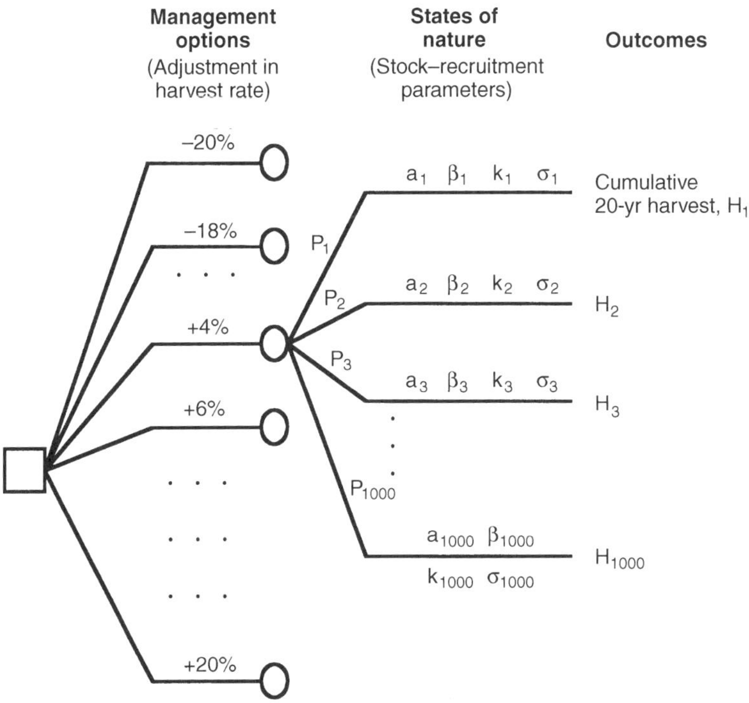

Fig. 29.4 Decision tree for Frederick and Peterman's (1995) analysis of optimal uncertainty adjustments to harvest rates for Atlantic menhaden. For simplicity, this figure shows only some of the adjustments and bootstrapped sets of Shepherd stock–recruitment parameters (a_i, β_i, k_i, and σ_i).

optimal safety margin depended on the degree of uncertainty in abundance estimates. When the annual abundance of the stock was assumed to be estimated with complete certainty ($CV = 0$), the optimal safety margin was a 4% reduction in the deterministic harvest rate of 72% (i.e. a harvest rate of 68%). In this situation, the safety margin thus reflected only uncertainty in stock–recruitment parameters. In contrast, when the *CV* of annual abundance estimates was 0.5, the optimal safety margin was a 20% reduction in the deterministic harvest rate, giving an optimal harvest rate of 52% (Fig. 29.5). For the latter *CV* case, the 20% reduction in the harvest rate resulted in an increase in the expected value of cumulative catch of 13.5% over that which would arise from using the deterministic optimal harvest rate. Additional sensitivity analyses showed that optimal decisions depended also on the management goal and on assumptions about the productivity of the stock at low abundances. For

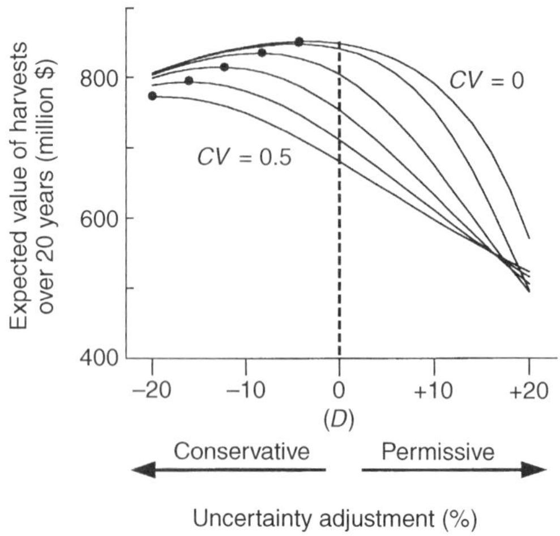

Fig. 29.5 Expected present economic value of cumulative catches for various uncertainty adjustments to the deterministic harvest rate. The six curves represent results for an annual discount rate of 5% and different *CVs* of the annual stock abundance estimates (see text for details). Solid circles on each curve indicate the optimal uncertainty adjustment (i.e. the one that maximizes the expected present value of catch) for each *CV*. Each uncertainty adjustment reflects an increase or decrease from the deterministic harvest rate (*D*) of 72%; e.g. applying an uncertainty adjustment of −20% resulted in an exploitation rate of 52%. Adapted from Frederick and Peterman (1995).

example, when the model was modified to include a spawning biomass below which recruitment failed (critical depensation), the optimal safety margin changed from 8% to 24% (i.e. optimal harvest rate changed from 64% to 48%) for $CV = 0.3$. The expected value of cumulative catch using that safety margin was more than double the value expected from using the deterministic optimal harvest rate. Thus, fisheries managers should not arbitrarily adjust their harvesting strategies to account for uncertainty, but rather should quantitatively derive the optimal uncertainty adjustment for each situation (Frederick and Peterman, 1995). It should be noted, however, that this example did not simulate the effect of various harvest rates on future information about the uncertain parameters. This effect might lead to higher harvest rates being optimal, if, for instance, a stock was historically held in a relatively narrow range of moderate or high abundance. In that case, a higher harvest rate might increase the contrast in the data by reducing abundance, thereby improving information about the uncertain parameters (Walters, 1986).

29.3 RECOMMENDATIONS

These applications of decision analysis to fisheries management suggest three general recommendations for fisheries scientists and managers. First, uncertainties should be explicitly and quantitatively considered in decision making processes whenever possible to avoid potential losses associated with using only the best point estimates of parameters or making arbitrary adjustments to actions to qualitatively reflect uncertainty. Variability in natural systems, which contributes to uncertainties, exists regardless of whether it is acknowledged, and ignoring it can lead to suboptimal decisions. The resulting losses depend on the specifics of each situation, e.g. the objectives of managers, the magnitude of the uncertainties, how those uncertainties enter into the equations describing the dynamics of the system, and the symmetry of the loss functions and probability distributions (Morgan and Henrion, 1990). Therefore, to determine the magnitude of losses, it is necessary to quantitatively compare forecasts of expected outcomes of decisions when uncertainties are considered, with the outcomes expected when uncertainties are ignored. Decision analyses make such comparisons explicitly.

Second, decisions should be judged according to how they were made – not on their outcomes. In the short term, ad hoc analyses that ignore uncertainties may produce favourable outcomes owing to chance or fortuitous changes in external factors, such as increased ocean productivity. However, decisions based on comprehensive analyses that quantitatively consider uncertainties will, in the long term, produce better results than decisions made using an ad hoc approach (von Winterfeldt and Edwards, 1986). Such comprehensive analyses should include peer review, full documentation of the decision-making process to promote future learning and consistency, and signed approval of the decision-making process by a qualified analyst before a decision is implemented. The last approach is analogous to the practice in some agencies of having biometricians approve research proposals before funds are granted to researchers.

Finally, fisheries managers are more likely to identify management strategies that will achieve sustainable harvests and conserve fish stocks if they acknowledge, quantify, and account for uncertainties in the ecological, economic, and sociological components of fisheries systems. However, managers should not use scientists' explicit quantitative descriptions of ecological uncertainties as a reason to put less weight on ecological concerns. This is because uncertainties also exist in economic and sociological components, and a quantitative statement of ecological uncertainties should not imply that ecological uncertainties are any greater than uncertainties in these other components of the fishery, which may simply be more difficult to quantify.

ACKNOWLEDGEMENTS

We thank Brian Pyper, Jeff Grout, Carl Walters, and Tony Pitcher for very useful comments on the manuscript. We also thank the following for financial support for the work described here: the Canadian Department of Fisheries and Oceans, the British Columbia Ministry of Environment (Fisheries Branch), the Natural Sciences and Engineering Research Council of Canada, and Simon Fraser University.

REFERENCES

Frederick, S.W. and Peterman, R.M. (1995) Choosing fisheries harvest policies: when does uncertainty matter? *Can. J. Fish. Aquat. Sci.* **52**, 291–306.

Keeney, R.L. (1982) Decision analysis: an overview. *Operations Res.* **30**, 803–838.

Kruse, G., Eggers, D.M., Marasco, R.J., Pautzke, C. and Quinn, T.J. (eds) (1993) *Proc. Int. Symp. Manage. Strategies Exploited Fish Populations*, Alaska Sea Grant College Program, Fairbanks, AK, 825 pp.

Morgan, M.G. and Henrion, M. (1990) *Uncertainty: A Guide to Dealing with Uncertainty in Quantitative Risk and Policy Analysis*, Cambridge, Cambridge Univ. Press, 332 pp.

Peters, C.N. (1996) Applying decision analysis to management of recreational fisheries: an evaluation of stocking rates for rainbow trout in Alleyne Lake, British Columbia, master's project in Resource and Environmental Management, Simon Fraser University, Burnaby, BC, 75 pp.

Robb, C.A. and Peterman, R.M. (1998) Application of Bayesian decision analysis to management of a sockeye salmon fishery. *Can. J. Fish. Aquat. Sci.* **55**, 86–98.

Shepherd, J.G. (1982) A versatile new stock–recruitment relationship for fisheries, and the construction of sustainable yield curves. *J. Cons. Int. Explor. Mer* **40**, 67–75.

von Winterfeldt, D. and Edwards, W. (1986) *Decision Analysis and Behavioural Research*. Cambridge, Cambridge Univ. Press, 604 pp.

Walters, C.J. (1986) *Adaptive Management of Renewable Resources*, MacMillan, New York, 374 pp.

Chapter thirty

A strategy for advancing stock assessment

Laura J. Richards and Jon T. Schnute

ABSTRACT

We propose a framework of shared data and analysis as a strategy for advancing stock assessment. Global access to documented data sets would allow investigators worldwide to address actual problems, such as the probability of stock collapse under a specified harvest regime. We argue that assessment science can mature only in the context of shared data. Consequently, lucid data provide an important legacy that fishery scientists now can leave to their successors.

30.1 INTRODUCTION

With the worldwide depletion of fishery resources (FAO, 1994), traditional approaches to stock assessment and fisheries management have come under challenge. Debate has focused on two extreme possibilities. The scientific assessment process could be badly flawed; alternatively, good scientific advice might be poorly implemented (Ludwig *et al.*, 1993; Rosenberg *et al.*, 1993). This debate has spawned a burgeoning literature on the implementation of fishery management decisions (Caddy, 1995; Clay *et al.*, 1995; Stephenson and Lane, 1995; Rice and Richards, 1996). Some proposed solutions seek fail-safe management plans, such as refugia in space and time, which supposedly reduce the dependency on scientific advice (Walters, Chapter 21, this volume).

Reinventing Fisheries Management. Edited by Tony J. Pitcher, Paul J.B. Hart and Daniel Pauly. Published in 1998 by Kluwer Academic Publishers, London. ISBN 0 412 83410 3.

We argue here that stock assessment will continue as an essential fisheries management tool. In fact, we foresee an expanding role for assessment. Trends indicate growing public concern for biodiversity, fish stock survival, and ecosystem health. Fisheries management alone cannot adequately address such concerns without studies on the effects of fishing and other human interventions. Public participation in resource decisions broadens the scientific perspective from that addressed by traditional stock assessment. Not only do the old questions persist, but new ones have been added.

Even if assessment science has been badly flawed, we contend that significant advances can be achieved. We define stock assessment as the process of making intelligent inferences about fish population dynamics, based on available data. Therefore, advancements must come through improvements in both data and analysis.

Based on this definition, we propose a conceptually simple framework within which stock assessment can progress. Such a framework must begin with data and offer prospects for testing a variety of proposed analyses. We suggest compiling a number of documented benchmark data sets from the world's major fisheries and posting them on the Internet. Investigators worldwide can use these data retrospectively to predict the consequences of measured harvests, including the possibility of stock collapse. In general, data up to year t can be used to predict historically known outcomes in year $t + 1$. Can any investigator produce analyses that convincingly predict the major features of the world's fisheries? If so, what distinguishes the data sets and analyses that succeed from those that fail? Stock assessment can be deemed to have failed only if successful approaches cannot be found. Analogous criteria could be applied to questions of biodiversity and ecosystem health.

In many cases, data that have been collected may prove inadequate. Investigators should be encouraged to examine (and post on the Internet) other documented data sets, such as those on environmental conditions, in the search for better explanations of stock dynamics. Data from one fishery may provide useful predictive information for another fishery. We envisage a scientific environment in which analysts of particular stocks routinely study data from fisheries throughout the world.

30.2 SHARED DATA

International working groups who share data on particular species illustrate how this framework might operate. Unfortunately, results from such groups often lie buried in grey literature and protected data files, unavailable to the larger scientific community. We acknowledge the significant costs of data collection, the research publication rights of collecting agencies, and the potentially sensitive nature of certain data. Our proposal contemplates a global effort to archive data that can be released into the public domain, even if agencies must charge user fees to recover costs. Subject to such constraints, we envisage open data and analyses leading to primary publications. Within this framework, the world's collective scientific efforts could be brought to bear on reliable data sets. Analyses in published articles could be checked by readers with access to the same data.

Some prototype data sets already exist (Myers *et al.*, 1995b; see also 'recruitment table' in FishBase, Froese and Pauly, 1997) and have been used to demonstrate how assessments can benefit from comparisons among stocks (Myers *et al.*, 1995a). In general, however, fisheries data are much more complex than simple time series. Typically, the data have many potential levels of resolution in space and time. One important issue pertains to the level required for successful assessments. If analysts worldwide are to address this issue, data at a fine level of resolution must be publicly available.

Data from marked Pacific salmonids illustrate the complexities of data documentation and exchange. Over 670 million juvenile salmonids have been marked by numerous agencies during the last 25 years, and millions of these have been recovered as adults (Schnute *et al.*, 1996). Owing to the migratory behaviour of these species, fish released by one agency are often recovered by another. Because the collective data must be shared to be useful, an international committee has defined an exchange format (PSMFC, 1995). A key difficulty involves sample context, i.e. the stratum in which a particular marked fish has been recovered. This issue is best resolved within a relational database, where recovery records can readily be linked to corresponding sample records.

Long time series of data are essential for understanding stock dynamics. Ideally, data should include cycles of natural fluctuation and periods with and without significant human intervention. Thus, analogous to worldwide access, data must be shared across generations of analysts (Pitcher and Pauly, Chapter 24, this volume). Valuable historic data have already been lost for reasons ranging from simple lack of foresight to major political upheaval (Mathews, 1993). We believe that lucid data, carefully archived now, provide an important legacy that current fisheries scientists can leave to their successors (Schnute and Richards, 1994). With a sustained commitment to this goal, assessment science in 2096 should represent a dramatic advancement over that in 1996.

Physics offers a historical precedent for this approach to scientific advancement. Data on planetary motions collected by Tycho Brahe (1546–1601) provided the foundation on which Kepler (1571–1630) devised his theory of elliptical orbits. That theory, in turn, led Newton (1642–1727) to his much more general laws of dynamics. As an example from modern statistics, data sets in the text by Box and Jenkins (1970) led the way to progressive development of time series analysis. Unlike the physical sciences, where experiments can often be replicated in different laboratories, fisheries cannot be replicated. Consequently, data must be shared to enable objective, scientific development.

30.3 INTELLIGENT ANALYSIS

As a motto for the analyst, we quote Sherlock Holmes: 'It is a capital mistake to theorise before one has data. Insensibly, one begins to twist facts to suit theories, instead of theories to suit facts' (Conan Doyle, 1891). Of course, once the clues have been assembled, it is necessary to make sense of them. Similar to a detective, the analyst must extract the information from the noise.

Figure 30.1 illustrates the search for information in age structure data for a stock of Pacific ocean perch, *Sebastes alutus*, from the central coast of British Columbia (Richards and Olsen, 1996; Richards *et al.*, 1997). In spite of obvious variability and missing data for some years, the figure provides evidence of occasional strong year classes and high recruitment variability. Also, the position of modes along each column suggests that availability to the fishery increases with age for younger ages.

The figure also illustrates typical data collection problems. Two ageing methods have been used. Prior to 1977, true ages of fish older than about 15 years were underestimated. The opportunity to apply the current method to these historical age structures was lost when the samples were destroyed by fire. Consequently, we cannot determine from available data whether the absence of older fish in the early years resulted from biased age determination or high fishing mortality rates.

We have used Fig. 30.1 to demonstrate the first steps in a thoughtful analysis. Eventually, such data provide only one component of a comprehensive assessment model. At the core, almost all fishery models involve removal of the observed catch from the underlying fish population. To understand the population dynamics, removals must be related to subsequent population changes. The analysis depends critically on detecting the effects of these removals. Negligible effects can be explained in the model only by assuming that the actual population is much larger than the catch. Thus, models and data with inadequate sensitivity are likely to overestimate abundance and potential yield.

For example, if a simplistic model for the data in Fig. 30.1 assumed constant recruitment, then this model might have difficulty explaining

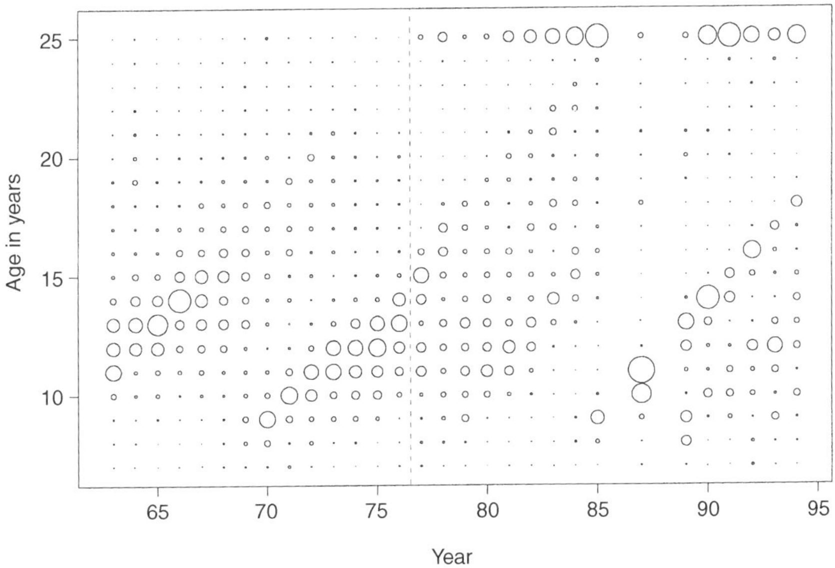

Fig. 30.1 Proportion at age sampled from the commercial catch of Pacific ocean perch, *Sebastes alutus*, from Queen Charlotte Sound, British Columbia, between 1963 and 1994. The radius of each circle shows the relative proportion at age in each year, where proportions sum to 1 in each column, and the age-25 proportion includes all fish of that age and older. Ages are determined from sagittal otoliths using the surface method (years earlier than the dashed vertical line) or the break-and-burn method.

large catches taken from cohorts with strong recruitment. As a result, the model might predict a very large population in which even the largest catches have little effect. An adequate model for the data in Fig. 30.1 must include greater complexity, such as a distinct recruitment parameter for each cohort. In our experience, oversimplified models can lead to unduly optimistic abundance estimates.

Data for a comprehensive model may not be self-consistent. In our analyses of a British Columbia lingcod, *Ophiodon elongatus*, stock, we found that a tagging experiment provided statistically distinct estimates of mortality rates from an analysis of the fishery using catch, effort, and fish size composition (Schnute *et al.*, 1989; Richards, 1991). Schnute and Hilborn (1993), Gove *et al.* (1996) and Schnute (1996) show that contradictory data can lead to bimodal likelihood functions. In such cases, the analyst may be forced to choose among extremes, in which one data set is presumed correct to the exclusion of others.

Contradictory data sometimes result from poorly designed abundance indices. For example, catch per unit effort (CPUE) is considered a poor index of abundance (Hilborn and Walters, 1992; Richards, 1994). Nevertheless, well-known problems with CPUE do not invalidate attempts to design intelligent indices from catch and effort data. Indeed, records of fishery catch and effort often represent the only data available. We anticipate that new technologies will provide greater opportunities in survey methods. Because assessment models tend to perform better when measurement errors are reduced (Schnute and Richards, 1995), improvements to data collection procedures should be a major goal of stock assessment research.

Problems with designing an intelligent index of abundance illustrate the fundamental relationship between models and data. Modelling suggests revised methods of data collection, and thoughtful data analysis indicates the need to change the model. Cycles of simulation and estimation can guide this process, prior to collecting expensive data. A simulation model, based on the analyst's concept of the ecosystem, produces hypothetical data. A corresponding estimation model then indicates how effectively parameters of interest could be estimated from these data. Such an exercise can at least guide the scope of data worth collecting and temper unrealistic expectations of what can be achieved.

We like to imagine a research context in which today's analyst could routinely use credible data dating back to the 1890s, when Conan Doyle crafted his stories of Sherlock Holmes. Given the changes to fisheries and the world's ecosystem that have occurred during the past century, data from that time period would surely enable more intelligent assessments than are realistically possible today. With such long series of data, the limitations of assessment would probably be more evident. Similarly, current data may provide important clues for assessments conducted a century from now. We see our strategy for advancing stock assessment as essential for this science to mature.

ACKNOWLEDGEMENT

We thank Jim Easley for drawing our attention to the quote from Sherlock Holmes.

REFERENCES

Box, G.E.P. and Jenkins, G.M. (1970) *Time Series Analysis: Forecasting and Control*, Holden-Day, San Franciso, 553 pp.

Caddy, J.F. (1995) Comment – Fisheries management science: a plea for conceptual change. *Can. J. Fish. Aquat. Sci.* **52**, 2057–2058.

Clay, P., Catanzano, J. and Sinclair, M. (eds) (1995) Fisheries science and management. *Aquat. Liv. Resour.* **8**, 201–288.

Conan Doyle, A. (1891) *The Adventures of Sherlock Holmes: A Scandal in Bohemia* (Reprinted in *The Illustrated Sherlock Holmes Treasury* (1986) Chatham River Press, New York, p. 3.)

FAO (Food and Agricultural Organization of the United Nations) (1994) Review of the state of world marine fishery resources. *Fish. Tech. Pap.* no. 335, 143 pp.

Froese, R. and Pauly, D. (1997) *FishBase 97: Concepts, Design and Data Sources*, ICLARM, Manila, 257 pp. [distributed with 2 CD-ROMs also including over 500 of R.A. Myers' series of recruitment and related items].

Gove, J.H., Linder, E. and Tzilkowski, W.M. (1996) Bimodality of the combined removal and signs-of-activities estimator for sampling closed populations. *Env. Ecol. Stat.* **3**, 65–78.

Hilborn, R. and Walters, C. (1992) *Quantitative Fisheries Stock Assessment: Choice, Dynamics and Uncertainty*, Chapman & Hall, New York, 570 pp.

Ludwig, D., Hilborn, R. and Walters, C. (1993) Uncertainty, resource exploitation, and conservation: lessons from history. *Science* **260**, 17, 36.

Mathews, C.P. (1993) On preservation of data. *NAGA, ICLARM Q.* **16**(2–3), 39–41.

Myers, R.A., Barrowman, N.J., Hutchings, J.A. and Rosenberg, A.A. (1995a) Population dynamics of exploited fish stocks at low population levels. *Science* **269**, 1106–1108.

Myers, R.A., Bridson, J. and Barrowman, N.J. (1995b) Summary of worldwide spawner and recruitment data. *Can. Tech. Rep. Fish. Aquat. Sci.* no. 2024, 327 pp.

PSMFC (Pacific States Marine Fisheries Commission) (1995) CWT data set definition, specification, and validation: Pacific Salmon Commission Format Ver. 3.1. 65 pp. (Obtain by anonymous ftp from *ftp.psmfc.org/pub/cwt/documents/psc_031.ps*).

Rice, J.C. and Richards, L.J. (1996) A framework for reducing implementation uncertainty in fisheries management. *N. Am. J. Fish. Manage.* **16**, 488–494.

Richards, L.J. (1991) Use of contradictory data sources in stock assessment. *Fish. Res.* **11**, 225–238.

Richards, L.J. (1994) Trip limits, catch, and effort in the British Columbia rockfish trawl fishery. *N. Am. J. Fish. Manage.* **14**, 742–750.

Richards, L.J. and Olsen, N. (1996) Slope rockfish stock assessment for the west coast of Canada in 1996 and recommended yield options for 1997. *Can. Tech. Rep. Fish. Aquat. Sci.* no. 2134, 91 pp.

Richards, L.J., Schnute, J.T. and Olsen, N. (1997) Visualizing catch-age analysis: a case study. *Can. J. Fish. Aquat. Sci.* **54**, 1646–1658.

Rosenberg, A.A., Fogarty, M.J., Sissenwine, M.P., Beddington, J.R. and Shepherd, J.G. (1993) Achieving sustainable use of renewable resources. *Science* **262**, 828–829.

Schnute, J.T. (1996) Discussion of Gove *et al.* 1996. *Env. Ecol. Stat.* **3**, 78–80.

Schnute, J.T. and Hilborn, R. (1993) Analysis of contradictory data sources in fish stock assessment. *Can. J. Fish. Aquat. Sci.* **50**, 1916–1923.

Schnute, J.T. and Richards, L.J. (1994) Stock assessment for the 21st century. *Fisheries* **19** (no. 11), 10–26.

Schnute, J.T. and Richards, L.J. (1995) The influence of error on population estimates from catch-age models. *Can. J. Fish. Aquat. Sci.* **52**, 2063–2077.

Schnute, J.T., Richards, L.J. and Cass, A.J. (1989) Fish survival and recruitment: investigations based on a size-structured model. *Can. J. Fish. Aquat. Sci.* **46**, 743–769.

Schnute, J.T., Wallace, C.G. and Boxwell, T.A. (1996) A relational database shell for marked Pacific salmonid data. *Can. Tech. Rep. Fish. Aquat. Sci.* no. 2090, 25 pp.

Stephenson, R.L. and Lane, D.E. (1995) Fisheries management science: a plea for conceptual change. *Can. J. Fish. Aquat. Sci.* **52**, 2051–2056.

Part Six

Overview and Synthesis

Speaking for themselves: new acts, new actors and a New Deal in a reinvented fisheries management

Daniel Pauly, Paul J.B. Hart and Tony J. Pitcher

ABSTRACT

Judged by a dismal record, the methodology, data, institutions and policy goals for fisheries stand worldwide in need of reinventing. Despite many new tools available for considering uncertainty, incorporating multiple data sources and listening to a variety of stakeholders, a reinvented fisheries management needs to adopt new, broader policy goals and to find new ways to examine, predict, shape, and ultimately mitigate the impact of fisheries on aquatic ecosystems. In parallel, new actors are asking for a voice in fisheries management, and for new legislative and other arrangements that will be required to accommodate their roles. The alternative to the inclusion of these new acts and actors, i.e. continuation of the present single-species focus, centralized arrangements, ineffective control of human fishing, and the resource exploitation patterns they induce, imply no less than a miserable, unrelenting erosion, and an eventual destruction of marine fisheries resources.

One of the articles in this book reminds us that 'fishers can speak for themselves' (Jentoft, Chapter 12, this volume). Trivial as it may seem, this

Reinventing Fisheries Management. Edited by Tony J. Pitcher, Paul J.B. Hart and Daniel Pauly. Published in 1998 by Kluwer Academic Publishers, London. ISBN 0 412 83410 3.

observation is deep, as it implies – because fish cannot speak – a very different relationship between social science and fisheries management than that between fishery biology and fisheries management. The book of which this is the concluding chapter contains, in different voices, many similar snippets, small truisms from which deeper questions are derived.

Emerging from the voices in this book are four clear themes and needs:

- reinventing the collection and handling of the data necessary to monitor marine resources;
- reinventing the methodology to analyse these data;
- reinventing the institutions designed to implement policy;
- reinventing policy goals for fisheries.

Therefore, we consider that:

1. fisheries globally are really in deep trouble (Buckworth, Chapter 1, this volume; Pitcher and Pauly, Chapter 24, this volume), far more so than is admitted to the public, to whom technical fixes are still being sold;
2. government institutions and stock assessment scientists charged with managing fisheries are still largely engaged in a quasi-exclusive partnership with the fishing industry – often with its least enlightened subsector – and further are engaged in denying the self-destructive streak that characterizes this subsector; and
3. addressing the issues in (1) and (2) requires a 'New Deal' in fisheries, involving new institutional arrangements that empower currently marginalized stakeholders to participate in fisheries management, both at strategic (who should have access to what still largely is, after all, a public resource) and at tactical levels (what gear should be used to extract what out of the multispecies assemblages constituting our resource base).

An important element of any New Deal must be the recognition that a central problem of fisheries management is human behaviour. Indeed, the heart of this book has no fewer than nine chapters addressing aspects of human behaviour in fisheries. The perspective from contemporary advances in social science is represented by Jentoft (Chapter 12, this volume) and by McGuire (Chapter 18, this volume). In the past, social scientists have felt ignored by fisheries managers, probably because of their different traditions and because social scientists have been much influenced by all-embracing theories through which they see the world. This often drives them to take stances that are at odds with the data-driven approach of biologists and the pragmatism of managers. On the other hand, rigorous social analysis of the politics and social beliefs that structure the bureaucratized science favoured by governments has been of considerable value in understanding convoluted processes leading to damaging decisions about fish resources (Finlayson, 1994, Hutchings *et al.*, 1997), and it may be good for all our souls to be deconstructed once in a while. But at the extreme, and fortunately not represented in this book, deconstructionists may regard the entire methodology, traditions and institutions of science as an arbitrary social construct (Latour, 1988), a view that is profoundly misguided (Sokal, 1996; Sokal and Bricmont, 1997). Other social science disciplines, such as economics, have attempted to look more at the empirical base of human behaviour, as shown by the two chapters by Scott (15) and by Hannesson (19), while a more anthropological view is taken by Harris (20), who emphasizes the link between the nature of the resource base and that of human social behaviours. Rice and Brown in their respective chapters (14, 13) show how the balance of power in fisheries management needs to be carefully monitored so that important groups are not left out, nor feel left out. A further perspective is provided in the two chapters by Hart and Pitcher (16) and by Hart (17), where an attempt is made to use an evolutionary perspective to understand the resource acquisition behaviour of fishers. This approach makes use of the extensive theory about the biology that underlies conflicts of interest among individuals, a science that has been developed by behavioural ecologists to make experimentally testable predictions about behaviour.

Central to all these views is that developing new relationships with fishers, fish workers and fish traders should be a central task of a reinvented fisheries management. Their lives are defined and given meaning by fishing, and they should be listened to and brought into the management system if they are to agree to be managed (Link and English, Chapter 10, this volume; Hart and Pitcher, Chapter 16, this volume).

The recent dialogue between fisheries planners and researchers from African, Caribbean and Pacific (ACP) countries and the European Union (EU) to develop the ACP–EU Fisheries Research Initiative is a case in point of an emerging new relationship (Anon., 1995, 1996, 1997; Chakalall *et al.*, 1997; Ruddle, 1997). This dialogue is founded on the twin concept of partnership and co-responsibility, which involves overcoming the outmoded donor–recipient relationships between industrialized and developing countries and their institutions (Williams, Chapter 11, this volume). Throughout the dialogue it was recognized that governments are increasingly unable to control and regulate economic activity through bureaucratic instruments, and to fund the required high levels of services and enforcement. On the other hand, non-governmental actors, e.g. in the corporate and cooperative sectors, are entering the space left by governments (Nauen *et al.*, 1996). This goes hand in hand with the need for new approaches in research, where a transition to transsectoral approaches is overdue, along with a shift from technological 'fixes' to a vision that takes cultural and social contexts into account.

It required a science-fiction author, Stanislaw Lem, to point out that the future is not realized as a choice between rosy Utopia and bleak Apocalypse, but through transmutations of old crises, through half-hearted fixes, into new crises (Lem, 1981). Thus, if he is right, and following on another theme of this book, one can easily envisage the crises that will occur globally, or in individual countries, if none of the adjustments in (1) to (3) above are implemented:

(a) fishing down the food webs of marine and freshwater ecosystems, as defined in Pauly *et al.* (1998) and in Pitcher and Pauly (Chapter 24, this volume), will continue, and will subvert scattered attempts at sustaining this or that single-species stock;
(b) the first cases of the global extinction of marine fish species will occur, complementing the ongoing extinction of many of their populations, and of freshwater fish species (Bruton, 1995; Baillie and Groombridge, 1996), and making a mockery of our governments' ratification of various treaties devoted to the protection of biodiversity; and
(c) we will still end up with subsidy-guzzling, overcapitalized fisheries, gradually pushed aside by mariculture, the next technological 'fix'.

On the other hand, there are powerful forces working against business as usual, notably the entry of new stakeholders into fisheries management, a process forming the most hopeful theme of this book. Thus Charles (Chapter 6, this volume), Walters (Chapter 21, this volume), and especially Sutton (Chapter 8 this volume) illustrate the potential of public involvement. Both Charles and Walters stress the public as the owner, and thus ultimate 'client' of fisheries resources management – in contrast to the often irresponsible resource users who operate at the public's tolerance.

Also, the public at large can be expected to affect fisheries management. This may happen indirectly, through the demand for legislation to protect the natural environment in general (e.g. the Magnuson Act, intended to protect sea mammals, among others, and which can be interpreted as also protecting the forage fish that the marine mammals require), through industrial standards (Sproul, Chapter 9, this volume), or through market mechanisms such as ecolabelling schemes (Sutton, Chapter 8, this volume) which allow for well-managed fisheries to be rewarded by purchase of their products. The involvement of this new actor in fisheries management, through non-governmental organizations or otherwise, will in many cases require new 'acts', i.e. new legislation to create new arrangements in the narrower context of co-management (Brown, Chapter 13, this volume), social 'collimation' (Dunsire, 1994) or broad participatory schemes (Hyden and Bratton, 1992; Putnam, 1993; Kooiman, 1994).

But the public is not willing to sacrifice short-term advantage for long-term gain. Like fishers, the consumers will act to protect their own interests. What is also required, therefore, are institutions that devise ways of incorporating the prospect of future benefits into the daily decision process so that some fish are left in the water (Hart, Chapter 17, this volume; Sutton, Chapter 8, this volume).

Perhaps paradoxically, it needs stating here that reinventing fisheries management does not include only changes in the socio-economic and institutional aspects of fisheries, though much emphasis must be given to this. Indeed, in addition to interdisciplinary work (Pitcher *et al.*, Chapter 3, this volume; Preikshot, Chapter 7, this volume), there is a role for good 'science' (Policansky, Chapter 4, this volume), i.e. for fisheries 'biology': data will have to be made and kept available for rigorous analysis (Richards and Schnute, Chapter 30, this volume), ocean patterns will have to be interpreted with the most advanced theory at hand (Bakun, Chapter 25, this volume), uncertainties will have to be accounted for (Bundy, Chapter 26, this volume; Peterman *et al.*, Chapter 29, this volume), trophic interactions between resource species and their preys and predators will have to be accounted for (Christensen and Pauly, 1993; Walters *et al.*, 1997; Neill, Chapter 22, this volume), and the optimal size and placement of marine protected areas will have to be researched, such as to enable them to play the beneficial role that is now widely expected of them (Sumaila, Chapter 23, this volume).

Our current inability to resolve these themes, except in a piecemeal and uncoordinated 'band-aid' fashion (Walters, 1995), is symptomatic of a profound lack of a coherent policy goal. The disaster that is the likely future of our seas is not what the responsible fishery scientists represented in this volume are diligently working towards; however, without change, we fear that this depressing scenario is what they will get. Accordingly, we argue strongly in this book for the adoption of an ecosystem-rebuilding goal for fisheries management (Pitcher and Pauly, Chapter 24, this volume). Thus, reinventing fisheries management also requires that we creatively expand our disciplines beyond their current boundaries. Business as usual, on the other hand, is bad science, and bad fisheries, such as those that now give us 'black greens', the very small (25–30 cm) but near fully grown cod that are now (illegally) landed in British and French ports by a North Sea fishery in denial of the need to change.

ACKNOWLEDGEMENT

We would like to thank Cornelia Nauen for reviewing the first draft, and suggesting a useful bit to add.

REFERENCES

Anonymous (1995) ACP–EU Fisheries Research Initiative. Proceedings of the First Dialogue Meeting, Eastern and Southern Africa, Indian Ocean and the European Union. Swakopmund, Namibia, 5–8 July, 1995. Brussels, *ACP-EU Fish. Res. Rep.* **1**, 144 pp.

Anonymous (1996) Initiative de Recherche Halieutique ACP–UE. Actes de la Deuxième Réunion de Dialogue sur l'Afrique centrale et occidentale et l'Union Européenne. Dakar, Sénégal, 22–26 avril 1996. Bruxelles, *Rapp. Rech. Halieut. ACP–UE* **2**, 178 pp.

Anonymous (1997) ACP–EU Fisheries Research Initiative. Proceedings of the Third Dialogue Meeting, Caribbean and Pacific and the European Union. Belize, Belize City, 5–10 December 1996. Brussels, *ACP–EU Fish. Res. Rep.* **3**, 180 pp.

Baillie, J. and Groombridge, B. (eds) (1996) *The 1996 IUCN Red List of Threatened Animals*, IUCN, Gland, Switzerland, 368 pp + appendices.

Bruton, M.N. (1995) Have fishes had their chips? The dilemma of threatened fishes. *Env. Biol. Fishes* **43**, 1–27.

Chakalall, B., Mahon, R. and McConney, P. (1997) Fisheries governance in the Caribbean, in *ACP–EU Fisheries Research Initiative. Proceedings of the Third Dialogue Meeting, Caribbean and Pacific and the European Union.* Belize, Belize City, 5–10 December 1996. Brussels, *ACP–EU Fish. Res. Rep.* **3**, 131–163, Annex 11.

Christensen, V. and Pauly, D. (eds) (1993) *Trophic Models of Aquatic Ecosystems. ICLARM Conf. Proc.* **26**, 390 pp.

Dunsire, A. (1994) Modes of governance, in *Modern Governance: New Government–society Interactions* (ed. J. Kooiman), Sage Publs, London, New Delhi, pp. 21–34.

Finlayson, A.C. (1994) *Fishing for Truth*, Inst. Social Econ. Res., Memorial Univ., Newfoundland, 176 pp.

Hutchings, J.A., Walters, C. and Haedrich, R.L. (1997) Is scientific inquiry incompatible with government information control? *Can. J. Fish Aquat. Sci.* **54**, 1198–1210.

Hyden, G. and Bratton, M. (eds) (1992) *Governance and Politics in Africa*, Lynne Rienner Publs, Boulder and London, 329 pp.

Kooiman, J. (ed.) (1994) *Modern Governance: New Government–society Interactions*, Sage Publs, London, New Delhi, 280 pp.

Latour, B. (1988) A relativistic account of Einstein's relativity. *Social Stud. Sci.* **18**, 3–44.

Lem, S. (1981) *Summa Technologiae*, Suhrkamp Taschenbuch 678, Frankfurt, 655 pp.

Nauen, C.E., Bangoura, N.S. and Sall, A. (1996) La 'gouvernance' des systèmes aquatiques en Afrique occidentale et centrale: leçons du passé, ouvertures pour l'avenir, in *Initiative de recherche halieutique ACP–UE. Comptes-rendu de la Deuxième Réunion de Dialogue, Afrique de l'Ouest et Centrale et Union Européenne. Dakar, Sénégal, 22–26 avril 1996* (ed. Anon.), Bruxelles, *Rapp. Rech. Halieut. ACP–UE* **2**, 157–177, Annexe 11.

Pauly, D., Christensen, V., Dalsgaard, J., Froese, R. and Torres, F. Jr (1998) Fishing down marine foodwebs. *Science* **279**, 860–863.

Putnam, R.D. (1993) Democracy, development, and the civic community: evidence from an Italian experiment, in *Culture and Development in Africa. Proceedings of an International Conference Held at the World Bank, Washington, DC, April 2 and 3, 1992* (*World Bank Env. Sustainable Develop. Proc. Ser.* **1**) (eds I. Serageldin and J. Tabaroff), World Bank, Washington, DC, pp. 33–73.

Ruddle, K. (1997) The role of local management and knowledge systems in ICAM in the Pacific region: a review, in *ACP–EU Fisheries Research Initiative. Proceedings of the Third Dialogue Meeting, Caribbean and Pacific and the European Union. Belize City, Belize, 5–10 December 1996* (ed. Anon.), *ACP–EU Fish. Res. Rep.* **3**, 121–130, Annex 10.

Sokal, A. (1996) A physicist's experiments with cultural studies. *Lingua Franca* **6**(4), 62–64.

Sokal, A. and Bricmont, J. (1997) *Impostures intellectuelles*, Editions Odile Jacob, Paris, 276 pp.

Walters, C.J. (1995) *Fish on the Line: the Future of Pacific Fisheries*, Suzuki Foundation Report, Vancouver, 87 pp.

Walters, C.[J.], Christensen, V. and Pauly, D. (1997) Structuring dynamic models of exploited ecosystems from mass-balance assessments. *Rev. Fish Biol. Fish.* **7**, 139–172.

Author Index

Species index

Bold page numbers refer to figures; *italic* page numbers refer to tables

SCIENTIFIC NAMES

COMMON NAMES

Subject index

Bold page numbers refer to figures; *italic* page numbers refer to tables